7 BIG PROJECTS
For a Better World

Written and Edited by Gordon Danby & James Powell,

the Inventors of Superconducting Maglev,

with Robert Coullahan, Ernest Fazio, Fletcher Griffis, George Maise, Charles Pellegrino, Jesse Powell, John Powell, John Rather, John Skaritka and James Jordan, Managing Editor

Also by Gordon Danby, James Powell

The Fight for Maglev, Making America the World Leader in 21st Century Transport

Maglev America, How Maglev Will Transform the World Economy

By James Powell, George Maise, and Charles Pelligrino

StarTram, The New Race to Space

By James Powell, Jesse Powell and James Jordan

Silent Earth Will Humans Give Up Fossil Fuels?

7 Big Projects

For a Better World

ISBN-13: 9781982076016

ISBN-10: 1982076011

Library of Congress Control Number: 2018901301

Printed in the United States of America

"Human beings are not wicked by nature. We have enough intelligence, goodwill, generosity, and enterprise to turn Earth into a paradise both for ourselves and for the biosphere that gave us birth. We can plausibly accomplish that goal, at least be well on the way, by the end of the present century. The problem holding everything up thus far is that Homo sapiens is an innately dysfunctional species. We are hampered by the Paleolithic curse: genetic adaptations that worked very well for millions of years of hunter-gatherer existence but are increasingly a hindrance in a globally urban and techno-scientific society. We seem unable to stabilize either economic policies or the means of governance higher than the level of a village. Further, the great majority of people worldwide remain in the thrall of tribal organized religions, led by men who claim supernatural power in order to compete for the obedience and resources of the faithful. We are addicted to tribal conflict, which is harmless and entertaining if sublimated into team sports, but deadly when expressed as real-world ethnic, religious, and ideological struggles. There are other heredity biases. Too paralyzed with self-absorption to protect the rest of life, we continue to tear down the natural environment, our species' irreplaceable and most precious heritage. ...

Our species dysfunction has produced the hereditary myopia of which we are all uncomfortably familiar. People find it hard to care about other people beyond their own tribe or country, and even then past one or two generations. ..."

From "The Meaning of Human Existence" by Edward O. Wilson, widely recognized as one of the World's preeminent biologists and naturalists. The author of more than 20 books and winner of 2 Pulitzer Prizes.

Dedication

To Our Children, Grandchildren and their Children and to the Grandchildren of the World.

Acknowledgements

The authors again wish to express their deep gratitude and thanks to their many colleagues for their valuable contributions that have advanced the technologies for the 7 big projects. They and their efforts are acknowledged in our previous books, "The Fight for Maglev", "Maglev America", and, "StarTram, The New Race to Space".

For the "7 Big Projects for a Better World" book, we want to again express our deep gratitude and thanks to the following colleagues for their valuable present efforts in the areas covered in the "7 Big Projects" book – Maglev Transport, Space Launch and Beamed Power, Synthetic Fuels and Large Ice Structures. Listed alphabetically, they are they are Tom Berg, Paul Earle, Frank Genadio, Chris Kempner, Josh Levin, John Mankins, Chris Miller, Thomas Wagner, and Bob Worthing.

The efforts of Douglas Rike and Barbara Roland in preparing the many wonderful drawings and the very attractive layout of the 7 Big Projects book have been outstanding, and are very gratefully appreciated.

Foreword

"For all the immediate challenges that we gather to address this week — terrorism, instability, inequality, disease — there's one issue that will define the contours of this century more dramatically than any other, and that is the urgent and growing threat of a changing climate."

President Barack Obama, UN General Assembly for Climate Week Sept, 2014

In the final chapter of this book the authors conclude that humankind has reached a Fork in the Road. Do we continue to power the World's seemingly successful path to growth and prosperity with the numerous inventions for transport based on the combustion of fuels refined from minerals formed in the Earth's crust more than 300 million years ago or do we come to grips with the realities and the future which we can see?

Inspired by the inventive minds of James Powell and Gordon Danby, the authors and contributors to this book define projects that can create a more sustainable future for our species.

Contents

FIGURE 1

Preface

AFTER US – THE EXTINCTION

"Après moi le deluge" – After me, the flood, was supposedly uttered by Louis XV (Figure 1), the Monarch of France from 1715 to 1774, though Madame de Pompadour was the more likely source. In any case, it sums up the conditions in France in the years prior to the 1789 storming of the Bastille and the Revolution's Reign of Terror (Figure 2).

Historians characterize Louis XV's reign as one of stagnation with lost wars, political fights between the monarchy and Parliament and religious feuds. "Things fall apart, the center will not hold", to quote from Yeats' poem, 'The Second Coming". "The aristocracy knew that the flood would come, hopefully, after they were gone unless they acted to stop it." They didn't try, just continued eating, drinking, and being merry.

Now, almost 300 years later, the World, not just France, faces collapse unless it acts and acts soon. The Sixth Mass Extinction is already underway, with projections that half of the World's species will be extinct by the end of the Century, 85 years from now. Our children and grandchildren will experience the Sixth Extinction.

The culprit? Us humans. In Pogo's words, "We have met the enemy and he is us." Earth's previous five mass extinctions were due to natural events like volcanic eruptions and an asteroid impact. The Sixth Extinction is a result of human activities – deforestation, habitat destruction, hunting, poisoning soil and water sources, and so on.

But the most important activity that is driving us to the Sixth Extinction is our massive consumption of fossil fuels. Their greenhouse gas emissions are rapidly warming the planet, far faster than any previous period in Earth's 4 Billion year history.

As a consequence, large areas of the planet are experiencing massive droughts and crop failures, worse wildfires, stronger storms, rising sea levels and coastal flooding, increased disease for plants, animals and humans, and the extinction of many species.

The oceans are rapidly acidifying as they absorb the increasing amount of carbon dioxide in Earth's atmosphere. As ocean acidity increases, it become harder and harder for shelled micro and macro organisms to form shells. At some point, projected to occur before the end of the 21st Century, the ocean will be so acidic that shells cannot form – no more oysters, mollusks, and the micro-organisms that support the ocean food chains. According to a recent New York Times story ("As Oysters Die, Climate Policy Goes on the Stump" by Coral Davenport Aug. 3, 2014) billions of baby oysters are already dying due to ocean acidity. When that happens, many of the ocean species dependent on the food chains will become extinct; as they did in the Permian-Triassic Extinction, 250 million years ago.

Making the oceans lifeless seas would be catastrophic. To prevent this from happening, humanity must act to stop consuming fossil fuels, and to be effective, must stop soon – no, "After us, the extinction."

It is possible that it's already too late to stop global warming and avoid the on-coming extinctions? Discussions of global warming tend to believe that humanity can continue to consume fossil fuels until things get so bad that the decision is made to switch to other energy sources, and that conditions will not get worse after that.

This belief is very wrong. In reality, at some point, the on-going global warming will pass a trigger point and runaway with no possibility of stopping, even if humanity completely stops consuming fossil fuels. There are 1,000's of billion tons of carbonaceous material locked up as methane hydrates in the sea beds and organic materials in the Arctic permafrost.

As the oceans and permafrost warm they are releasing methane and carbon into the atmosphere. Ocean methane boils have been observed in the Arctic, along with methane and carbon dioxide emissions from the warming permafrost.

In terms of its effect on global warming, methane is 20 times more potent a greenhouse gas than carbon dioxide. A recent study has concluded that the "release of 50 billion tonnes of predicted amount of hydrate is highly possible for abrupt release at any time". That would be equivalent in greenhouse effect to a doubling in the atmospheric level of CO_2, now at 400 parts per million, to 800 parts per million.

Have we already passed the trigger point for runaway global warming? Nobody knows. If humanity keeps consuming fossil fuels, will runaway global warming occur? Yes. If the trigger point has not already been passed, when will it occur? Nobody knows. Much more study of what will trigger runaway global warming and when it could happen is needed. But, it is very likely that it will not be possible to know how much time we do have before it does occur.

In this book, we describe 7 Big Projects that will greatly reduce humanity's need for fossil fuels. The projects are practical and can be developed and implemented in the near future. Not only will the projects help to reduce the damage caused by global warming and hopefully avoid triggering irreversible runaway warming, but their major economic and social benefits will provide a better quality of life and higher living standard for billions of people.

Humans will survive the Sixth Extinction. The real question is will modern human civilization survive, grow, and improve the World's environment, or will it collapse under the stresses brought on by continuing its dependence on fossil fuels, bringing on a new Dark Age. The choice is up to us – continue on our present path, or switch to a sustainable future.

FIGURE 2

Prologue

The Seven Big Projects and Why We Need Them

Modern civilization is supported by 5 fundamental technology pillars:

- **Transportation**
- **Energy**
- **Food and Water**
- **Manufacturing**
- **Communications**

Without them, modern civilization would not exist. Humans would be back in the hunter-gatherer World, walking around, looking for plants and animals for food, and exchanging grunts when they meet another human.

The 5 technology pillars are not isolated entities, but are critically dependent on each other in complex ways. To get our food and water, we depend on transportation, which in turn depends on energy. Extracting fossil fuel energy from the Earth requires transport of manufactured equipment. Making the equipment takes energy and communications about what to make and were to transport it. Growing food and obtaining water takes energy and manufactured equipment and communications, ordering the energy and equipment, and finding out where to transport the food and water.

One could go on in detail about the complex interactions between the 5 technology pillars, and how they affect governments and social behavior, but it would take many more books than just this one to do so.

In this book, we focus on 7 Big Projects in the areas of energy, food and water, and transportation that will support modern society and reduce their costs to society, both in economic terms and in their damage to Earth's environment.

The 7 Big Projects are:

Chapter 1: The 29,000 mile 300 mph United States National Maglev Network that will interconnect all of the metropolitan areas in the Lower 48 US states, transporting passengers, trucks, freight and, autos with passengers at low cost with no pollution and emissions of greenhouse gases.

Chapter 2: The Global Maglev Network that will interconnect North America, South America, Asia, Europe, and Africa. Passengers and freight will be able to travel to any point in the Global Maglev Network at low cost.

Chapter 3: The StarTram Maglev Launch System that will launch payloads into space at $1/100^{th}$ the cost of rocket launch. StarTram will enable: very low cost beamed electric power to Earth from space solar power satellites, protection of Earth from asteroids and comet impacts, asteroid mining, enhanced worldwide broadband communications, and exploration and colonization of the Solar System.

Chapter 4: Beamed Solar Power to Earth. Space solar power satellites launched by StarTram will enable continuous 24/7 beaming of millions of megawatts of clean, sustainable electric power to Earth at lower cost than presently possible. All points on the Earth will be able to use StarTram electric power.

Chapter 5: Synthetic Gasoline, Diesel and Jet Fuel from air and water. Using low cost electric power to extract carbon dioxide from the atmosphere and generate hydrogen by electrolyzing water as raw materials, synthetic gasoline, diesel, and jet fuel can be produced using existing chemical processes. Burning the synthetic fuels for transport does not generate net carbon dioxide emissions into the atmosphere, in contrast to fossil fuel oil and gas, which do. The carbon dioxide release from synthetic fuels is balanced by the carbon dioxide extracted from the atmosphere to manufacture the fuels.

Chapter 6: LISA Islands (Large Ice Structure Assemblies) for OTEC Electric Power Generation and Fresh Water Production with large thermally insulated floating ice islands as sites for Ocean Thermal Energy Conversion (OTEC) large amounts of low cost, clean, sustainable electric power can be generated using the temperature difference between warm ocean surface water and the cold deep ocean water. Fresh water can also be produced by desalinization of the ocean water. One LISA OTEC island can generate 2,000 megawatts of electric power and 1.9 Billion gallons of fresh water daily.

Chapter 7: LISA Coastal Barriers against Storm Surges, Tsunamis, and Rising Sea Levels. Thermally insulated large LISA ice barriers can be erected to protect vulnerable coastal areas from storm surges, tsunamis, and rising sea levels. The capital and refrigeration costs for the LISA are very low. Had a LISA barrier been in place to protect New York from Hurricane Sandy, New Orleans from Hurricane Katrina, and Fukushima from the tsunami, hundreds of Billions of dollars of damage would have been avoided and many lives would have been saved.

In Transportation, we describe 3 Big Projects based on Maglev, the first new mode of transport since the airplane. Maglev vehicles do not physically contact the ground or rails. Instead, they are magnetically levitated and propelled along a guideway, with air drag the only frictional force acting on them. In vacuum tunnels or tubes, Maglev vehicles can reach orbital speeds of 18,000 mph. In air, they can travel at 300 mph with energy requirements that are much lower per passenger mile and ton mile of truck freight than existing transport modes. In contrast to autos, trucks, airplanes and diesel powered trains that emit many billions of tons of carbon dioxide per year, a major cause of global warming, Maglev uses clean electric power, which if it supplied from non-fossil-fuel sources, emits zero greenhouse gases and does not contribute to global warming.

Chapter 1 describes how Maglev can enable a National Maglev Network for the United States, transporting passengers, highway trucks, freight and autos at 300 mph from any point in the 48 continental states to any other point. 74 percent of the US population would live within 15 miles of a Maglev Station on the 29,000 mile Maglev Network, from which they could travel to any other station in the 48 States. Figure 1 shows a map of the US National Maglev Network. Chapter 1 also describes how Maglev can be adapted to use existing commuter rail and subway tracks for local transit at much faster speeds, with cleaner, healthier, lower cost and more comfortable travel.

Chapter 2 describes how Maglev can be extended into a Global Network, enabling the entire World to experience its benefits. Freight containers can be transported thousands of miles by Maglev in much shorter time and at lower cost than by ships, greatly reducing oil consumption and health damage from the pollutants presently emitted by ships. Similarly, passengers will be able to travel by Maglev all over the World, more conveniently, and in shorter time, than by airplane. The chapter describes how Maglev can connect Europe to Africa across the Strait of Gibraltar and Asia to North American across the Bering Strait.

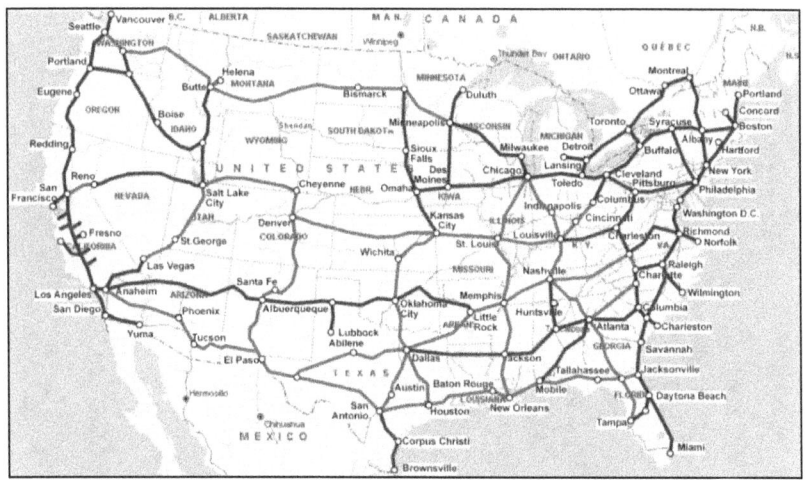

Figure 1: US National Maglev Network
Table 1: Population and States Served by the Network

Maglev Network	States In Network	Population of States in Network (millions)	Population Living Within 15 Miles of Stations (millions)	Route Miles in Network
First, Second and Third Waves Completed	48 plus Toronto, Montreal & Vancouver	315 includes Toronto, Montreal & Vancouver	232 includes Toronto, Montreal & Vancouver	29,000
74% of population in States live within 15 Miles of a Station				

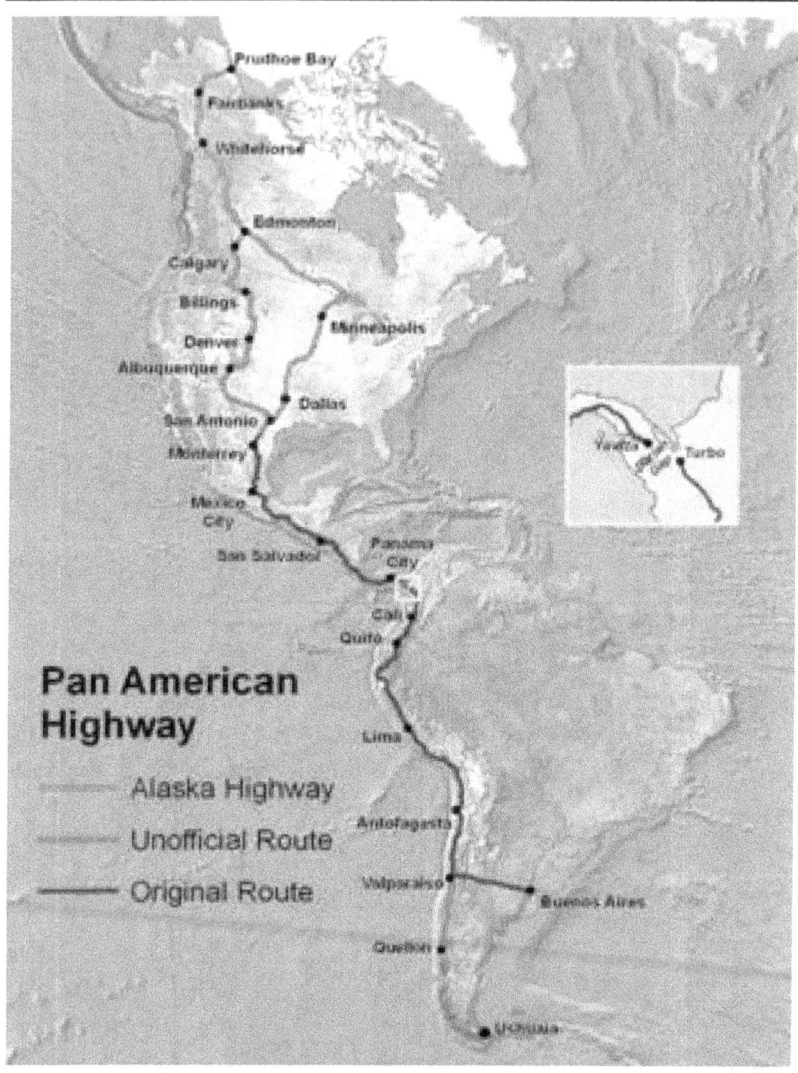

Figure 2

Figure 2 shows a map of part of the Global Maglev Network linking North and South America.

Chapter 3 describes how Maglev can launch cargo into space at much lower cost than presently done by rockets. The StarTram system magnetically accelerates cargo spacecraft to orbital speed in a vacuum tunnel. After reaching orbital speed the cargo craft transitions through a portal into the atmosphere, coasting upwards to orbit. Payloads launched into space by rockets have extremely high launch costs, on the order of $5,000 per kilogram. The launch cost using StarTram is only $50 per kilogram, a factor of 100 cheaper. The World presently launches approximately 200 tons of payload into space per year. With Maglev, a single StarTram launch facility could economically launch over 100,000 tons of payload per year, a factor of 500 greater. Figure 3 shows an artist's drawing of a StarTram spacecraft launched into the upper atmosphere from the end of its vacuum acceleration tunnel.

The very low-cost capability of StarTram will bring enormous benefits to humanity.

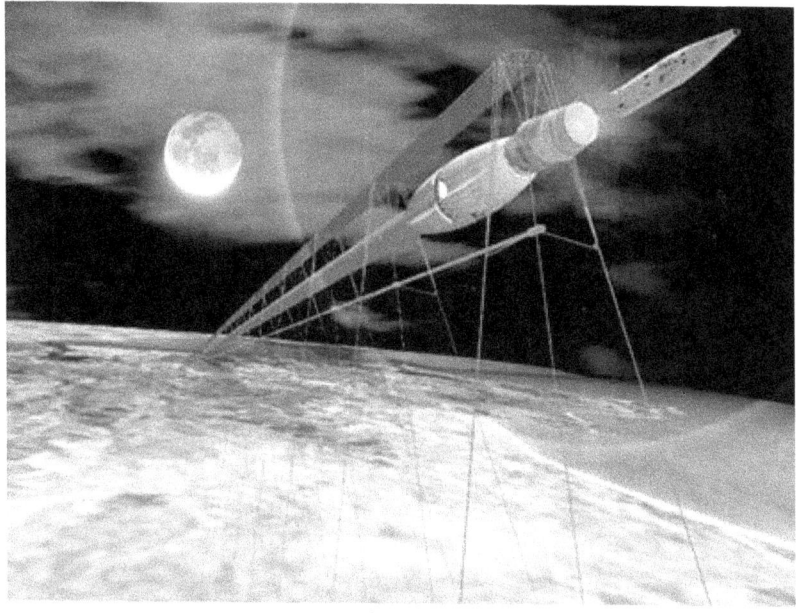

Figure 3

Chapter 4 describes how StarTram makes it practical to generate low cost clean electric power from solar power satellites in Geosynchronous Orbit (GEO) and beam it down to Earth. With beamed solar electric power, there will be no need to burn fossil fuels for electric power. There are many other benefits from the StarTram Maglev Launch system, including robust defense against asteroids and comets impacting the Earth, exploration and colonization of the Solar system, much improved communications and environmental surveillance by satellites in Earth orbit, mining of asteroids for scarce and costly resources, and so on.

In the area of Energy, Maglev will play a vital role as discussed in Chapters 4 and 5. Chapter 4 describes how solar power satellites placed in orbit by StarTram (Chapter 3) can beam down enormous amounts of clean electric power to Earth, greatly reducing fossil fuel consumption. Figure 4 shows an artist drawing of a solar power satellite system beaming power to Earth.

Chapter 5 describes how carbon based fuels – gasoline, diesel fuel and jet fuel for airplanes – can be produced for transportation at locations not served by Maglev transport, using beamed solar power to manufacture synthetic carbon based fuels, from carbon dioxide extracted from the atmosphere. This eliminates the need for fossil oil and gas to fuel transport systems. The synthetic fuels Big Project enables the continued use of combustion powered transport systems where needed, without adding new carbon dioxide to the atmosphere.

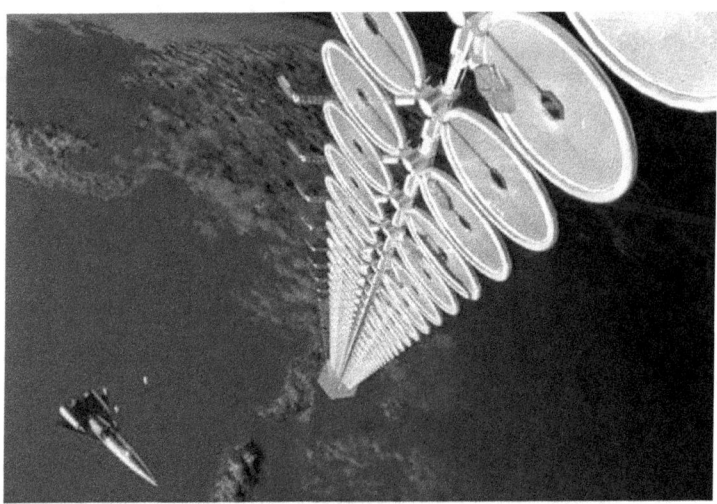

Figure 4

Today, fossil fuels are critically important for the viability of modern society. Without oil, natural gas and coal, we would be back in the Middle Ages, or even worse with horses and wagons and almost everybody living as peasants. We could still build pyramids and cathedrals using enormous amounts of human labor. We could still travel across the oceans in sailing ships, or float down rivers on rafts. We also would have periodic plagues – new "Black Deaths". We could still fight wars with bows and arrows and spears using metal armor smelted by charcoal. Heating? Wood and charcoal. Water? Lakes and springs for drinking, with a bath every year.

However, there are 2 fundamental problems with fossil fuels, which will prevent them from providing long-term sustainability of Modern Civilization:

- Global warming
- Depletion of reserves

At some point, probably well before the end of the 21st Century, these problems will require that the World transition from fossil fuel energy to non-fossil fuel energy sources, if modern civilization is not going to collapse and fall back to the Middle Ages.

We have previously discussed the problem of the effect of greenhouse gas emissions from fossil fuel on global warming and ocean acidification. It appears very likely that acidification will turn the World's oceans into lifeless seas, well before 2100 AD, based on the effects already observed on marine organisms. It also appears very likely that sometime in the coming decades we will pass the trigger point for runaway global warming, based on already observed methane boils in the Arctic Ocean and methane & carbon dioxide emissions from the warming permafrost.

The second problem for fossil fuels, depletion of reserves, is very likely to occur before 2100. As fossil fuels run out, production will decrease, and other sources of energy will have to be implemented if modern society is not to collapse.

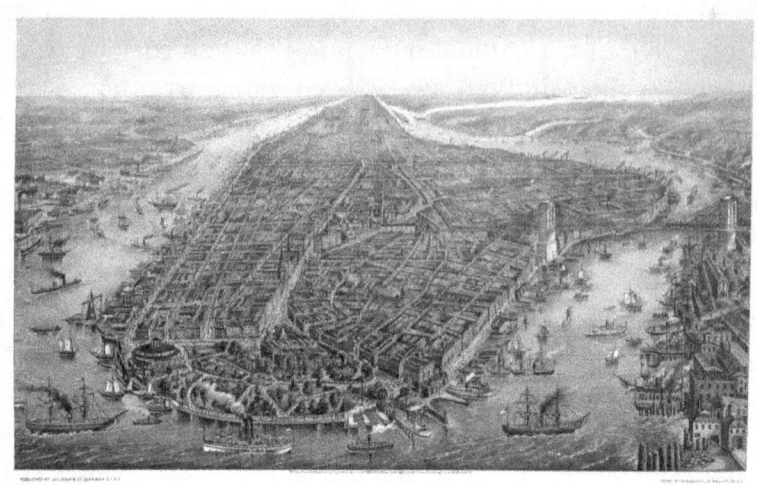

Figure 5

The amounts of fossil fuels that the World consumes are staggering -- 1,000 barrels of oil and 250 tons of coal per second. To illustrate this visually, rather than just giving numbers, which are difficult to grasp, imagine stacking fossil fuels on Manhattan Island. Figure 5 shows a lithograph of Manhattan Island in 1873, before all of those tall buildings were built on top of it. It was a lot less crowded then.

Manhattan Island has a surface area of 22.6 square miles. That's very big. Now imagine stacking 1 year of World consumption of: oil, natural gas, and coal on top of the 22.6 square mile Manhattan Island. Figure 6, overleaf, compares the height of the coal, oil and natural gas layers from 1 year of World production with the height of the 1378 foot tall World Trade Center building #1, before it was destroyed in the 9/11 attack of 2001. The oil and coal layers are at normal density. The natural gas layer is depicted as Liquefied Natural Gas (LNG) with its normal density of 450 kilograms per cubic meter. The annual World production rates for oil and natural gas are taken from the BP Statistical Review of World Energy. June 2014 (bp.com/statistical review). The annual World production rate for coal is taken from the Wikipedia article on coal (en.wikipedia.org/wiki/coal).

The thickness of the oil, gas, and coal layers are stunning, with a combined height of 337 meters (1,110 feet) compared to the 417 meter (1,368 feet) height of the WTC building. Imagine – a pile of *one year's fossil fuel production* that is 22.6 square miles in area and 1,110 feet high.

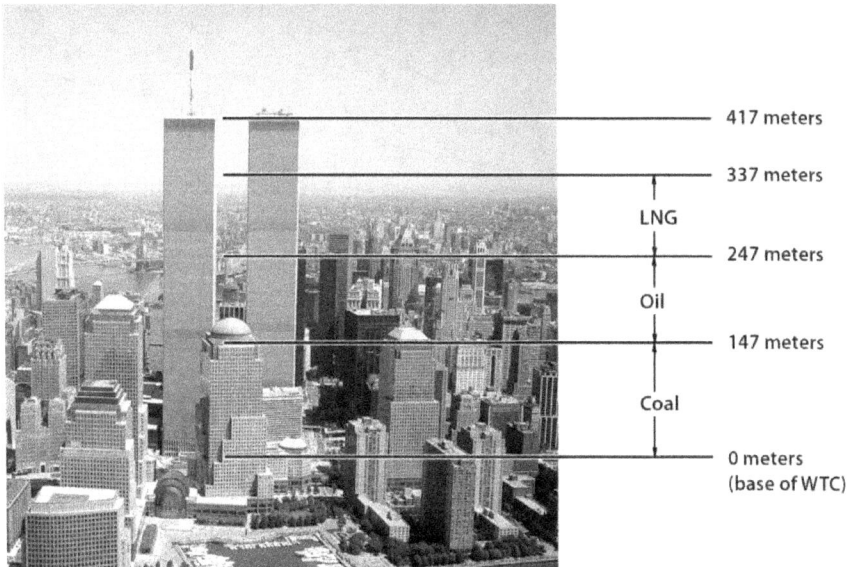

Figure 6

Even more stunning is Figure 7, which shows the thickness of the Manhattan Island fuel pile as a function of time, starting with the year 2015. The production rates of oil, gas, and coal are assumed to stay constant at present value over the period 2015 to 2100 AD.

In 2040, the Manhattan fuel pile reaches a height of 5.5 miles, the height of Mt. Everest, the highest point on Earth. By 2070 when the known reserves of oil and gas (BP Statistical Review) are exhausted, the combined height of the fuels is almost 12 miles, twice as high as Everest. Assuming new reserves of oil and gas are discovered – which is likely, though the new deposits will be deeper and more difficult to extract, and much increased fracking for natural gas will be necessary

– by 2100 the Manhattan Island pile would be 18 miles high, 3 times the height of Everest.

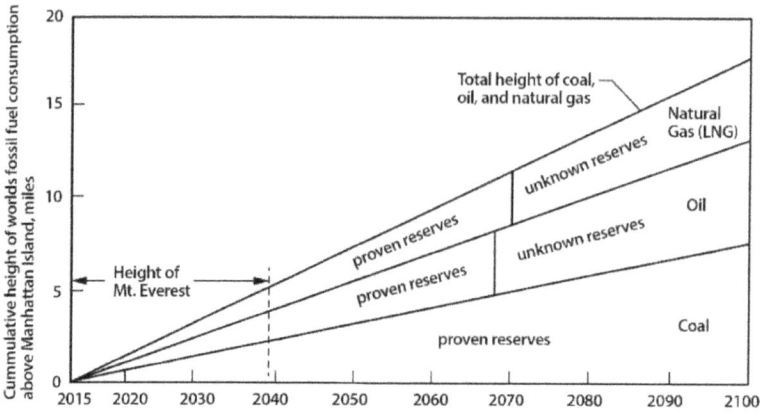

Basis: 22.6 square miles area of Manhattan Island, 2013 world consumption rates for oil, coal, and natural gas

Figure 7

And that is just assuming that the coal, oil, and gas consumption rates will remain constant at today's values. That certainly has not been the case in the past. Below are given the carbon dioxide emissions rates from fossil fuels, based on EPA (US Environmental Protection Agency) data.

Year	World CO_2 Emissions, Millions Tonnes per Year
1900	2,000
1950	5,000
1960	10,000
1970	15,000
1980	20,000
1990	23,000
2000	25,000
2008	32,000

In 2008, World CO_2 emissions were more than 6 times greater than in 1950, 32,000 million tonnes compared to 5,000 million tonnes. This is a consequence of the ever increasing World industrialization and better quality of life. This process will continue in the coming decades. The underdeveloped World wants to, and will, catch up to the developed World.

A measure of how far the World has to go to catch up and what it means in terms of fossil fuel use can be obtained by comparing US per capita consumption of fossil fuels with the World average per capita consumption.

Per capita, Americans presently consume 25 barrels of oil per year, compared to the World average of 4.5 barrels of oil per capita per year for the 7 Billion people in the World, a factor of 6 lower . Does the rest of the World want to catch up to America? Yes! The number of automobiles and trucks in the World is rapidly growing as is World shipping and air travel.

US per capita consumption of coal is presently 3.2 tonnes per year. The World average per capita consumption is 1.1 tonnes per year, a factor of 3 lower. US per capita consumption of natural gas is 2,330 cubic meters per year, compared to the World average of 478 cubic meters per year (BP Statistical Review), a factor of 5 lower for the World population. Does the rest of the World want a standard of living more like America's? Yes!

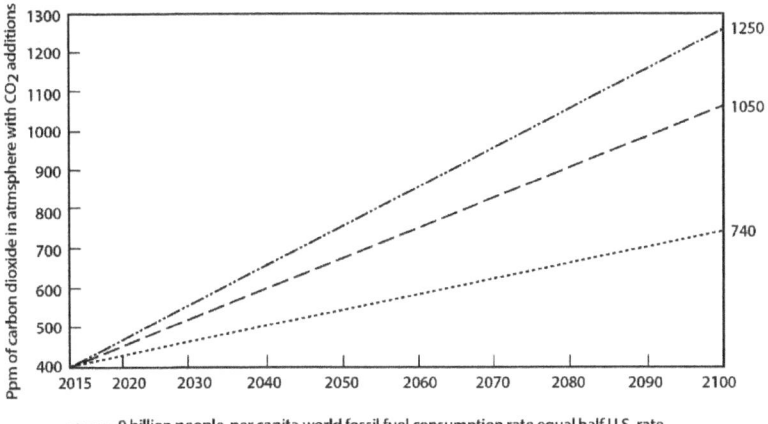

Figure 8

As a way of visualizing the effect of increasing World industrialization and the resulting increased consumption of fossil fuels, consider the following 2 scenarios.

1. For the present World population of 7 Billion people, the per capita consumption of oil, gas, and coal increases to be ½ of America's present per capita consumption rates for the various fuels.

2. For the projected World population of 9 Billion people in 2050 AD, the per capita consumptions of oil, gas and coal increases to be ½ of America's present per capita consumption rates for the various fuels.

In scenarios #1 and #2, the known reserves of oil, gas, and coal are exhausted in a much shorter time. Based on the known reserves of oil (1,688 Billion Barrels), gas (186,000 Billion cubic meters), taken from the BP Statistical Review, and coal (861 Billion tonnes) taken from the Wikipedia article on coal, we can calculate how many years of consumption are left before the know reserves run out.

R, Years Until Known Reserves Run Out

Fuel Type	Present World Consumption Rate 7 Billion People (2015)	Scenario #1 World Avg Per Capita=1/2 Present US Value 7 Billion People	Scenario #2 World Avg Per Capita = 1/2 US Value 9 Billion People
Oil	53	22	17
Gas	55	22	17
Coal	112	78	60

The number of years remaining before the reserves are exhausted can be expressed as values of R, the ratio of known Reserves/Annual Production. Below are listed the R values for oil, gas and coal for today's rates of consumption, and for scenarios #1 and #2. Figure 9 Shows the years that the known reserves of oil, gas, and coal will run out for present World consumption rates, Scenario #1 Consumption rates, and Scenario #2 Consumption rates. Given the increasing World consumption rates, it is likely that the known reserves of oil and gas will be gone by 2050, and coal will be gone by 2100.

If World fossil fuel consumption stays at its present rate, CO_2 concentration in 2100 AD will be 740 ppm, almost double today's (2015 AD) concentration. In scenario #1 it would be 1050 ppm, a factor of 2.5 increase, and in scenario#2, 1,250 ppm, a factor of 3 increase. World civilization would not survive such increases in CO_2 concentration. Climate scientists state that for the long term, 350 ppm, 50 ppm lower than the present concentration of 400 ppm, is the maximum Earth can have if it is to avoid environmental devastation. By 2050, the CO_2 concentration will probably be on the order of 650 ppm unless we act decisively, and very soon, to transition from fossil fuels.

Figure 9

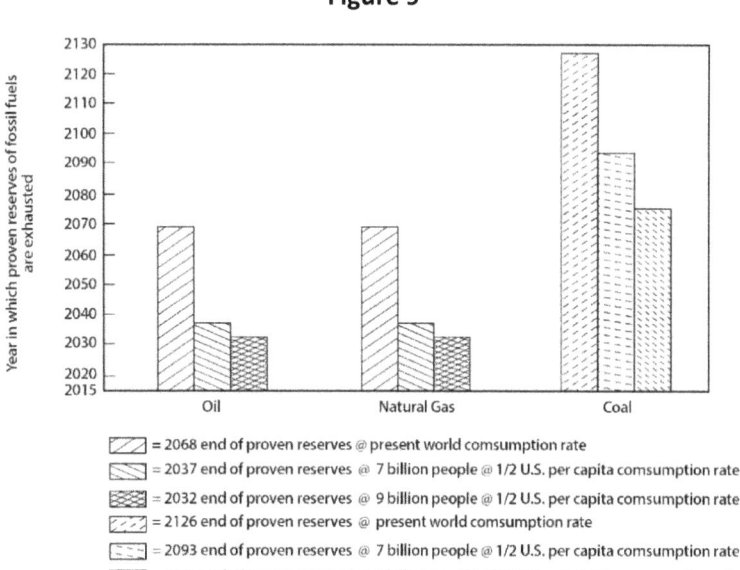

We hope our readers are not bored by the numbers described above, because they are not boring. They are frightening. They show that unless the World transitions from fossil fuels soon, modern civilization will collapse, probably well before the end of the Century.

Those who deny that the carbon dioxide emissions from the consumption of fossil fuels is causing global warming should talk to sophomore chemical engineering students who have to calculate the temperature of a hot radiating body inside a furnace as a function of the carbon dioxide concentration in the furnace atmosphere. The higher the CO_2 concentration, the higher the temperature of the radiating body. Just look it up in Perry's Chemical Engineering Handbook.

In the area of food and water, Chapter 6 describes how the temperature difference between the warm surface ocean water and the deep cold water underneath the surface can be used by Ocean Thermal Energy Conversion (OTEC) plant ships to generate very large quantities of low desalinated fresh water and electric power. The fresh water can be transported by pipeline or fresh water tankers to regions suffering from serious drought such as California, to help farmers obtain sufficient water to maintain their crop outputs, so that the populations depending on the food that they produce can continue to eat. Five large OTEC plants would produce 10 Billion gallons of fresh water per day, 30 percent of California's daily consumption.

The World has a major problem with not enough fresh water in drought areas, which the Big Project described in Chapter 6 helps to mitigate. The World also has a major problem with too much seawater in coastal areas, with storm surges and rising sea levels that flood low lying coasts and islands, causing hundreds of Billions of dollars in property damage, and hundreds of thousands of deaths and injuries. Chapter 7 describes how low cost barriers can be built along coast lines that will fully protect their inhabitants from storm surges, tsunamis, and rising sea levels. Figure 10 shows a photo of New Orleans after Hurricane Katrina. 80% of the city was flooded when its levees failed. Chapter 7 describes how better coastal barriers can provide complete protection for

World Coasts against storm surges, tsunamis and rising sea levels.

Bottom Line. If modern civilization is to continue, the World must transition to energy sources that do not depend on fossil fuels which are running out and causing irreversible environmental catastrophe. The transition must begin very soon, and will require a number of very big projects, 7 of which are described in this book.

Which path will humanity choose? The "After Us – The Extinction" path or the "Transition to a Sustainable Future". We don't know, nobody knows. All we know is that the choice must be made soon.

Figure 10 Flooding of 80% of New Orleans by Hurricane Katrina

1

What is a Big Project? Why Do We Need Big Projects?

The traditional definition of a Big Project is an undertaking that spends a lot of money, regardless of whether it results in just something that's impressive, but doesn't do anything that is useful and important.

Big spending is not our definition of a Big Project. To us, Big Projects create much greater capabilities and a better life for the society that undertakes them – safer, higher living standards, long term sustainability, better working conditions, more leisure time, better education opportunities, and a less stressful life.

v

Figure 11 Pyramids and Sphinx in Egypt

Figure 12 Pyramid del Sol in Mexico

Historically, there have been many Big Spending Projects that did nothing useful and important, but consumed enormous amounts of labor and were a drag on society. Examples include the Egyptian pyramids (Figure 11) whose only purpose was to glorify the Pharaohs, and temples like the Pyramid of the Sun (Figure 12) in Teotihuacán, Mexico, whose purpose was to glorify the local religion.

History has examples of Big Projects that benefited the society that undertook them. The Great Wall of China (Figure 13) was built to keep out the Mongol hordes from invading and ravaging Chinese society. While not completely successful, it did provide significant protection.

Figure 13 Great Wall of China

Over the last 200 years, America has undertaken a number of Big Projects that have greatly benefited the nation and enabled it to grow into the powerful modern society it is today.

In transport, the first Big Project was the Erie Canal (Figure 14). 363 miles in length, it connected New England to the Midwest, enabling horse drawn barges to transport goods between the 2 regions. 11 million cubic yards of digging, all by hand, twice as much as the English "Chunnel". 1,000 men died of swamp fever during the dig. Economic historians credit the Erie Canal with the economic growth of New York and it economy surpassing Philadelphia, Baltimore and other cities on the East Coast.

Then came the Transcontinental Railroad (Figure 15) that connected California to New York in 1869. In 1876, the Transcontinental Express traveled from San Francisco to New York in 84 hours, the same length of time it takes today on Amtrak. The Transcontinental Railroad unified America – before it was built it took months to cross America by wagon trains, or sail 14,000 miles around Cape Horn.

The next Big American Project in transportation was the Panama Canal (Figure 16). The French tried to dig the Canal but gave up after 22,000 of its workers died. President Theodore Roosevelt restarted the project in 1904. Completed in 1914, it is absolutely essential for World shipping.

America started the World revolution in transport by mass producing autos and trucks with Henry Ford, along with the development of airplane transport, invented by the Wright brothers.

The Interstate Highway System (Figure 17) was America's next Big Project, created by President Eisenhower. Its 47,000 miles of high speed highways carry 24 percent of America's auto and truck transport, connecting all of the continental United States into an efficient network for the movement of people and goods. Without it, America's economy and standard of living would be poorer.

Then came America's Big Project in space, the Apollo Program. Created by President Kennedy in 1962, it not only landed men on the Moon in 1969 (Figure 18), it made America the world leader in transport to space, with the ability to launch satellites for World-wide communications and surveillance.

In Energy, America has also been the world leader in the development of nuclear power, through its Manhattan Project. Today, 10 percent of America's electric power needs are met by 1,000 megawatt nuclear reactors (Figure 19), a product of the Manhattan Project.

In the area of food and water, the US has also been a world leader with Big Projects – the Grand Coulee and Hoover Dams (Figure 20), the California Aqueduct System (Figure 21), and others to provide water to grow America's crops. Without these projects, our food and water supplies would be much smaller and our living standards would suffer.

So, Big Projects have made America what it is today – prosperous, a world power, an efficient, safe nation. If they had not been undertaken, America would have been very different and much poorer.

Sadly, today, America seems to have lost its thrust for Big Projects that will dramatically enhance our quality of life in the areas of transportation, energy, and food & water. We talk a lot about innovation, but in these areas, mostly talk, not action.

Have we lost the talent and ability to take on innovative Big Projects? No. Have we lost the willpower to do so? It appears that we have. Today's investors and politicians don't want new projects hurting their existing operations and assets. As a result, the highways are deteriorating, there is no real move towards reducing the consumption of fossil fuels – which will lead to environmental and economic disaster, if we do not reduce their use, as discussed in the Prologue, and our food and water sources are endangered from global warming.

The purpose of this Book is to layout 7 Big Projects that can help us meet the oncoming environmental and economic challenges and produce a better, more sustainable life for all humans. All of the projects are practical, both technically and economically, and can be carried out soon – they are not far-out fantasies. What is required for their implementation is decisiveness and courage of the type exemplified by past American leaders – Governor DeWitt Clinton for the Erie Canal, President Abraham Lincoln for the Transcontinental Railroad, President Theodore Roosevelt for the Panama Canal, President Franklin D. Roosevelt for the Hoover and Grand Coulee Dams and the Manhattan Project, President Dwight D. Eisenhower for the Interstate Highway System, and President John F. Kennedy for the Apollo Program.

In deciding on which future big projects to undertake, society's leaders should evaluate their benefits and impacts objectively on how well they meet the following criteria:

- Improve the economy, job opportunities, living standards, and quality of life for the entire nation and its citizens

- Reduce deaths, injuries, and health damage from accidents, pollution, natural disasters, and other causes

- Reduce environmental damage to localities and the world from greenhouse gas emissions, toxic wastes, industrial and mining operations, and other sources

- Provide safety, security, and operational reliability for the nation and its citizens from natural disasters, breakdown of unreliable systems, and attacks.

How well do the 7 Big Projects meet these criteria?

Very well!

The National and Global Maglev Network described in Chapters 1 & 2 will substantially reduce the cost of transporting passengers and goods, eliminate greenhouse gas emissions, be very safe and reliable, reduce highway deaths and injuries, be very quiet, comfortable, and much faster, and avoid delays due to congestion and bad weather.

The StarTram Launch to Space and Space Solar Power Beaming projects described in Chapters 3 & 4 will eliminate the need for fossil fuels for generating electric power, preventing global environmental catastrophe from greenhouse gas emissions. The much lower cost of electric power enabled by the StarTram and Space Power Beaming projects will result in a sustainable higher standard of living for all of the world's inhabitants. The much lower cost of Launch to Space using StarTram will result in much lower cost communications, as well as improved monitoring and forecasting of weather conditions, including storms, droughts, crop conditions and yield, etc. StarTram will also provide complete protection from asteroids and comets impacting the earth.

The Synthetic Fuel from Air and Water projects described in Chapter 5 will enable transport systems that require hydrocarbon fuel to operate, such as airplanes, to continue operation without greenhouse gas emissions. Using low-cost electrical power beamed from space solar power satellites, we will be able to generate hydrogen, which is chemically combined with carbon dioxide extracted from the atmosphere, to produce synthetic gasoline and jet and diesel fuel at acceptable cost. This eliminates the need to extract fossil fuels, which when combusted, release large amounts of carbon dioxide into the atmosphere, further warming the planet.

Finally, Chapters 6 & 7 describe two very promising big projects based on the construction of large, thermally insulated, low cost ice structures. Using ocean thermal energy conversion (OTEC) plants operating on large ice islands, in the world oceans, as described in Chapter 6, can produce thousands of megawatts of low-cost electric power and billions of gallons of fresh water per day.

OTEC power generation using LISA ice islands, in combination with electric power from space solar power satellites, can provide virtually all of the world's electric power needs, cleanly, without greenhouse gas emissions from fossil fuels, preventing global environmental catastrophe and reducing damage to public health from pollution.

Large amounts of clean freshwater provided by OTEC LISA ice islands will help to maintain adequate water supplies for food crops and inhabitants in regions that are suffering from major drought. Access to adequate amounts of clean fresh water is vital for a good standard of living.

The seventh big project describes the use of thermally insulated LISA ice structures to protect the world's coastlines from increasingly severe storm surges tsunamis and rising sea levels. Storms like Hurricane Sandy and Katrina, together with Fukushima and other tsunami disasters, have cost hundreds of billions of dollars of damage and hundreds of thousands of deaths.

Researchers at the National Academy of Science have forecasted that by 2100 AD, 400 million people could be flooded annually by rising sea levels, with an expected annual economic loss of as much as 9.6 percent of global GDP - many trillions of dollars per year.

Environmentally acceptable LISA ice barriers that will prevent these enormous economic losses, and many thousands of deaths and injuries, and can be built quickly at very low cost.

When there are competing technologies which might accomplish the objectives of a Big Project, it is important that the competing technologies be tested and evaluated on a level playing field, and not judged on the basis of political or commercial pressures. In the face of climate change and economic challenges, we believe these 7 Big Projects have the potential to secure a bright future for the United States and the World as a whole, but only if our leaders have the vision to pursue them.

Figure 14 View of Erie Canal by John William Hill, 1829

Figure 15 Linking of Transcontinental Railroad at Promontory Point, Utah 1869

Figure 16 Panama Canal Scenes-President Theodore Roosevelt at Controls of Steam Shovel, 1906

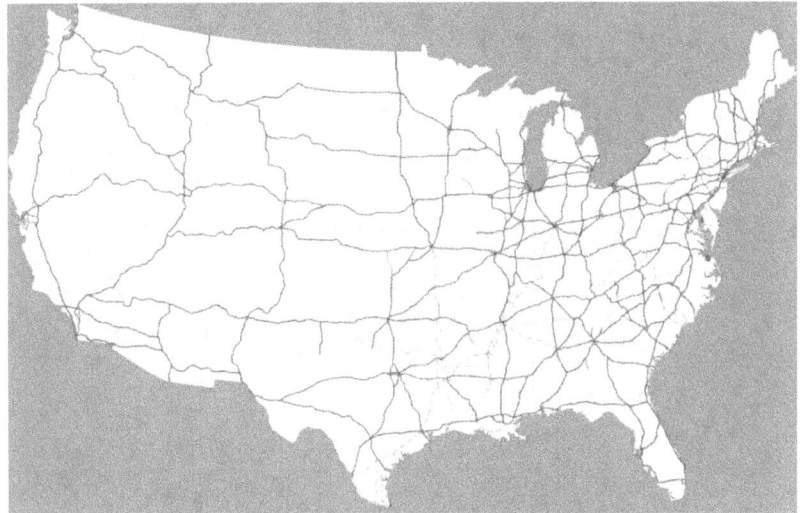
Figure 17 Map of US Interstate Highway System

Figure 18 Astronaut and US Flag on the Moon

Figure 19 Idaho National Laboratory's Advanced Test Reactor (ATR). Powered up, the fuel plates can be seen glowing bright blue. The core is submerged in water for cooling.

Figure 20 Hoover dam from air
From Wikimedia Commons, the free media repository

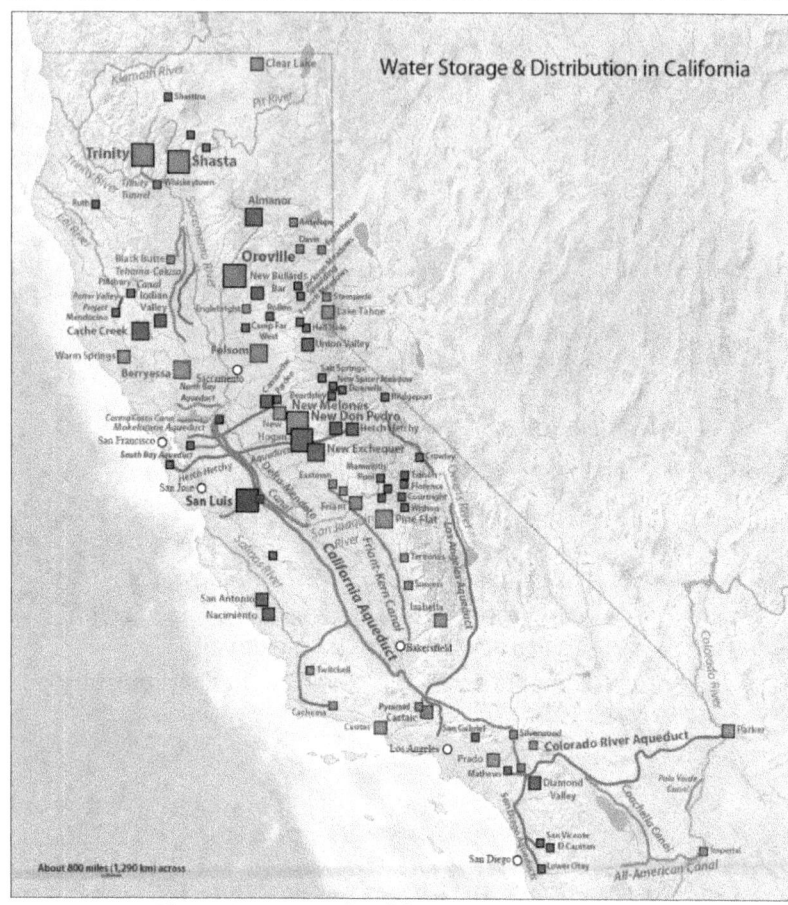

Figure 21 Map of California State Water Project

California's interconnected water system serves over 30 million people and irrigates over 5,680,000 acres (2,300,000 ha) of farmland. As the world's largest, most productive, and most controversial water system, it manages over 40,000,000-acre feet (49 km3) of water per year. From Wikipedia

Chapter One

The National Maglev Network

James Powell, Gordon Danby, Robert Coullahan, Fletcher Griffis and James Jordan

"This land is your land

This land is my land

From California to the New York Island

From the Redwood Forests to the Gulf Stream Waters

This Land was made for you and me" -- Woody Guthrie

Why American Needs the National Maglev Network

For many reasons, America's present transport systems, which primarily rely on fossil fuel, are not sustainable and will not be able to meet our future needs. Within this Century, known oil reserves will be exhausted and the increasing carbon dioxide concentration in the atmosphere will result in global environmental disaster, unless humanity eases its dependence on fossil fuels, as discussed in the Prologue.

The realities of America's present transport systems are described in the next section of the chapter – the increasing amounts of transport required to sustain our economy and standard of living, the incredible numbers of horrible deaths and injuries on our Nation's highways, the trillions of dollars we spend annually on transport, our growing congestion and travel delays, and so on.

The solutions proposed for long term sustainable transport – biofuels, hydrogen cars, electric autos, and high speed rail are not practical and will not meet America's future transport needs. Biofuels compete with food production and could only satisfy a tiny fraction of our transport fuel needs. Hydrogen cars would require enormous amounts of electric power to manufacture the hydrogen, and would be extremely dangerous to drive on the highways. Electric cars are limited in range, very expensive, and pose reliability problems for large-scale operations on highways. High speed rail (HSR) systems operate in other countries, but travel costs are very high, and require major government subsidies for construction and operation. Even in countries with HSR systems, they only supply a small fraction, less than 10%, of the countries' transport needs. In America, because of its much larger geographic size and much lower population density than in countries like Japan, China, France and the rest of Europe, HSR would play an even smaller role, and with only a few systems such as Boston to Washington, DC and Los Angeles to San Francisco that would provide much less than 1% of US passenger mile transport.

America's future will be the National Maglev Network (Figure1.1). Built along 29,000 miles of the U.S. Interstate Highway System, it will transport passengers, highway trucks, and autos at 300 mph between Americas 174 metropolitan areas having 250,000 or more people. Over 70 percent of America's 312 million population will live within 15 miles or less from a Maglev station, from which they can travel to any other Maglev station in the continental United States.

The Senate Environment and Public Works Committee, chaired by the late Senator Daniel Patrick Moynihan, proposed an idea of a new committee member, Senator Harry Reid of Nevada, to use the rights-of-way of the Interstate Highway System to erect low cost elevated Maglev guideways (This concept illustrated in (Figure 1.2).

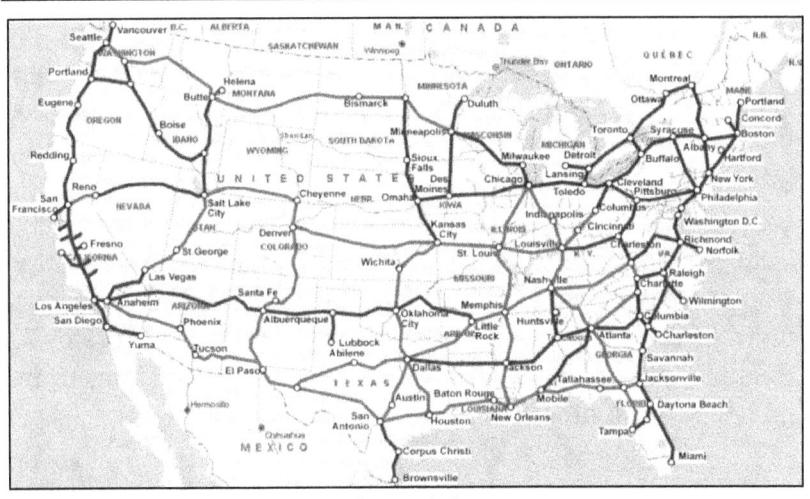

Figure 1.1
National Maglev Network
Population and States Served

Senator Reid became the Majority Leader of the U.S. Senate and he agreed to meeting with Dr. Powell in the Summer of 2008 and seized the moment of the 2008 elections to alert Candidate Barack Obama to the potential of using the

Maglev Network	States In Network	Population of States in Network (millions)	Population Living Within 15 Miles of Stations (millions)	Route Miles in Network
First, Second and Third Waves Completed	48 plus Toronto, Montreal & Vancouver	315 includes Toronto, Montreal & Vancouver	232 includes Toronto, Montreal & Vancouver	29,000
74% of population in States live within 15 Miles of a Station				

Powell and Danby Maglev to increase the efficiency and capacity of the Interstate Highway System to upgrade the economic efficiency of the national logistics system by deploying the 2nd generation superconducting Maglev for passengers, Maglev vehicles for roll-on, roll-off highway trucks (Figure 1.3) and Maglev vehicles for autos (Figure 1.4) with their passengers. All travel on the same guideway at 300 mph. If a Maglev vehicle is scheduled to stop at an off-line station to load/unload passengers, autos, or trucks, it will electronically switch at high speed from the main line to a secondary guideway that leads to the station. After loading /unloading, the Maglev vehicle then accelerates, switches back to the mainline at high speed and heads for its next stop. No cumbersome, low speed mechanical switches. Maglev vehicles can by-pass stations at full speed that they are not scheduled to stop at. Maglev vehicles can travel at full speed as individual units directly to their destination, without having to stop at multiple stations along the route, a necessity for long trains pulled by locomotives.

Figure 1.2
Depiction of Interstate Highway Rights-of-Way being used for Maglev guideways operating with both a freight carrier to left and passenger vehicle to the right side.

Figure 1.3
Artist's depiction of loading a roll-on, roll-off Maglev vehicle carrier

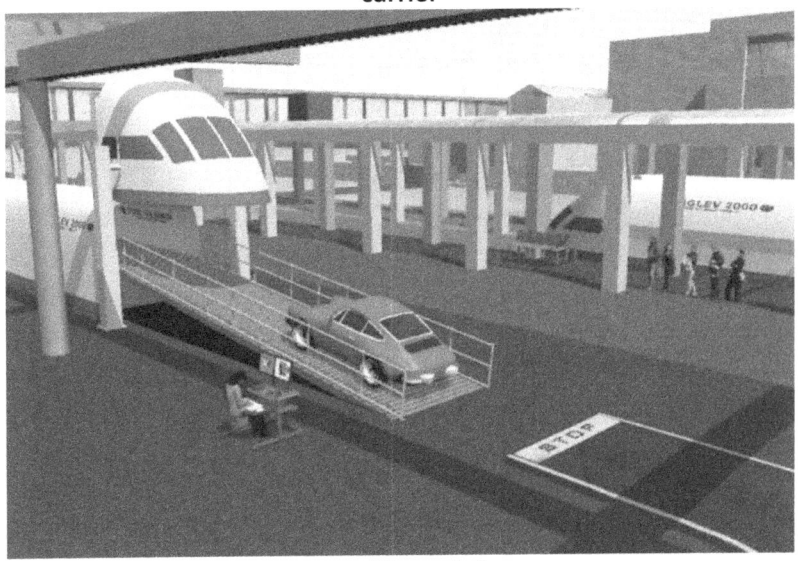

Figure 1.4
Artist's depiction of a Maglev auto carrier loading a double-deck carrier.

The cost for travel on the National Maglev Network Network? Considerably less than travel on highways, using autos and freight trucks, and less than by air or passenger rail.

Freight rail remains the cheapest in cents per ton-mile, but present revenues for highway freight trucks are much greater than for freight rail, 100s of Billions of dollars per year for truck freight compared to 10's of Billions for rail freight. The reason why truck freight revenues are much greater? Trucks deliver their loads in much shorter times and pickup/deliver loads much more conveniently. Today, high value freight goes by highway truck – 10.5 Trillion dollars' worth per year – while low value freight – coal, iron ore goes slowly by rail. Tomorrow, intercity highway trucks will go 300 mph on Maglev, instead of 50 mph on highways, at considerably lower costs.

The safety and environmental benefits of Maglev? Enormous! The National Maglev Network will save many 1,000's of lives and many 100's of 1,000's of injuries now happening in accidents (Figure 1.5) on America's highways every year. It will greatly reduce the 5 Billion barrels of fossil fuel we now use for transport and the 1.8 Billion tons of CO_2 greenhouse gas emitted from our tailpipes and jet engines. It will also greatly reduce the damage to our hearts and lungs from the pollutants and microparticles emitted by our 230 million automobiles and 10 million trucks.

The benefits to the economy and our quality of life will be tremendous. By taking trucks and autos off the road, not only will the National Maglev Network substantially reduce highway deaths and injuries, it also will greatly reduce highway traffic congestion (Figure 1.6) and delays, which today cost the US economy 100 Billion dollars annually. This, plus reducing the 500 Billion dollars now spent on medical expenses, insurance, health damage and lost income from highway accidents, plus Maglev's considerably lower cost per passenger mile and truck ton mile, will greatly benefit the US economy, saving each of us more than $1,000 per year.

Figure 1.5

Figure 1.6

And the reduction in everyday stress on all of us? Much shorter trip times, less worry about accidents, more comfortable and convenient travel, less damage to personal health – it's difficult to quantify these benefits in dollars, but they will be enormous.

Working in tandem with the National Maglev Network, there will be many Maglev Public Transit Systems that operate in America's metropolitan areas. The 2nd Generation Maglev-2000 System described later in the chapter not only is the basis for the National Maglev Network, but also can provide faster, better, cheaper Public Transit.

The 2nd Gen Maglev System has the unique capability to be adapted at very low cost to **existing** railroad and subway tracks for travel by magnetic levitated and propelled passenger vehicles. The adaptation can be done at night and low travel periods, without interfering with conventional train and subway schedules, during times when travel by the existing equipment is infrequent. The adaptation for Maglev travel is easy and quick. Simple panels containing loops of ordinary aluminum conductor wire are attached to the crossties of the railroad and subway tracks. Conventional trains and subway cars can continue to use the tracks after they have been adapted for Maglev, if desired.

The Benefits of Maglev Public Transit? Lower operating costs – much less maintenance required for tracks and vehicles, more energy efficient, increased employee efficiency and productivity, more convenient and more frequent service, shorter trip times, much lower government subsidy requirement, low fares, and much more comfortable and healthier travel.

As an example, Figure 1.7 shows a map of the Long Island Railroad (LIRR) System, the largest commuter rail system in the United States. It carries 280,000 passengers per day on weekdays, with a total of 81 million passengers per year – 3 times the total Amtrak ridership for all America. The average LIRR fare cost paid by passengers is 26 cents per passenger mile; the actual average cost per passenger mile is 80 cents, with the difference of 54 cents per passenger mile paid by government subsides. With the Maglev LIRR, the government subsidy will be much less.

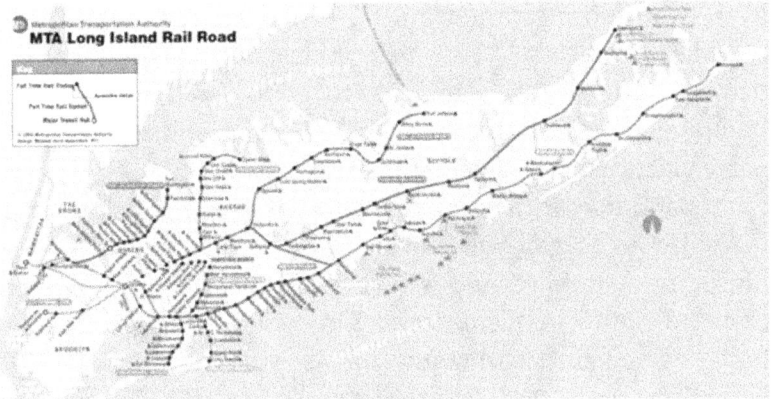

Figure 1.7

With a Maglev LIRR, trip times will be much shorter. The average speed of LIRR trains is about 30 mph – a result of the slow acceleration and deceleration of conventional long trains of many cars, and the requirement that the train stop at many stations along its route. Maglev LIRR vehicles will travel as individual units, able to accelerate and decelerate much faster, like ordinary automobiles, and able to travel past stations at full speed that they do not have passengers for.

Riders on the Maglev LIRR will love it. Trip times a factor of 2 shorter. Babylon to Montauk, a distance of 79 miles, today's travel time is 2 hours 22 minutes an average of 33 mph. On Maglev LIRR, it would be 1 hour 11 minutes, an average of 66 mph. There are presently 6 long trains per day on the Babylon – Montauk Branch. With Maglev LIRR, it could be 20 or more vehicles per day for the trip, much more convenient service. And, no more noisy, bumpy, and swaying rides. Just quiet, comfortable, no vibration – like sitting in a chair in the living room.

Another example. Figure 1.8 shows a New York City Subway car. The NYC Subway System is a marvel. It transports 6.5 million passengers daily. NYC's annual ridership is 2.4 Billion, 1/4th of the 10.4 Billion total annual US transit ridership for all modes – commuter rail, subways, and buses.

However, as anyone who has ridden the NYC subway knows, it is not the most pleasant ride. Noise levels are astronomic, reaching 100 decibels at some stations, with possible hearing damage. Riders are jammed together in very crowded, bumping and swaying cars, breathing in steel dust and other particulates from erosion of rails and brakes.

Adaptation of the NYC Subway, and other transit systems in the US for Maglev will provide much better ride quality – no noise, no bumping and swaying of the transit cars, less crowded, more frequent service, and much cleaner air – no brake or rail dust to breathe in. As with the Maglev LIRR, operations will be cheaper and more efficient, and maintenance will be much less, enabling substantial reductions in government subsidies for public transit.

Figure 1.8

The next section, The Realities of US Transport, shows the sad state of American transport and why it is not sustainable in the long term. Following it is the description of the National Maglev Network and Maglev Public Transit Systems – and how they work, their performance and cost, and the schedule for their implementation.

The Realities of US Transport

Today, US transport is a big mess, and it's only going to get worse, much worse, in the years ahead if we continue on our present path. The realities can be seen in the list of DOT statistics (1) given below.

We spend an enormous amount on transporting people and goods – 1.5 Trillion dollars per year, 10% of US Gross Domestic Product, $8,300 per household, as much as we spend on food plus clothing.

On average, Americans travel 14,400 miles per year, more than halfway around the World. Sadly, it's not "See the World" travel. 88 percent (12,600 miles) is on congested bumpy highways with lots of potholes, or jammed together in noisy public transit buses, subways, and commuter rail cars. 12 percent (1,730 miles), on crowded airplanes that are often late. And travel on slow, bumpy intercity trains? 20 miles per year per person.

And the highways are very dangerous – 32,367 persons died on America's highways in 2011, with 2.2 million injured. The medical, insurance, and health damage cost? $500 Billion per year.

Americans own 230 million cars, with an average of 0.83 cars per person in our population of 312 million. On average, each American travels 11,500 miles per year on our highways. Today, congestion delays are estimated to cost the US $100 Billion dollars per year. In 2040 AD the DOT projects that high congestion will increase by 366 percent. In 2035 on the 1,381 mile I-5 highway from San Diego to the Canadian border north of Seattle, 95% of the 550 miles of urban segments will be congested, with 85% of the rural segments congested. Traffic flow on the I-5 Highway will be enormous, with a maximum of 600,000 vehicles and 70,000 trucks per day.

Highway trucks are a vital part of America's transport network. In 2011, highway trucks moved 11 Billion tons of goods (35 tons per capita) worth 10.5 Trillion dollars (66% of US GDP), at a cost of 500 Billion dollars annually for truck operations. And truck transport will almost double by 2040,

Average Daily Long-Haul Freight Traffic on the National Highway System 2002

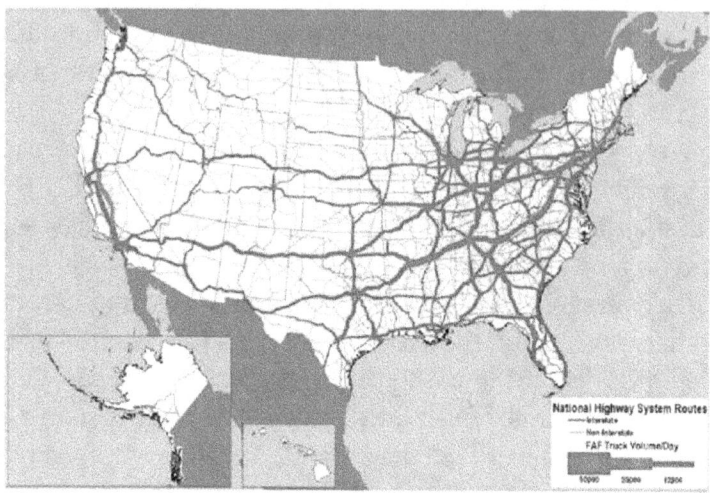

Average Daily Long-Haul Freight Traffic on the National Highway System 2035

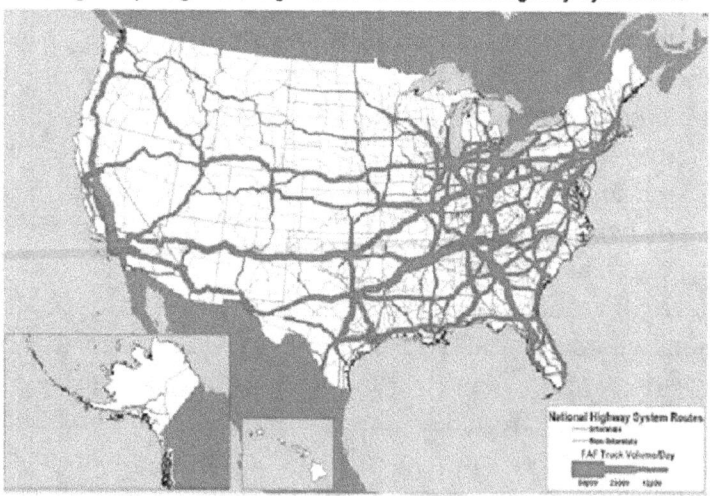

Figure 1.9

with projected movement of 17 Billion tons worth 21 Trillion dollars. Figure 1.9 compares the present US truck traffic flow with truck traffic flow in 2035.

Overhanging the problems of increasing congestion, greater travel, deteriorating highways, bridges, and tunnels and the many thousands of highway deaths and many billions of injuries that will occur in the next 20 years, is the inevitable prospect of less fossil fuel for America's transport system. The US consumes 7 Billion barrels of oil per year, 5 Billion of which goes for transport.

As discussed in the Prologue, the known reserves of oil will run out in only 50 years, at the present rate of World oil consumption. If, as is very likely, developing countries like China and India demand more oil and the consumption rate increases, oil reserves could run out in as little as 20 years.

What has been proposed to meet our future transport needs?

America needs to transform its future transport network to systems that do not depend on oil, and that enable safer less deadly, lower cost, faster, less congested, and less stressful travel for people and goods.

Sadly, in contrast to the big transport projects of the past, America now seems to have lost its courage to undertake new big transport projects. So far, we have no real plans that will achieve a better future transport network. Rather, highways continue to deteriorate, congestion gets worse, more people die and are injured in accidents, costs continue to climb and we remain completely dependent on fossil oil fuel.

What has been proposed to meet our future transport needs? Not much. And much of what has been proposed is not workable.

Four technology fixes have been proposed to meet future US transport needs.

- Biofuels as a substitute for fossil fuels for motor vehicles
- Hydrogen powered motor vehicles
- Electric powered motor vehicles
- High Speed Rail (HSR)

As we will show below, Biofuels are not practical because they conflict with food supply, hydrogen powered cars have major safety issues, electric powered vehicles are practical but are limited in capability, and High Speed Rail, while technically practical, would have only an extremely tiny and expensive role in America's transport future.

In the area of vehicle fuel, a few years ago, biofuels were all the rage. To think that biofuels will solve our dependence on fossil oil and help to mitigate global warming is silly. Follow the numbers. 40% of the US corn crop was sold in 2013 to make 13.9 Billion gallons of ethanol fuel (2, 3 Chapter 5 ref.9). It takes 1.5 gallons of ethanol to have the same fuel energy as 1 gallon of gasoline, so that 13.9 Billion gallons of ethanol actually equals only 9.3 Billion gallons of gasoline. The US consumes 200 Billion gallons of gasoline, diesel, and jet fuel annually, so 40% of our corn crop yields less than 5% of our transport fuel needs.

Hmm. How much farmland would it take to produce the equivalent of 200 Billion gallons of oil based fuel? A lot. At 350 gallons of ethanol per acre (2), equivalent to 230 gallons of gasoline, it would take 200 Billion/230, or 870 million acres. Total US farmland is 300 million acres, only 1/3rd of the required 870 million acres. Well, if we want to use ethanol for vehicle fuel, we can stop eating. However, we'll only be able to travel 1/3rd of the distance we travel today.

Clearly, biofuels cannot meet our and the World's transport needs. Cellulosic ethanol made from corn stalks, plants, trees, etc. has been invoked as a way to obtain ethanol, but it is still experimental. However, even if cellulosic ethanol proves possible, it still takes land to grow the cellulosic material, land that could be used to grow food. In a World of more than a Billion starving and malnourished people who need food, it is heartless to use food to make fuel for automobiles.

Along with the Biofuels hype a few years ago, there was a lot of excitement about hydrogen cars. "Only water comes out of the tailpipe!" "Hydrogen fuel cells are very efficient!" and so on. Unfortunately, there are 2 fundamental problems with hydrogen cars – where do we get the hydrogen, and safety. First, free hydrogen doesn't exist in nature. It either has to be manufactured from fossil fuels, i.e., natural gas, as in the production of hydrogen for ammonia fertilizer, or produced by electrolyzing water using massive amounts of electric power.

To use fossil fuels to make hydrogen for autos would be incredibly stupid. To make it by electrolyzing water using electric power from fossil fueled power plants, which today generate more than 80 percent of America's power is also incredibly stupid. It would greatly increase fossil fuel consumption and greenhouse gas emissions.

The electric power for hydrogen fuel for motor vehicles would have to come from nuclear and renewable sources – wind, solar, hydro. And, a hydrogen transport economy would need a lot of power. Based on EPA analyses electric cars use approximately 0.36 Kwh of electric power per mile of travel from their battery packs (4 same as chapter 5 ref. 11) with power transmission and distribution losses and battery inefficiencies, to extract 0.36 Kwh per mile from our electric car battery pack would require generating about 0.5 Kwh/mile at the electric power plant.

The electric power per mile for hydrogen cars would considerably greater due to power losses in electrolyzing water to make the hydrogen, compressing the hydrogen to 5,000 psi for on-board storage in cars or liquefying it to very low temperatures, 20 degrees Kelvin, for storage of liquid hydrogen in your car tank, plus the losses in the hydrogen fuel cells on the car that generate electric power for operating it. All told, the primary electric power generated for a hydrogen car will be approximately 1.0 Kwh/mile, about 2 times that for an electric car.

Total passenger vehicle miles in the US (autos, SUV's, light trucks in 2011 were 2.6 Trillion. Total US electric generation was 3.9 Trillion Kwh. If all 230 million cars were hydrogen powered, American would have to boost its power generation by 2.6 Trillion Kwh, a 66% increase – equivalent to adding 370,000 megawatts of wind, solar, or nuclear power – that's 370 new 1,000 nuclear reactor power plants.

Plus the safety issue. Visualize 230 million hydrogen cars driving 70 mph on America's congested highways, with each car holding either a tank of compressed hydrogen gas at 5,000 psi, or a cryogenic tank of liquid hydrogen at 20 degrees Kelvin. In a collision, if the hydrogen escapes into the atmosphere, it could explode. For a hydrogen fueled car with a 300 mile range, the explosive force would be equivalent to many pounds of TNT.

For the reasons given above, electric cars will probably be the choice for much of America's motor vehicle transport needs in a non-fossil-fuel future. However, they do have significant limitations.

- limited range
- run out of power on highway
- long charging times
- availability of charging stations
- high cost

The Chevy Volt has an EPA range of 35 miles and the Nissan Leaf, 75 miles. The Tesla Model S EPA range is 265 miles with an 85 Kwh battery pack. However, these ranges relate to ideal conditions – not in cold weather where heating of the vehicle interior would be necessary, not in hot weather where air conditioning is necessary, or on congested highways where there are long delays. Non-ideal conditions will substantially reduce electric car range. And battery capacity diminishes with age and recharges.

If an electric car runs out of power on a highway before it can find a charging station to recharge, it is a problem. Gasoline fueled cars can easily get a can of gas. An electric car would have to be towed to a charging station at considerable expense, and disrupt normal traffic flow.

There are on the order of 250,000 gasoline stations in America. It is likely that there will be a far smaller number of charging stations, making them harder to find and access. And charging times will be much longer. It only takes 2 or 3 minutes to fill one's gas tank, but even so, there often are multiple cars at a gas station at the same time. With charging times of 20 or 30 minutes and fewer charging stations, there will be many more cars waiting to get charged.

Finally, electric cars are more expensive than gasoline fueled cars, a burden for lower income families and individuals.

The National Maglev Network will enable fast, very low cost non-highway, long distance travel for passengers, trucks, and autos. For long distance trips, electric cars and trucks could travel on Maglev vehicles, instead of needing a very large battery pack, or charging at multiple stations on the journey. Plus, they could be electrically charged while on the Maglev car carrying vehicle. Electric cars and trucks could then operate locally with smaller, less expensive battery packs and more convenient charging locations.

Second, the Maglev Public Transit systems will substantially reduce the need for electric cars. Today, about 75% of people going to work get there in their own cars, because the public transit systems are slow, crowded, expensive and inconvenient, with poor service frequency. With Maglev Public transit, transport will be much more attractive, with a much greater percentage of workers using it rather than driving to work.

For locations where Maglev service is not available, motor vehicle transport will be necessary, either electric, or gasoline/diesel fueled. For the fueled vehicles the fuel would not have to use fossil fuels, however. Instead, as described in Chapter 5, using low cost beamed solar power from space, synthetic gasoline/diesel fuel can be manufactured from carbon dioxide extracted from the atmosphere and hydrogen generated by electrolyzing water. The cost per gallon of synthetic fuel will be comparable to the present cost of gasoline/diesel fuel from fossil oil.

While electric cars will be significant in America's transport future, High Speed Rail (HSR) will not. It is difficult to understand why politicians and policymakers think that HSR will benefit America, when in fact it will hurt it, by the government spending huge sums on HSR routes that would do virtually nothing for US transport.

At the present rate of 14,400 passenger miles per capita, during an 80-year lifespan, Americans travel the equivalent of 1.1 million miles, 46 times around our 25,000 mile circumference World. Traveling on intercity rail (Amtrak) at the present, an average of 20 miles per capita per year, would be equivalent to 1,600 miles, one round trip between Chicago and New York City, during a person's 80 year lifespan.

Building a few HSR corridors would increase intercity passenger rail traffic, but it would still be insignificant. Today, 8 million passengers per year fly between San Francisco and Los Angeles, while 3 million fly between Boston and Washington, DC. If they all were to switch from air to HSR, a very unlikely scenario, intercity rail passenger traffic would increase by 11 million passengers per year, from the present 26 million passengers to 37 million. The average rail travel for Americans would increase to about 30 miles per capital per year. Wow! One round trip between New York City and Miami, Florida in an 80 year lifetime.

The HSR construction expense would be enormous. A projected (before cost overruns) of $150 Billion for Boston to Washington, DC and 70 Billion for San Francisco to Los Angeles. Many Billions more in operating subsidies over the years. Almost every HSR route in the World requires substantial government subsidies.

And the cost per passenger mile. Riders on the Acela between NY City and Washington, DC pay $1 per passenger mile. At that rate a San Francisco – Los Angeles round trip would cost $800. Not many could afford it – just the top 1 percent – the 99 percent would have to travel some other way. But, they would still have to pay taxes to subsidize the HSR line that carried the 1%.

One wonders. Why don't our politicians and policy makers do the simple math, to see how useless HSR would be for America? We are a very big country in size, with a very low population density compared to countries that have HSR routes. Japan has 337 persons per square kilometer, a factor of 10 higher than the United States 32 persons/square kilometer, Germany has 229 persons/square kilometer, a factor of 7 higher. China has 142 persons/square kilometer, a factor of 4 higher. Even with countries with much higher population densities, HSR accounts for only a small percentage of total passenger travel – e.g., in France, average per capita HSR travel distance is 400 miles, compared to 7,600 miles on the highway.

Rather than the government spending hundreds of Billions of dollars to build and subsidize a few unconnected HSR lines, US politicians and policy makers should have the government test and certify the 2nd Generation Maglev System instead of spending hundreds of millions of dollars on HSR. Testing and Certifying 2nd Generation Maglev would cost much less, only 600 million dollars. Once certified, private investors would build and operate the National Maglev Network and Maglev Public Transit Systems.

The 600 million dollars spent for test and certification of the 2nd Generation Maglev System, the basis for the National Maglev Network and Maglev Public Transit Systems is only $2 per capita for America's 312 million population. Spent over a 5 year test and certification period, that's 40 cents per year – about 1/2 the cost of a small candy bar. Over the 5 year period, a total of 2 candy bars. The annual savings in transport costs per American will be greater than $1,000 dollars per year, amounting to well over $80,000 for an 80 year life span. Plus all of the other benefits – much safer, much faster, more comfortable, and much less pollution. A real bargain. Worth giving up two candy bars for 5 years.

Figure 1.10

The National Maglev Network and Maglev Public Transit Systems: Performance, Cost, and Schedules

Detailed descriptions of the 2nd Generation Maglev 2000 System, the National Maglev Network, and Maglev Public Transit are given in two books, "The Fight for Maglev" and "Maglev America," by James Powell, Gordon Danby, and James Jordan. The books are available at Amazon.com.

Maglev transport was a far out dream of Robert Goddard, the American pioneer of rocket transport, and other scientists and engineers, until James Powell and Gordon Danby published their 1966 breakthrough paper on Superconducting Maglev. Their paper sparked Maglev development programs in many countries, in particular, Japan and Germany.

Japan has developed Powell and Danby's 1966 Maglev inventions in their 1st Generation Maglev System, now operating in Yamanashi, Japan (Figure1.10). The demonstration route has carried over 100,000 passengers, at speeds up to 360 mph with an estimated running distance of hundreds of thousands of kilometers.

Japan plans to extend the Yamanashi line to become a 300-mile route between Tokyo and Osaka, which will carry 100,000 passengers daily with a trip time of 1 hour.

Powell and Danby have continued working to evolve 1st Generation Maglev to an even better, more capable, 2nd Generation Maglev System much like airplanes went from the 1st Generation propeller driven Ford Tri-Motor and DC-3 in the 1930's to today's modern jet airliners – Boeing's 767, Airbus' 380, etc.

The 2nd Generation Maglev System has the following unique capabilities not possessed by 1st Generation Maglev, that are critically important for the National Maglev Network and Maglev Public Transit for congested urban areas.

The 2nd Gen Maglev can:

1. Carry trucks and autos as well as passengers. Intercity truck revenues are much greater than intercity rail passenger revenues, more than $300 Billion annually, compared to 2 Billion annually for Amtrak rail

passengers. The much greater revenues will enable private financing, with no government subsidies.

2. Operate in levitated mode on **existing** rail and subway tracks that have been adapted at very low cost for Maglev – a capability not possible for 1st Generation Maglev Systems. This capability enables 2nd Generation Maglev vehicles to transition from 300 mph guideways to operate at lower speed on **existing** rail tracks when they come into suburban area. This eliminates the need to construct new guideways in suburban and urban areas – an expensive and disruptive process, like Boston's Big Dig. Conventional trains and subway cars can still use the existing tracks after the adaptation process with appropriate scheduling.

3. Electronically switch at high speed from the main guideway line to offline stations for loading/unloading. This eliminates mechanical switches, the method used in 1st generation Maglev. Mechanical switching at low speed significantly reduces the average speed of Maglev vehicles on the main line.

4. Erect elevated high speed guideways at much lower cost than 1st Generation Maglev. The 2nd Gen Maglev monorail guideway beams and piers can be mass produced in factories with all their attached equipment, and then transported at very low cost by highway trucks to the construction site, to be rapidly erected by conventional cranes on pre-poured footings. Very little field construction work is required for 2nd generation Maglev, which is typically much more expensive than factory production. 1st Gen Maglev systems require much more field construction work.

With cost efficient, mass production of Maglev components – beams, piers, aluminum loop panels, propulsion loops, and electronic as integrated units in factories that can be trucked to sites, the construction cost per 2-way mile of the 2nd Generation Maglev-2000 elevated guideways is only $30 million per 2-way mile. This does not include the cost of Maglev vehicles, stations, and any required land modifications.

The cost of adapting existing rail and subway tracks for Maglev operations is very low cost, about 5 million dollars per 2-way mile. The panels containing aluminum loops for levitation and propulsion are attached to the crossties of the existing trackage during periods of low travel by the existing conventional trains or subway cars. Field construction is not required.

The minimization of field construction, besides reducing costs, also enables both Maglev high speed elevated guideways for intercity transport and moderate speed Maglev public transport to be installed much more rapidly that systems that require substantial field construction. This is very important in rapidly implementing the National Maglev Network and Maglev Public Transit.

To illustrate how rapidly the 29,00 mile National Maglev Network could be implemented we assume a start date of January, 2019. The start date will depend on when and how aggressively the US Government initiates a testing and certification program, but it could be carried out in only 4 years, with an aggressive effort by the government starting in January 2015.

The National Maglev Network would be built in 3 waves, each wave taking 5 years. The First Wave shown in Figure 1.11 and Table 1.1 would be constructed on the US East and West Coasts where most Americans live.

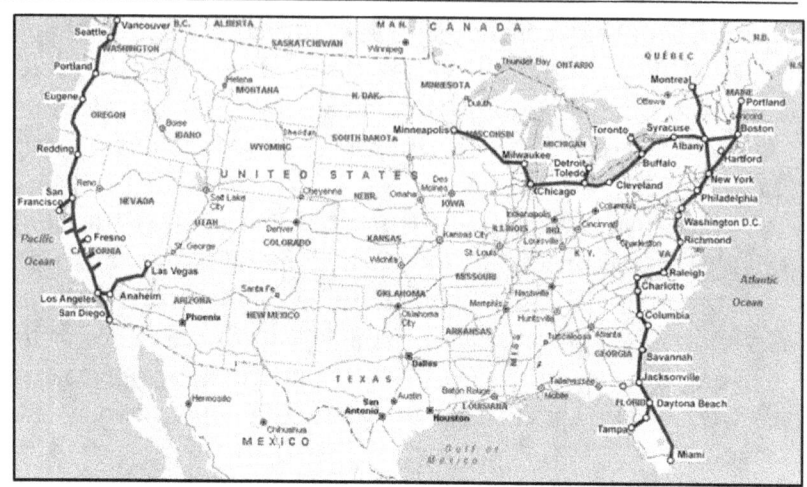

Figure 1.11: First Maglev Wave to Be Built 10 years from Start

Table 1.1 Population and States Served in First Wave

Maglev Network	States In Network	Population of States in Network (millions)	Population Living Within 15 Miles of Maglev Stations (millions)	Route Miles in Network
East Coast/Midwest Network	45 MN, WI, IL, IN, OH, PA, NY, MA, VT, NH, MN, ME, RI, DE, MD, VA, DC, NC, SC, GA, FL +Toronto & Montreal	175.8 (includes Toronto, Montreal)	102.9 (includes Toronto, Montreal)	4,224
West Coast Maglev Network	CA, NV, OR, WA & Vancouver, Canada	50.9 (includes Vancouver)	43.5 (includes Vancouver)	2006
Total for First Maglev Wave (Both Networks)	26 States Plus Toronto, Montreal & Vancouver	226.7	146.4	6230
65 % of population in States Served by the Networks live within 15 Miles of a Maglev Station				

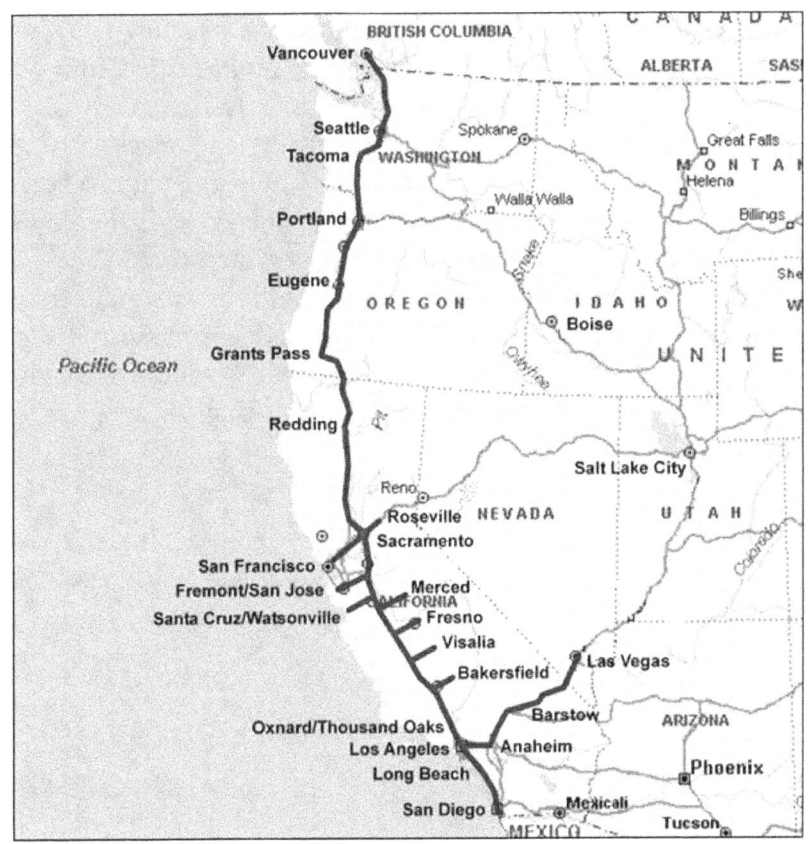
Figure 1.12

Figure 1.12 shows a more detailed map of the West Coast portion of the First Wave. It would connect 51 million persons to the West Coast Maglev System – 36.5 million in California, 3.8 million in Oregon, 6.5 million in Washington State, 2.6 million in Nevada, and 1.3 million in Vancouver, British Columbia. 85% of the population would live within 15 miles of a Maglev station. The San Diego to Vancouver route along Interstate 5 would be 1,380 miles in length. Side Maglev routes along I-15 to Nevada plus side routes in California would increase the West Coast Systems to a total of 2,000 route Miles,

Table 1.2 compares trip times for Maglev, air and highway travel for 4 illustrative trips – San Diego to San Francisco, and

Los Angeles to Las Vegas. Maglev has the shortest trip time except for air travel between San Diego and Seattle. However, Maglev is very competitive with air travel for those trips, and will be even faster when airplane boarding and check-in times are longer than 1 hour. Taking into account the very frequent Maglev service – along with flight delays and the lower frequency of airline service, plus lower fares on Maglev, it appears very likely that Maglev would be the choice of travelers. Table 1.3 shows the projected cost using Maglev transport for the 4 illustrative trips on the West Coast Maglev System. The costs are substantially lower than by air , rail, or highway.

Table 1.2

Compares trip times for Maglev to trip times by air, rail and highway.

Illustrative Trip	West Coast Maglev Passenger, Auto , or Truck	Air [1] Passenger	Rail Passengers [2]		Highway, Auto & Truck
			Conventional Rail	High Speed Rail	
San Diego to Seattle	4 Hrs 30 min	4 Hrs	21 Hrs	9 Hrs 40 min	25 Hrs 15 min
San Francisco to Los Angeles	1 Hr 45 min	2 Hr 30 min	6 Hrs	3 Hrs 45 min	9 Hrs 40 min
Portland to San Francisco	2 Hrs 30 min	2 Hrs 45 min	10 Hrs 45 min	4 Hrs 50 min	12 Hrs 45 min
Los Angeles to Las Vegas	1 Hr	2 Hr	4 Hrs 40 min	2 Hrs 10 min	5 Hrs 30 min

1. Includes 1 Hr Pre-Boarding Time at Airport for Check-In
2. Average Speed of 60 mph by Conventional Trains (Amtrak)
3. Average speed of 130 mph by High Speed Rail (French TGV)
4. Average Highway Speed of 50 mph (Including Congestion Delays & Rest Stops

The West Coast Maglev System will substantially reduce highway congestion along the I-5 Corridor. In 2035, traffic flow is projected to increase to 2 times the flow in 2007. Without Maglev, most urban plus rural segments of I-5, will be highly congested, greatly increasing trip times and congestion costs.

Table 1.3
Trip Times & Costs on the West Coast Maglev Network

Illustrative Trip	Trip Miles	Trip Time	One Way Maglev Trip Cost		
			Passenger	Auto w/ Passengers	Highway Truck *
San Diego to Seattle	1260	4 Hr 30 Min	$37.80	$403	$128
San Francisco to Los Angeles	480	1 Hr 45 Min	$14.40	$154	$49
Portland to San Francisco	640	2 Hr 30 Min	$19.20	$205	$65
Los Angeles to Las Vegas	275	1 Hr	$8.00	$88	$28

Basis

280 mph Avg Speed,

Daily Avg Traffic

30,000 Passengers

20,000 Autos w/Passengers

5,000 Highway Trucks

10 Cents/KWh

Unit Capital Cost

30 M$/2-Way Mile for Guideway

5M$/Maglev Vehicle

*$ Per Ton of Load, 30 Tons Load per Truck

Table 1.4

Vehicle Flows and Congestion Along the I-5 Corridor

	Traffic Flow on Corridor			
	2007		2035	
	Avg	Max	Avg	Max
Vehicles/Day	71,000	300,000	150,000	~600,000
Trucks/Day	10,000	35,000	22,000	~70,000
Urban Segments,* % Congestion	65%		95%	
Rural Segments, % Congestion	31%		85%	

*(550 miles of 1381 mile total length are urban segments)

Tables 1.5A and 1.5B show the metropolitan and micropolitan area in California, Oregon, Washington State, Nevada and British Columbia served by the West Coast Maglev System, assuming the following daily traffic flow on the West Coast System with their corresponding fraction of the 2007 traffic flow on the I-5 corridor (Table1.4) would be:

- 30,000 passengers daily (100 per maglev vehicle) (42% of auto/passenger traffic on 2007 I-5)
- 5,000 trucks daily (50% of 10,000 truck traffic on 2007 I-5)
- 20,000 person autos w/passengers daily (10 per vehicle) (71,000 28% of auto traffic on 2007 I-5)

In terms of 2007 traffic flow on I-5, Maglev would reduce auto traffic by 70% and truck traffic by 50%. 80% of the 50 million people served by the I-5 corridor would live within 15 miles of a Maglev station.

Table 1.5A

Metropolitan and Micropolitan Areas in California Served by West Coast Maglev Network.

- California I-5 Corridor Served By Maglev Network
- Plus Side Routes from I-5 (Side Routes Marked by *)

Metro/Micro Area	2007 Population (in millions)
San Diego - Carlsbad - San Marcos	2.98
Santa Ana - Anaheim-Irvine	3.00
Los Angeles - Long Beach - Glendale	9.88
Oxnard - Thousand Oaks -Ventura	0.80
Bakersfield*	0.79
Hanford – Corcoran*	0.15
Visalia – Parkersville*	0.42
Fresno*	0.90
Merced*	0.25
Santa Cruz – Watsonville*	1.80
San Jose – Sunnyvale – Santa Clara*	0.08
Tracy - Paterson	0.51
Modesto*	2.04
Sacramento – Arden – Arcad - Roseville*	2.48
San Francisco – Oakland – Fremont*	1.72
Red Bluff	0.06
Redding	0.18
Yreka	0.10
Total	28.44
California I-15 Corridor	
Riverside – San Bernardino - Ontario	4.08
Victorville – Apple Valley - Hesperia	0.20
Total	4.28
Total California Population Served by West Coast Maglev Network	32.70
Total California Population	36.6
% Served by WC Maglev Network	89%

Table 1.5B
Metropolitan and Micropolitan Areas in Nevada, Oregon, Washington & British Columbia served by West Coast Maglev Network

Nevada I-15 Corridor Served by West Coast Maglev Network

Metro/Micro Area	2007 Population (millions)
Las Vegas	1.84
Boulder City	0.15
Total	1.99
Total Nevada Population	2.56
% Served by WC Maglev Network	78%
Oregon I-5 Corridor Served By West Coast Maglev Network	
Medford	0.20
Eugene	0.34
Corvallis	0.08
Albany	0.11
Salem	0.39
Portland	1.75
Total	2.87
Total Oregon Population	3.75
% Served by WC Maglev Network	76%
Washington State I-5 Corridor to Canadian Border Served by Maglev Network	
Vancouver	0.43
Longview	0.10
Olympia	0.24
Tacoma	0.77
Seattle – Bellevue - Everett	2.54
Mount Vernon	0.12
Bellingham	0.19
Total	4.34
Total Washington State Population	6.47
% Served by WC Maglev Network	68%
British Columbia I-5 Corridor Served By West Coast Maglev Network	
Vancouver	0.54
Burnaby	0.19
Richmond, Surrey, Langley, White Dock	0.57
N. Vancouver	0.04
Total	1.52

Using the unit costs for Maglev transport of passengers, highway trucks and autos (cents per passenger mile, cents per auto mile, and cents per ton mile) shown in Table 1.6, along with existing transport costs, assuming the above Maglev transport volumes, annual savings would be 21 Billion dollars (Table 1.7) in terms of the 2007 AD traffic flow. The projected savings, of course, cannot occur, because it now is 2015, not 2007. When the West Coast Maglev System is operating, with the 2035 projected traffic (Table 1.4), 2 times that in 2007, the Maglev annual transport savings would be 42 Billion dollars (2x21 Billion/year).

Table 1.6
Comparison of Maglev Travel Costs with Costs for other Modes

	Unit Cost	Maglev [1]	Highway [2]	Air [3]	High Speed Rail [4]
Passengers	Cents/passenger mile	3	---	15	50
Autos	Cents/mile	32	40	---	---
Trucks	Cents/ton mile	10.2	30	---	---
Seattle to San Diego Trip (1,260 miles one way) Dollars					
Passengers		$37.80	Travel w/Auto	$190	$630
Auto w/Passengers		$403	$500	---	---
Trucks		$128	$480	---	---

1. Costs include amortization and maintenance of guideway & vehicles
2. Avg cost of operating a car (depreciation, fuel, tires, etc)
3. Avg cost per passenger mile (US Statistical Abstract)
4. Cost/passenger mile for High Speed Rail in Europe

For the 2007 Reference Case and the unit travel costs from Table 1.4 Maglev travel savings are given in Table 1.7.

Table 1.7
Annual Savings in 2007 Transport Cost Enabled by West Coast Maglev Network

	Maglev Travel (B$ per year)	Existing Mode of Travel (B$ per year)
Passengers	0.64 B$	6.6 B$ (Highway & Air)
Autos w/Passengers	4.7 B$	5.8 B$ (Highway)
Highway Trucks	7 B$	21 B$
Total	12.3 B$	33.4 B$
Savings by Maglev = 21B$ per year		

The West Coast Maglev Network is extremely attractive in terms of faster trip times, lower costs, increased safety, greater energy efficiency, and decreased greenhouse gas emissions. A more detailed description of the West Coast Network is given in "Maglev America", available at Amazon.com.

Turning now to the East Coast Maglev Network the other part of the First Wave of the National Maglev Network, Figure 1.13 shows a map of the Northeast-Midwest Section of the East Coast Maglev Network, down to Richmond, Virginia. The Richmond to Miami section is shown in Figure 1.11, the map of the First Wave.

While the Chicago to Albany, New York Maglev route is not geographically part of America's East Coast, it is very important to include it in the Maglev East Coast Network, because of the enormous truck traffic from the Chicago region to New York and other cities on the East Coast.

The Metro/Micropolitan areas served by the Midwest section of the East Coast Maglev Network are shown in Table 1.8 for New York State and Table 1.9 for the other states and Canada served by the Northeast-Midwest Maglev Network. Table 1.10 shows the Metro/Micropolitan areas served by the Southern section of the East Coast Network.

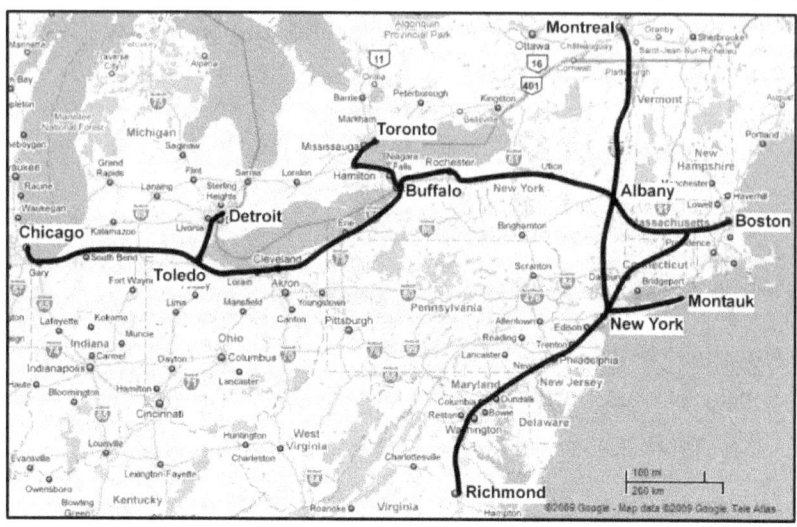

Figure 1.13

Summarizing, the West Coast Maglev Network has 2,108 miles of guideway, costing 63 Billion dollars. 43 million Americans and Canadians live within 15 miles of a Maglev Station on the West Coast Maglev Network. The East Coast Maglev Network has 4,224 miles of Maglev guideway. Cost is 126 Billion dollars. 103 million Americans and Canadians live within 15 miles of a Maglev Station on the East Coast Network. Total population living within 15 miles of a Maglev Station for the First Wave of the National Maglev Network, to be completed in 10 years from Start, would be 146 million.

It's interesting to compare the First Wave with the proposed High Speed Rail lines from Boston to Washington (estimated cost of 115 Billion dollars) and San Francisco to Los Angeles (estimated cost of 68 Billion). Cost for the First Maglev Wave? Comparable. 190 Billion dollars. Population served by the HSR lines? About 20 million riders. Population served by the First Maglev Wave? 146 million. Passenger fare by HSR, about $1 per passenger mile. Passenger fare by Maglev? About 10 cents per passenger mile, enabling a much larger ridership.

Table 1.8
The New York Portion of the Northeast-Midwest Section of the East Coast Maglev Network Routes and Metropolitan/Micropolitan/Cities Served by I-87 and I-90

Segment #	Maglev Route	Route Mileage	Metro/Micropolitan Areas & Cities	Population Served (millions)
1.	I-87, NY Thruway Major Dugan Expressway to Albany	142 miles	NY Metropolitan	
			Bronx County	1.37
			King County	2.53
			NY County	1.62
			Queens County	2.27
			Richmond County	0.48
			Rockland County	0.30
			Westchester County	0.95
			Subtotal	**9.52**
			Newburgh-Beacon	0.10
			Kingston	0.18
			Albany-Schenectady	0.85
			Total	**10.65**
2.	I-90, NY Thruway Albany to Buffalo to Pennsylvania Border	354 miles	Amsterdam	0.05
			Utica Rome	0.30
			Syracuse	0.64
			Auburn	0.08
			Seneca Falls	0.03
			Rochester	1.03
			Batavia	0.06
			Buffalo-Niagara	1.13
			Subtotal	**3.32**
3.	I-87, Albany to Canadian Border	200 miles	Plattsburgh	0.08
			Glens Falls	0.13
4.	I-90 NY Thruway	24		
	Total	**720**		**14.15**

Table 1.8 (continued)

The New York Maglev Network – Nassau and Suffolk Counties Served by I-495 and LIRR Adapted for Maglev			
Elevated Monorail along I-495 from NYC to Riverhead plus travel on LIRR Tracks	150 miles including LIRR Tracks	Nassau-Suffolk	2.76
Total NY Population (2007)			19.4
Total NY Population Directly Served by NY Maglev Network			16.9
% Directly Served by NY Maglev Network (Living within 15 miles of a Maglev Station			87%
Total Route Length	870 miles		

Table 1.9 Extension of the Maglev Network to Regions and States
Table 1.10

Seg ment #	Maglev Route Extension	Route Mileage	Metro/Micropo litan Areas & Cities	Popula tion (millio ns)
		Table 1.9 Continued Beyond New York		
5.	NY/PA Border to Erie (PA), Cleveland & Toledo (OH), Detroit (MI), South Bend & Gary (IN), Chicago (IL) Maglev	I-90 410 miles I-75 60 miles Total 470 Miles	Erie, PA	0.28
			Cleveland, OH	2.10
			Akron, OH	0.70
			Toledo, OH	0.65
			Detroit, MI	4.47
			South Bend, IN	0.32
			Gary, IN	0.70
			Chicago-	7.45
			Lake &	0.87
			Total	18.04
6.	Route I-95 from NY City to Richmond, VA	231 Miles to Washington, DC plus 106 miles from Washington, DC to Richmond, VA = 343 miles	New	0.36
			Princeton (NJ)	6.14
			Trenton (NJ)	0.36
			Camden (NJ)	1.25
			Philadelphia	3.89
			Wilmington	0.69
			Baltimore-	2.67
			Alexandria,	5.31
			Richmond, VA	1.21
			Total	16.1
	Total Additional Mileage for Maglev Extension Routes			1238
	Total Additional Population Served by Maglev			53.8
			Total	5.65

Metro/Micropolitan areas served by the Southern section of the East Coast Network.

State Served by East Coast Network	Maglev Route Miles in State	Metro and Micropolitan Areas Served	State Population Served by Maglev (in millions)	Total State Population, (in millions)
North Carolina	190	Fayetteville (I-95) Raleigh-Cary Durham, Greensboro, Winston-Salem Total	0.36 1.09 0.49 0.71 0.47 3.12	9.22 Total Pop % Served 34%
South Carolina	325	Columbia, Charleston Total	0.73 0.65 1.38	4.48 Total Pop % Served 31%
Georgia	130	Savannah	0.33	9.69 % served 3.4%
Florida	506	Cape Coral-Ft. Meyers	0.41	18.3 Total Pop % Served 79%
		Deltona-Deltona Beach-Ormand	0.50	
		Gainesville	0.26	
		Jacksonville	0.51	
		Lakeland—Winter Haven	0.58	
		Miami - Ft. Lauderdale	5.42	
		Naples-Marico Island		
		Orlando-Kissimmee	2.06	
		Palm Bay-Melbourne, Titusville	0.54	
		Port St. Lucie	0.40	
		Tampa-St. Pete, Clearwater	2.73	
		Sarasota-Brandon	0.69	
		Total	14.43	

Figure 1.14 shows a map of the Second Wave of the National Maglev Network, to be built from year 11 of the National Maglev Network project through year 15. It would connect America's West and East Coast Maglev Networks built in the First Wave, with 3 Transcontinental Maglev routes, located in the Northern part of the US, the Middle part of the US, and the Southern part along the Gulf Coast. Additional routes would be built to connect inland cities to the Second Wave network. Total route mileage would increase from the 6,230 miles in the First Wave to 18,630 miles when the Second wave was completed (Table 1.11). The population living within 15 miles from a Maglev station would grow from the 146 million in the First Wave to 210 million in the Second Wave.

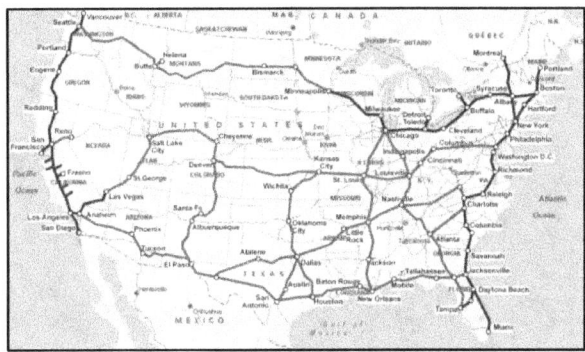

Figure 1.14

Table 1.11: Population and States served in Second Wave

Maglev Network	States In Network	Population of States in Network (millions)	Population Living Within 15 Miles of Maglev Stations (millions)	Route Miles in Network
First Wave Plus Second Wave	45 (Iowa, Nebraska & S. Dakota not in Network) plus Toronto, Montreal & Vancouver	310 (includes Toronto, Montreal & Vancouver)	210 (includes Toronto, Montreal & Vancouver)	18,630

Figure 1.15 shows a map of the Third Wave of the National Maglev Network to be built from Year 16 year 20 of the 20 year long project. North-South Maglev routes would be built to connect America's metro/micropolitan areas into an even closer knit Network, with shorter trips between the North and South of the US. Total route mileage would increase from the 18,630 miles the Second Wave to 29,000 miles when the Third wave was completed. The population living within 15 miles of a Maglev station would increase from 210 million to 230 million (Table 1.11).

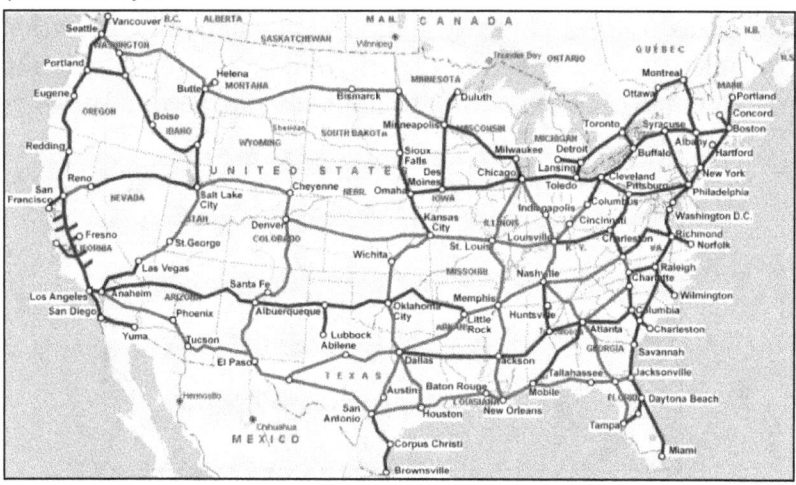

Figure 1.15: Map of Third Wave of National Maglev Network

Table 1.11: Population and States Served in Third Wave

Maglev Network	States In Network	Population of States in Network (millions)	Population Living Within 15 Miles of Stations (millions)	Route Miles in Network
First, Second and Third Waves Completed	48 plus Toronto, Montreal & Vancouver	315 includes Toronto, Montreal & Vancouver	232 includes Toronto, Montreal & Vancouver	29,000
74% of population in States live within 15 Miles of a Station				

Maglev Public Transit

We now turn to Maglev Public transit – application to heavy rail, commuter rail and light rail systems. Heavy rail systems include the New York City subway, San Francisco's BART and others. Table 1.12 lists the 10 largest US heavy rail systems, which together have an annual ridership of 3.5 Billion passengers (Table 1.13), 30 % of America's total annual ridership of 10.4 Billion passengers on all transit modes – heavy rail, commuter rail, light rail, and buses. Of the 3.5 Billion heavy rail passengers, the New York City Subway accounts for 2.4 Billion per year, (6.5 million daily), 70 percent of the total.

Table 1.12 Selected Service Indicators, Ten Largest U.S. Heavy Rail Operators, 2009

City (Transit System)	Operating Expense (in millions)	Active Vehicles	Average Fleet Age*	Daily Passenger Miles	Daily Passenger Trips	Daily Miles of Service	Daily Hours of Service
New York City (NYCT)	$3,313.1	6,360 (1)	17.4 (4)	27,322,683 (1)	6,461,133 (1)	965,821 (1)	52,934 (1)
Washington D.C. (WMATA)	804.8	1,120 (3)	18.9 (5)	4,569,588 (2)	813,307 (2)	196,721 (2)	7,737 (3)
San Francisco (BART)	484.2	669 (4)	11.7 (1)	3,951,025 (3)	314,122 (5)	185,872 (4)	5,320 (4)
Chicago (CTA)	462.0	1,186 (2)	25.7 (8)	3,290,783 (4)	554,984 (3)	187,924 (3)	10,226 (2)
Boston (MBTA)	298.5	442 (5)	21.8 (7)	1,558,839 (5)	407,354 (4)	61,575 (6)	3,533 (5)
New York/New Jersey (PATH)	233.0	369 (6)	29.4 (10)	934,784 (8)	219,329 (8)	33,433 (8)	1,813 (8)
Atlanta (MARTA)	167.0	270 (8)	19.2 (6)	1,443,898 (6)	228,347 (7)	67,304 (5)	2,523 (6)
Philadelphia (SEPTA)	158.0	369 (6)	16.7 (3)	1,158,870 (7)	260,576 (6)	46,267 (7)	2,375 (7)
Los Angeles (LACMTA)	88.8	104 (10)	13.0 (2)	623,717 (9)	128,469 (9)	16,651 (10)	736 (10)
Miami-Dade (MDT)	78.4	136 (9)	27.0 (9)	363,753 (10)	49,985 (10)	18,333 (9)	806 (9)

(Ranking among all heavy rail transit systems, 1 = most/newest, 10 = least/oldest)

*Age based on end of year 2009 data.

Source: National Transit Database 2009, Federal Transit Administration (Daily numbers calculated from Annual Averages in NTD.)

Figure 1.16 shows a map of the NYC Subway System, with Table 1.14 listing its parameters. In terms of government subsidies, it performs very well compared to other public transit systems. The average passenger fare per trip is $1.05, with an actual operating cost per trip of $1.40. The average fare per passenger trip for all US transit modes is $1.18, while the actual cost is $3.54 per trip, 3 times the fare cost. The $2.26 subsidy per trip is paid by taxpayers.

Adapting the NYC Subway System to Maglev will result in many benefits:

- Reduced subsidies from taxpayers

- Faster, much more comfortable trips – no bumping and swaying
- Quiet trips – no 100 decibels noise, which causes hearing loss
- No breathing in steel dust and other health harming microparticles generated by braking on steel rails
- Greater energy efficiency

Similar benefits will result from adaptation of Maglev to the other US heavy rail systems listed in Table 1.14.

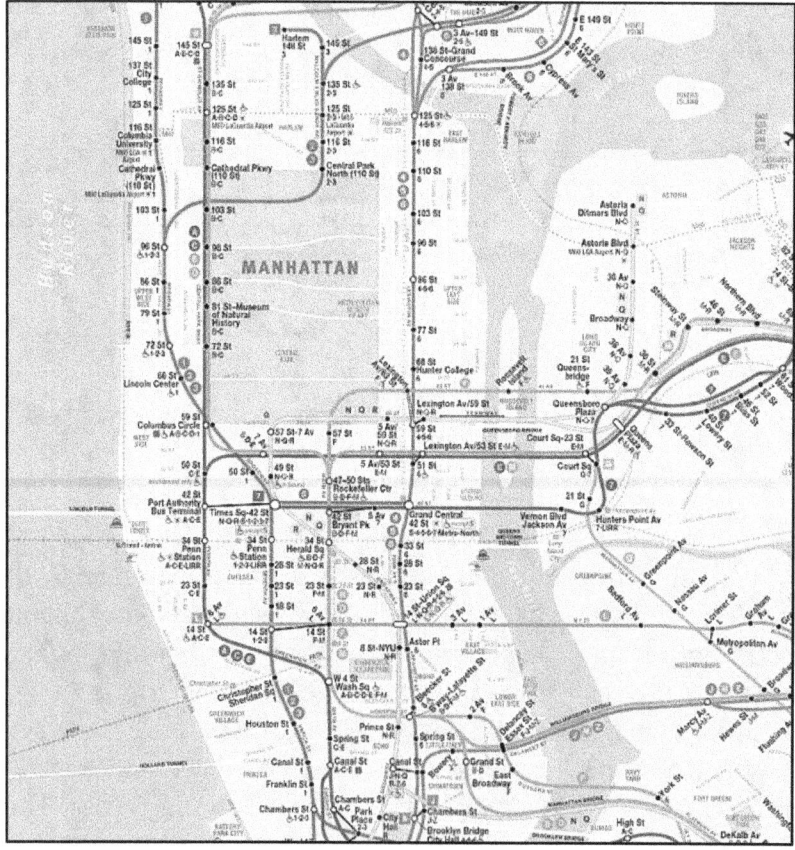

Figure: 1.16 NYC Subway System

Table 1.13
Comparison of National US Passenger Transit Parameters with New York City Subway Parameters (2009 Data)

Item	Parameter	US National Transit All Modes (1)	US National Transit Heavy Rail (1,2)	New York City Subway (3)
1.	Annual Ridership	10.4 Billion	3.5 Billion	2.4 Billion
2.	Annual Vehicle Miles	5.2 Billion	0.68 Billion	0.35 Billion[2]
3.	Revenue Vehicles	173,000	11,460	6,760[2]
4.	Total Annual Operating Expenses (Fares + Subsidies)	$37.2 Billion	$6.1 Billion	$3.3 Billion[2]
5.	Annual Capital Expenditures	$18.0 Billion	N.A.	N.A.
6.0	Passenger Fare Revenues	$12.3 Billion	N.A.	$2.5 Billion[3]
7.0	Rate, Fares/Operating Expense	0.33	N.A.	0.75[3]
8.0	Average Fare Per Trip	$1.18	N.A.	$1.05[3
9.0	Actual Average Operating Cost Per Trip	$3.54	$2.14	$1.40[3]
10.	Subsidies per trip	$2.36	N.A.	$0.35[3]
11.	Annual Passenger Miles	N.A.	16.5 Billion	10.0 Billion
12.	Actual Average Operating Costs Per Passenger Mile	N.A.	$0.39	$0.33[2]
13.	Average Trip Length	N.A.	4.7 miles	4.2 miles

(1) Statistical Abstracts of the United States (Table 1115)
(2) Benchmarking Efficiency for the Metropolitan Transportation Authority's Service (April 2011)
(3) MTA New York City Transit

Table 1.14
New York Subway System Parameters (5),(8)

21 Interconnected Subway Routes Serving Manhattan, Queens, Brooklyn, The Bronx and Staten Island through MTA Staten Island Railway

- Numbered Routes: 1, 2, 3, 4, 5, 6, 7
- Letter Routes: A, B, C, D, E, F, G, J, L, M, N,Q, R,S, T

209 Route Miles

- 656 Revenue track miles, 842 track miles including non-revenue trackage
- Longest train Route, 31 miles on A train from 207th St in Manhattan to Far Rockaway in Queens

468 Stations

- 60% underground; 40% other (elevated, embankment, open-cut)
- Longest distance between stations is 3.5 miles on the A train between Howard Bench/JFK airport and Broad Chalked in Queens

7775 Weekday Train Trips

- 343 Million Miles of Subway Car Service per Year
- 173,000 miles between repairs for a subway car
- 6,280 subway cars

Annual Ridership at 10 Busiest Stations is 308 million (2011)

- Busiest station is 60.6 million riders at Times Square/42nd St
- 10 Busiest Stations account for 20% of total system ridership

Subway Cars powered by 625 Volt DC Third Rail

- Maximum speed between station is
- 8 to 11 cars in train
- Cost of 2 million dollars per new subway car (2012)
- R160 cars maximum capacity of 246 passengers (44 seated, 202 standing)
- R160 cars weight 85,000 lbs.
- R142/R-188 New Cars weight 70,000 lobs
- BMD cars 60 or 75 feet in length, 10 feet wide.
- IRT Cars 8 Feet 9" wide, 51 feet long

Details of the adaptation process for the NYC Subway System for Maglev operation is described in *"Maglev America"*. Summarizing, the capital cost of the installation of the aluminum loop panels on the cross ties of the subway track (Figure 1.17) plus the capital cost of the Maglev vehicles and their superconducting Magnets is projected to be 10 Billion dollars. Amortized over 30 years that would be 330 million dollars per year, 10 percent of the NYC subways present operating budget of 3.3 Billion dollars per year. The savings in operating costs made possible with Maglev would more than offset the adaptation cost. The adaptation process could be carried out in as little as 2 years, given adequate funding and high priority for the program.

Figure 1.17

Maglev can also be adapted to commuter rail systems. Details of the adaptation process are given in **"Maglev America"**. The Long Island Railroad (LIRR) System shown in Figure 1.7 is America's busiest commuter rail system. It carries 280,000 passengers daily on weekdays, 81 million annually. It connects New York City to many stations on Long Island located on its 12 branches. It has 315 route miles and 700 miles of one-way track.

Table 1.15 shows the annual ridership on the 12 branches, together with the fares and actual operating cost per passenger for each branch. The ridership and actual costs per passenger vary widely, while the fares that passengers pay are relatively constant. On the Greenport branch, with only 70,000 passengers annually, it actually costs $85.91 per passenger trip, while the passenger fare is only $10.38. On the Ronkonkoma Branch with 14 million passengers annually, a passenger pays $7.53, while the actual cost is $11.96.

For the whole LIRR System, the average passenger fare is $6.46, 45% of the $14.68 actual cost. The difference is government subsidies. Adaptation of the LIRR for Maglev service would result in major benefits to passengers, taxpayers, and people living near the LIRR tracks, including:

- Much lower taxpayer subsidies
- Much shorter trip times, a factor of 2 to 3 shorter using higher speed, faster accelerating Maglev vehicles.
- Lower passenger fares
- Very quiet operation, no rail or locomotive noise for passengers and people living near the LIRR tracks
- More frequent service – individual Maglev vehicles, no infrequent long trains of many cars pulled by locomotives
- More comfortable rides: no vibration, bumping and swaying of RR cars, less crowded passenger seating
- No diesel emissions of greenhouse gases and microparticulates

- Safer operation – no 3rd rail, able to stop much faster in emergencies

Table 1.15

Long Island Railroad Ridership, Cost, Revenue and Fares by Branch

Source: 2010 MTA Financial Plan

LIRR Branch	Annual Customers (millions)	Average (1) Direct Cost Per Passenger	Total (2) Average Cost per passenger	2011 Fare per Passenger
Greenport	0.07	$33.82	$85.91	$10.38
Montauk	2.6	$22.22	$56.43	$9.58
Port Jefferson	3.4	$11.29	$28.63	$6.54
Oyster Bay	2.1	$10.59	$26.91	$6.49
West Hempstead	0.85	$8.18	$20.79	$5.72
Far Rockaway	4.2	$5.94	$15.08	$5.66
Hempstead	4.3	$5.70	$14.48	$5.77
Babylon	19.7	$5.22	$13.25	$6.81
Huntington	12.2	$4.92	$12.51	$6.74
Long Beach	6.5	$4.70	$11.95	$6.46
Ronkonkoma	14.00	$4.71	$11.96	$7.53
Port Washington	13.3	$3.34	$8.49	$5.72
LIRR System Wide	83.3	$5.78	$14.68 (4)	$6.46 (3)

(1) Direct variable costs include train crew, fuel & propulsion, rolling stock maintenance, running repairs, cleaning.

(2) Total cost includes all operating costs, including Direct Costs (#1), overhauls, administrative & support, etc.

(3) Total Fare Revenues are 538 million dollars

(4) Total cost is 1.22 Billion dollars

Cost and schedule for adapting the LIRR to Maglev? For 700 miles of one-way track, the capital cost at 4 million dollars per one-way mile would be 2.8 Billion dollars, about 93 million dollars annually over a 30 year amortization period. The track adaptation annual cost would be approximately 5% of the annual LIRR budget. Put in another way, the 2.8 Billion dollars to adapt 700 miles of LIRR track is about 1/4 of the 10 plus Billion dollars the LIRR is now spending to dig a tunnel under the East River to connect the LIRR to Grand Central Station in New York City.

At 5 million dollars per Maglev vehicle, the cost of 300 vehicles to transport the LIRR's 280.000 daily riders would be approximately 1.5 Billion dollars, about 1/2 of the track adaptation cost. With mass production of Maglev vehicles, the unit cost will probably be less than 5 million dollars.

Maglev can be adapted to other commuter rail systems in the US, like Metro North in New York State. We have considered adapting Maglev to US light rail systems; however, light rail ridership generally appears too low to be cost effective, and adaptation would be more difficult and expensive than for heavy rail and commuter rail.

In summary, the National Maglev Network and Maglev Public Transit will be of great benefit to America in its capability for:

- Much lower cost of transport
- Faster and more comfortable travel with shorter trip times
- Safer, less congested highways, with substantial reductions in deaths and injuries
- Greater energy efficiency and reduced pollution
- Increased economic productivity.

Chapter Two
On the Road to Global Maglev
James Powell, Gordon Danby, and James Jordan

"In a spatially inhomogeneous system, living things are much favored by mobility. A couple of billion years ago bacteria were already equipped with rotating flagella, stirred by electric micromotors of the kind physicists call step motors, and even capable of traveling in reverse."

"The 21st century will come to be dominated by the maglev— from the office to the moon, to offer a motto. Consistent with basic instincts, maglevs can serve as the pinnacle of a supersystem for green mobility. ... Even the bacteria should be impressed with the mobility humans can achieve."

From Jesse H. Ausubel and Cesare Marchetti— *"The Evolution of Transport"* in The Industrial Physicist a publication of the American Institute of Physics

"The World is a Book, and those who do not travel read only a page."

St. Augustine

The Global Maglev Network – Why It is Needed. By 2050 AD, only 35 years from now, World leaders and experts project that the World economy and its population will grow greatly. Compared to now (2014), by 2050:

- World population will grow from today's 7 Billion to 9 Billion. (1)
- GDP will almost triple from approximately 70 Trillion US$ Annually to more than 190 Trillion US$.(2)
- Electricity generation will double, from 20 Trillion Kilowatt Hours per year, to 39 Trillion Kilowatt Hours.(3)
- Number of automobiles will increase from 1 Billion to 2.5 Billion, a factor of 2.5 (4)
- Passenger miles will increase by a factor of 2.6

- Tonne miles of freight will increase by a factor of 2
- Global carbon dioxide emissions from transport will increase by a factor of 2.1, assuming that transport continues to be based on fossil fuels.

The World economy and population will not be sustainable at these levels if we continue to depend on fossil fuels for energy. As described earlier in the Prologue, global warming, ocean acidification, and depletion of fossil fuel reserves will lead to the collapse of modern civilization if humanity continues to use fossil fuels as its primary source of energy.

Transport of passengers and freight is critically important to modern civilization, today, and will continue to be in the future. However, transport must transition from fossil fuels to a different and sustainable source of energy, and must do soon.

As discussed in *Chapter One, The National Maglev Network*, biofuels and hydrogen fuel are not practical options because of the serious problems they pose. Electrification of transport, i.e., electric cars and rail, together with reductions in air travel and water-borne shipping, which cannot be electrified, is the only solution.

Maglev will make the electrification of global transport practical in 3 ways. First, by transporting passengers and freight, at high speed, high efficiency and low-cost, on Maglev guideways built between metropolitan areas, greatly reducing intercity travel by highway and airplanes. Second, by adapting already existing railroad and subway tracks at very low-cost for Maglev public transit, in urban and suburban areas. Maglev public transit will be much faster, cheaper, more comfortable, more frequent in service, and better environmentally. Subsidies from taxpayers will be much less. These advantages will help shift travel by highway vehicles to travel on public transit systems.

Third, with an intercontinental global Maglev Network, transport of freight by container and other ships will be much less, greatly reducing oil consumption and greenhouse gas emissions from shipping. It takes 35 days to ship containers from Asia by water. On Maglev it will be only a day, and shipping cost will be less.

Chapter 1 points out that for Maglev to be a major mode of transport in the United States, it must function as a National Network. Building isolated routes between a few city pairs, like the proposed Los Angeles to San Francisco High Speed Rail (HSR) System or the Boston-New York City, Washington, DC HSR route, will do very little for US transport needs.

To be effective, transport systems must be a connected network that serves the whole country like the US Highway system, the US railroad system, and the US airplane/airport system. Isolated routes between 2 points, with no connection to other points in the country, would be of little use.

Similarly, to be greatly effective globally, Maglev must be a Global Maglev Network. Described in more detail later in the Chapter, here we list its highlights. The Global Maglev Network will:

- Interconnect the World's principal continents – Asia, Africa, Europe, North America and South America
- Connect to the principal cities in each continent
- Adapt existing public transit rail systems for Maglev service where appropriate
- Transport passengers, trucks, and autos on the same guideway or railroad trackage adapted for Maglev
- Transport a major portion of the containers and freight presently transported by ship.

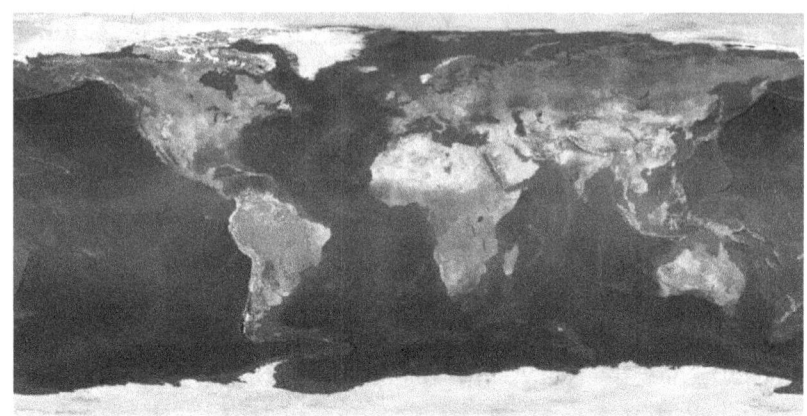

Figure 2.1

The World is very, very big (Figure 2.1) 25,000 miles for a trip around the World, 197 million square miles in area, 3/4th of it water. Very high mountains, immense oceans, very large deserts – all major challenges for creating a Global Maglev Network. The US National Maglev Network described in Chapter 1 has 29,000 miles of high speed Maglev intercity routes. The Global Maglev Network will require many more miles.

Starting with the Maglev intercontinental connections, the Europe to Africa Maglev connections would be made through the Strait of Gibraltar (Figure 2.2) using either a tunnel underneath the seabed, similar to the English Chunnel from Britain to France, or a floating undersea tube anchored to the sea floor. The tube could be at a sufficient depth, on the order of 100 meters, that wave movements would be very small. Using an undersea tube, the crossing length would be only 14 miles, much shorter than the 31 mile long Chunnel.

Figure 2.2 The Strait of Gibraltar

Europe and Asia are actually one continent with the distinction being cultural, not geographical. The great distance between the societies in Europe and those in the Far East, e.g., China and Japan, makes them seem like separate continents. Historically, there has been land transport of humans and goods between Europe and Asia for centuries – Marco Polo, the Silk Road, Genghis Khan, and so on.

The Europe to Asia Maglev Connection will in general follow the same path as the existing Trans-Siberian Railway that runs 5,753 miles across Russia from Moscow to Vladivostok on the Pacific Ocean (Figure 2.3). It takes 6 days from Moscow to Vladivostok and Beijing on the Trans-Siberian Railway. By 300 mph Maglev, it will take only 1 day. It takes 35 days by container ship from China to Europe. With Maglev it will be only 1 day, and it will be cheaper.

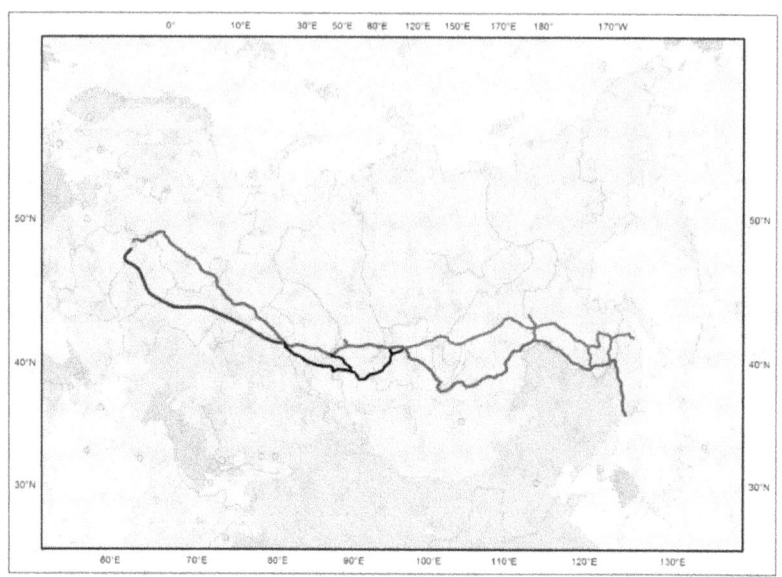

Figure 2.3 The Trans-Siberian Railway

The Asia to North America Maglev connection would be made across the Bering Strait (Figure 2.4) through either a tunnel beneath the seabed or a tube resting on the sea floor. The concept of a bridge or tunnel across the Bering Strait is over 120 years old, and has been proposed many times over the years. The Bering Strait crossing would connect Russia's Chukchi Peninsula to America's Diomedes Islands, located midway in the 50-mile crossing, and then. from the Diomedes Islands to Alaska's Seward Peninsula. The Bering Strait crossing is described in more detail later in the chapter.

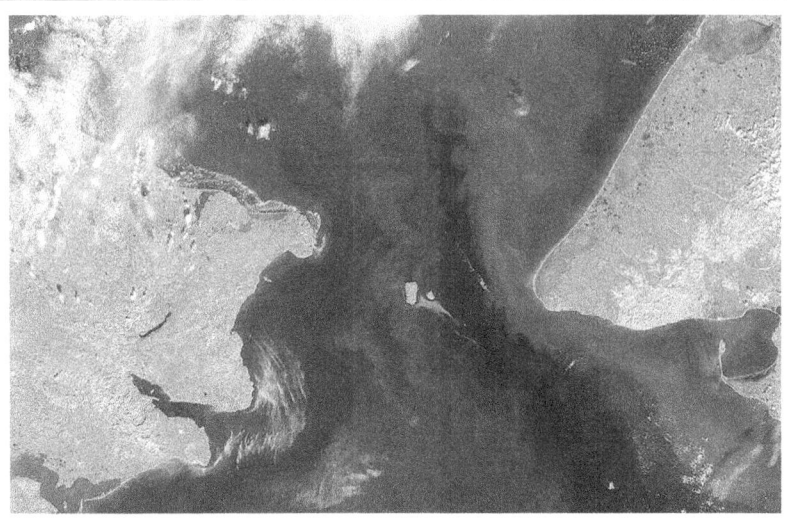

Figure 2.4 The Bering Strait, Separating Siberia from Alaska in the North Pacific. NASA image, taken by MISR satellite

A Maglev System would then connect Alaska through Canada to the US National Maglev Network described in Chapter 1. From the US, Maglev routes would run South through Mexico, Central America and Panama, connecting to Maglev routes along the West Coast of South America to form the Pan American Maglev Highway (Figure 2.5), extending all the way to the bottom of South America. The Pan American Maglev Highway would directly serve Columbia, Ecuador, Peru, and Chile. Additional Maglev Systems (not shown) would connect the other countries in South America, i.e., Venezuela, Brazil, Argentina, Bolivia, Paraguay and Uruguay to the Pan American Maglev Highway.

That's 5 continents connected together. What about the 6th and 7th, Australia and Antarctica? Only a few scientists reside in Antarctica, so there is no point to connecting them to the Global Maglev Network. It would be very desirable to connect Australia, but unfortunately, the ocean crossing distance is too great to be practical. Australia can still have its own National Maglev Network.

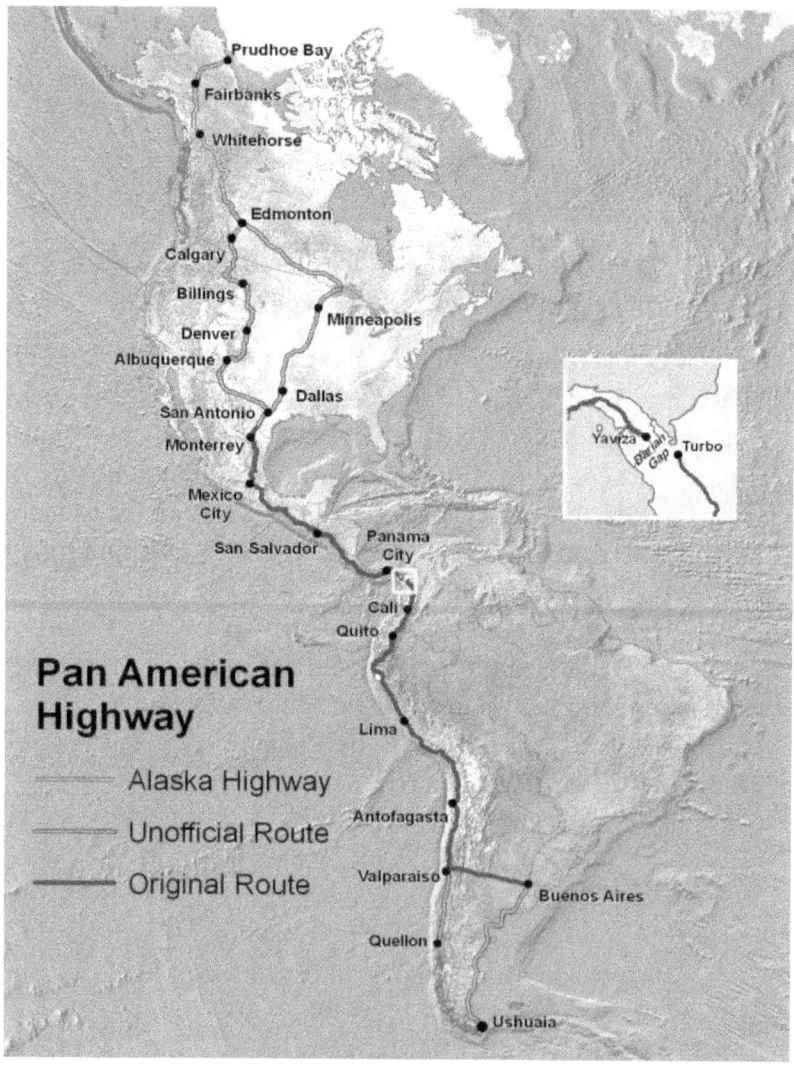

Figure 2.5 Map of Pan American Highway

China and India will be of special importance in the Global Maglev Network. Their combined population is 30% of total World population. As described below, China's annual GDP (Gross Domestic Product) is projected to grow by a factor of 5 from today's (2014) value by 2050 AD, making it the #1 ranking in 2050 GDP. India's GDP is projected to grow by a factor of 10 by 2050 making it #3 in the ranking of 2050 national GDP's.

China and India already have extensive national railway systems (Figures 2.6 and 2.7). Adapting them for Maglev travel, where practical, is a very attractive possibility, particularly in China, which has in the past few years, built many thousands of miles of High Speed Rail (HSR) lines. Where appropriate, new Maglev guideway systems would be built.

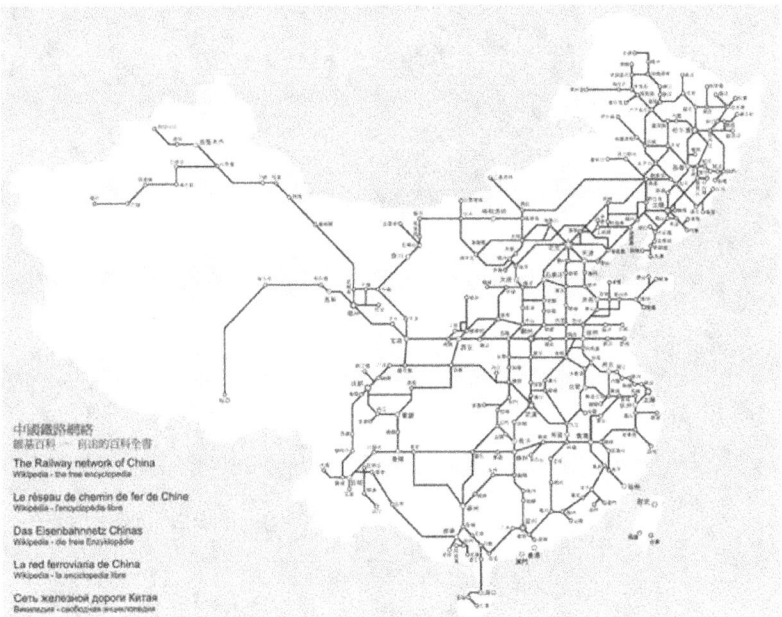

Figure 2.6 Map of Railway Network in China

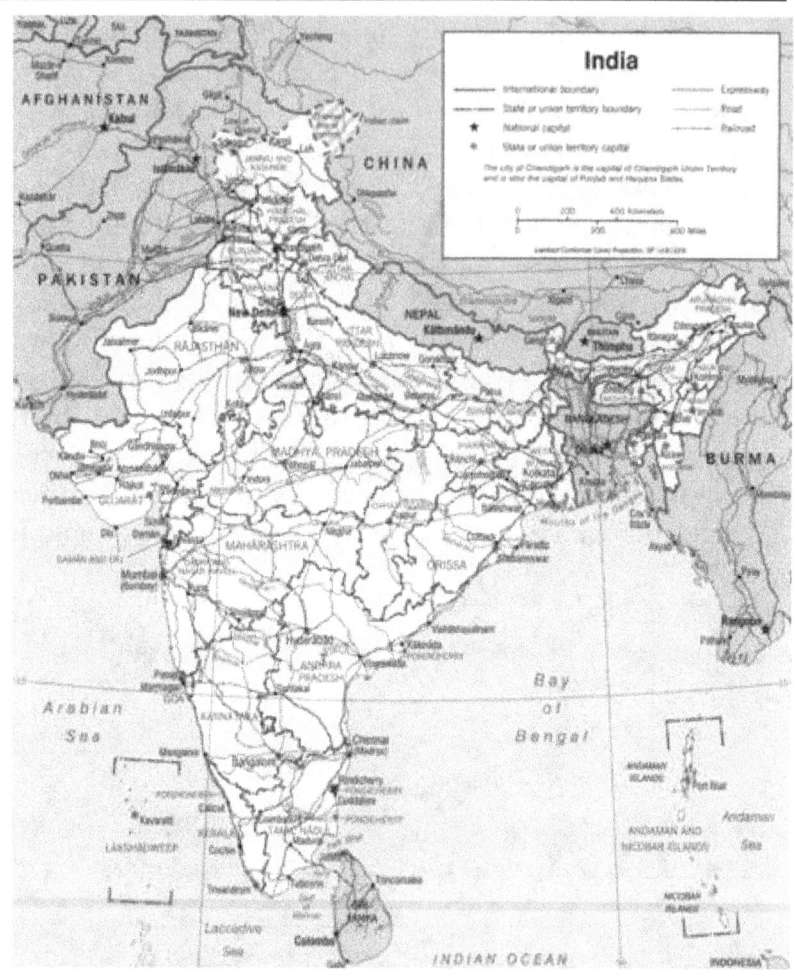

Figure 2.7 Map of India's Transport Networks

Before describing the Global Maglev Network in more detail, we now describe the present and future status of global transport, if we continue on the present path of transport that is based on fossil fuels.

Table 2.1

Index of Global Transport Activity, 2000 AD to 2050 AD

Reference: Transport Outlook 2011 (5)

Year	Index		
	Global Passenger Transport Activity (Passenger Kilometers)	Global Freight Transport Activity (tonne Km)	Global CO2 Emissions from Transport
2000	100	100	100
2014	140	150	130
2020	170	170	150
2030	210	200	170
2040	270	250	230
2050	360	300	270
Ratio 2050/2014	2.6/1	2/1	2.1/1

Global freight transport activity in 2050 will be 2 times greater than today (2014) and global passenger transport activity will be 2.6 times greater than today.

Along with the large increase in global transport activity will come a large increase in global carbon dioxide. Global CO_2 emissions from fossil fuels in 2014 were 32 Billion tonnes (6). Of that, approximately 27% came from transport (7), about 7.4 Billion tonnes. For an R value of 2.1 for CO_2 emissions, 2050/2014, the global carbon dioxide emissions in 2050 would be 2.1 x 7.4 = 15.5 Billion tonnes, almost 1/2 of the present (2014) emissions.

This is very important. One can shift from fossil fuels used for electrical generation to wind, solar, and especially beamed solar power as described in Chapter 4, and use the electrical energy from them for the industrial and residential sectors.

However, it is also very important to electrify transport if runaway global warming and ocean death are to be averted. As discussed light duty vehicles (LDVs)—that is, cars, SUVs, light trucks, together with highway freight trucks, buses and ships account for approximately 85% of transport CO_2 emissions, as shown in Table 2.2 for 2,000, 2030, and 2050 (5). Airplanes account for only about 12% of transport emissions.

Table 2.2

Modal Composition of Global CO2 Emissions from Transport Vehicles Use (6)

Mode	2000 AD	2030AD	2050AD
Freight & Passenger rail	2.3	1.9	1.5
Buses	6.3	4.3	3.0
Air	12.4	13.8	12.0
Freight Trucks	23.5	23.3	21.6
LDVs	42.5	45.2	52.1
2-3 Wheelers	2.4	2.2	2.0
Waterborne	10.6	9.2	7.8
	100%	100%	100%

The biggest contributor to CO2 emissions from transport, approximately 1/2 of the total is Light Duty Vehicles – autos, SUVs and light trucks. This is not surprising. People love LDVs. They can go when they want, where they want, without other people crowded around them, at reasonable cost. If they want to travel long distances, hundreds of miles, they will consider taking airplanes or rail. If parking is not available or too expensive, they will take busses and other modes of public transit, but reluctantly.

The higher the per capita income in a country, the more motor vehicles per capita. With more money, people are more likely to buy one. Figure 2.8 illustrates the correlation between per capita GDP income of a country (8) and motor vehicles per 1,000 passengers (9).

A list of the 20 countries' GDP per capita and motor vehicles per capita is given in Table 2.3, together with their population and land area. It is clear that motor vehicles per capita only correlates with GDP per capita and not with population or land area. Russia, with a land area of 17 million square kilometers, had 293 motor vehicles per capita compared to Mexico, which has 275 motor vehicles per capita, and an area of only 2 million square kilometers. Their GDP per capita's are close, $17,887 for Russia, and $15, 563 for Mexico. Similarly, India has a population of 1170 million and 41 motor vehicles per capita while Vietnam has 23 motor vehicles per capita and a population of only 90 million.

Of particular interest are how the World average motor vehicles per capita also correlate with World average GDP per capita. In 2011, World average GDP per capita was $10,000 per capita, with a World population of 7 Billion. In 2050 AD with 9 Billion people, would average per capita is projected to be $22,000, a factor of 2.2 increase. With 1 Billion motor vehicles in 2014 and 2.5 Billion projected for 2050, the motor vehicle per capita will increase from 143 vehicles per 1,000 persons to 278 vehicles per 1,000 persons, slightly less than a factor of 2.

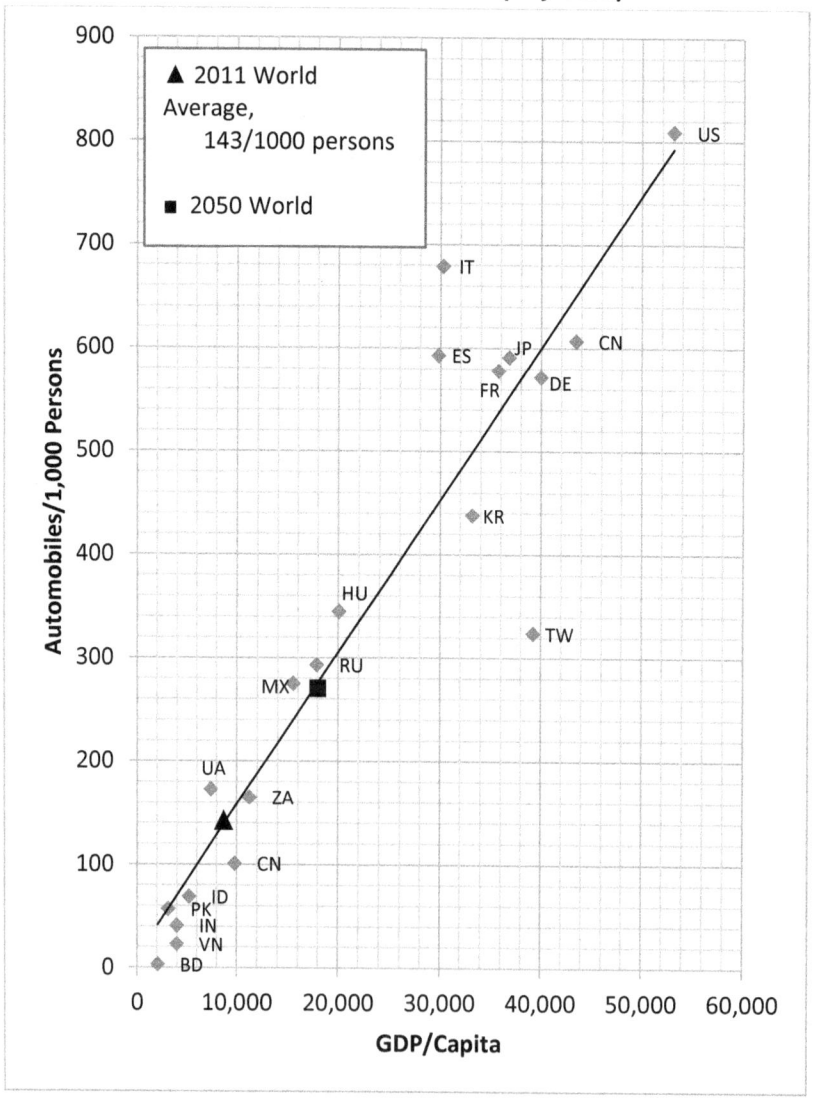

Figure 2.8

Automobiles Per 1,000 Persons as a Function of GDP Per Capita for Different Countries

Basis: 2010 – 2013 Data PPP (Adjusted)

Table 2.3

GDP per Capita, automobiles per 1000 Persons, Population, and Area for Different Countries

Basis 2010-2013 Data (PPP Adjusted)

Rank	Country	Autos per Thousand Persons	GDP Per Capita in US$	Population in Millions	Area in Millions of square KM
1	US	809	53,101	308	9.5
2	Italy	679	30,289	61	0.30
3	Canada	607	43,472	34	10.1
4	Spain	593	29,851	47	0.50
5	Japan	591	36,849	128	0.38
6	France	578	35,784	65	0.64
7	Germany	572	40,007	82	0.35
8	S. Korea	438	33,188	49	0.10
9	Hungary	345	20,065	10	0.09
10	Taiwan	324	39,267	23	0.04
11	Russia	293	17,887	139	17.1
12	Mexico	275	15,563	112	2.0
13	Ukraine	173	7,423	45	0.60
14	S. Africa	165	11,259	49	1.2
15	China	101	9,844	1,330	9.6
16	Indonesia	69	5,214	243	1.9
17	Pakistan	57	3,149	184	0.85
18	India	41	4,007	1,170	3.2
19	Vietnam	23	4,012	90	0.33
20	Bangladesh	3	2,080	156	0.15

In this chapter, we examine how Maglev can meet the challenge of the quickly growing passenger and freight transport activities in countries with rapidly expanding economies. We also examine how Maglev can improve passenger and freight transport in already industrialized countries to be low in cost, more energy efficient, and more environmentally acceptable.

Table 2.4
Gross Domestic Product (GDP) of Words 10 Largest Economies in 2050 AD, Compared to 2014 GDP Levels
Basis: Nominal GDP (Non PPP)(2)

2050 Rank	Country	2050 GDP in Trillion US$	2014 GDP in Trillion US$	Ratio 2050GDP/ 2014GDP
1	China	50.8	10.0	5.1
2	United States	38.0	17.5	2.2
3	India	21.0	2.0	10.5
4	Japan	7.8	4.8	1.6
5	Brazil	7.5	2.2	3.4
6	Russia	6.2	2.1	3.0
7	Germany	6.2	3.9	1.6
8	United Kingdom	5.8	2.8	2.1
9	Mexico	5.8	1.3	4.4
10	France	5.7	2.9	2.0

Table 2.4 lists the GDP's of the 10 Countries projected to have the largest economies in 2050 AD(2), and their present (2014) GDP's, together with the ratio of their 2050 GDP/2014 GDP.

- India has the largest growth ratio, R=21.0/2.0=10.5
- China is the next largest, with R=50.8/10.0=5.1
- The US, now the largest economy, is projected to slow, with R=38.0/17.5=2.2
- Japan has the lowest growth rate, with R=7.8/4.9=1.6
- Along with Germany, R=6.2/2.9=2.0, France is a bit better, with R=5.7/2.9=2.0
- Russian and Brazil are in-between , with R values on the order of 3, with Mexico at R=5.8/1.3=4.4, slight less than China

China and India are the biggest in growth and also the biggest in national populations (10).

	China	India	China + India	Total World	% of Total World
2015 Populations in Billions	1.36	1.25	2.71	7	39%
2050 Population in Billions	1.30	1.66	2.96	9	33%

Their very large populations and large future economies make them especially significant in terms of future World transport. Later in the chapter, we examine in detail how Maglev will help meet their transport needs.

In summary, the projections for global transport and Global GDP in 2050 predict growth factors of 2 to 3 in travel activity, GDP per capita, numbers of motor vehicles, and carbon dioxide emissions from transport.

Such growth factors are not sustainable if transport, which is now primarily based on fossil fuels, continues to depend on the combustion of fossil fuels. Global warming, ocean acidification, and depletion of fossil fuel reserves will act, causing environmental and economic disaster.

For a sustainable, prosperous World with a better life for humanity, it is absolutely necessary to transform to electric transport and cease its dependence on fossil fuels, and to make the transition soon. Maglev will be a necessity in this inevitable global transition to electric transport.

The Global Maglev Network is described in the following 4 sections:
- Maglev Intercontinental Connections
- Maglev Systems for Countries in Each Continent
- Maglev Alternative to Ocean Shipping
- Global Maglev Schedule and Cost

Turning to the first section, Maglev Intercontinental connections, there are 4 intercontinental Maglev connections.

- Europe to Africa across the Strait of Gibraltar
- Asia to North America across the Bering Strait
- Europe to Asia through Siberia
- North America to South America through the Isthmus of Panama

The Europe to Africa connection is across the Strait of Gibraltar. Figure 2.2 shows a satellite view of the Strait of Gibraltar crossing, but so far none have been attempted. At its narrowest part, the strait is only 9 mile across, where in 1986, Professor Ty Lin proposed building a gigantic bridge from Europe to Africa (11). The towers of the Bridge were 3,000 feet high, twice the height of the Empire State Building, with two spans, each 4.5 miles long, and 10 lanes of traffic.

A more realistic approach is an underwater tunnel, similar to the English "Chunnel". Spain and Morocco appointed a commission in 2003 to study the feasibility of a tunnel under the Strait of Gibraltar's sea floor (12). The proposed tunnel would be 25 miles long and 980 feet deep. It would be very expensive. The English Chunnel cost approximately 10 Billion dollars, when it was built in 1985 (13), and would probably be twice that if built today. Moreover, it would be much deeper than the Chunnel, with less favorable geology and earthquakes.

The Maglev Europe to Africa crossing could be made in 2 one-way underwater tunnels like the Chunnel or more likely, in underwater floating tubes of reinforced polymer concrete, which is much stronger than ordinary concrete. The tubes would have positive buoyancy, and be anchored to the sea floor beneath by cables. They would be located at a depth of approximately 100 meters (300 feet) to minimize wave action.

The Maglev crossing would be approximately 14 miles long, running from Tanya in Spain to Tangier in Morocco. The location chosen reduces the length of the anchor cables from the maximum water depth of 1,000 meters to the Seafloor for the 9 mile wide crossing to about 500 meters. Capital cost would be in the range of 1 to 2 Billion dollars.

The English Chunnel transports approximately 18 million passengers and 10 million tons of freight per year (13). The Maglev Gibraltar crossing could carry much greater traffic. With a 4 tube Maglev system, 2 tubes for passengers and 2 tubes for freight, Maglev vehicles travel from Gibraltar to Morocco in one to tube and return in the other tube of the pair.

Using consists of 3 Maglev vehicles coupled together, each carrying 100 passengers with 1 minute headings between consists, the Maglev Gibraltar crossing could transport 320 million passengers per year, 20 times the 16 million passengers traveling through the English Chunnel. Clearly, this passenger capacity far exceeds the likely traffic, but it does allow for large growth in the Gibraltar crossing passenger traffic.

Similarly, with a freight loading of 40 tons per Maglev vehicles, 3 vehicles per consist, and 1 minute heading between consists, the 2 freight tubes in the Gibraltar crossing could transport 120 million tons of freight per year, 12 times greater than the 10 million tons transported by the English Chunnel.

Only a small number of Maglev vehicles will be required for the Gibraltar crossing. At 200 mph, the Maglev vehicles consist would take only 4 minutes to travel the 14 mile tube system between the 2 terminals in Spain and Morocco. Allowing 8 minutes at each terminal for unloading/loading operations, the total round trip time for the consist would be 2 (4+8)=24 minutes. With a 50 percent load factor, i.e., 150 passengers out of the 300 passengers maximum, a single 3 vehicle consist could transport 6 million passengers per year, about 1/3 of the Chunnel's passenger traffic.

Similarly, a single 3 vehicle consist carrying 40 tons of freight per vehicle with a round trip time of 24 minutes, could transport 5 million tons of cargo annually, about 1/2 of the tonnage transported through the Chunnel. Allowing for peak demand periods and maintenance, a fleet of 30 Maglev passenger vehicles and 20 Maglev vehicles could easily handle traffic flow volumes well in excess of the passenger and freight volumes currently transported through the English Chunnel.

The cost of the 50 Maglev vehicle fleet would be well under 1 Billion dollars.

The Asia to North America Connection across the Bering Strait. (Figure 2.4) This crossing is longer, 50 miles in length, compared to 14 miles for the Gibraltar crossing, but the maximum water depth is much less, 180 feet versus 3,000 feet for the Gibraltar crossing.

First proposed in 1890 as a Bridge across the Strait by William Gilpin, the first Governor of Colorado Territory, it has been advocated by many since then. Joseph Straus, the future designer of the Golden Gate Bridge, proposed a Bering Strait Bridge in 1892 in his senior thesis (14). In 1905 Tsar Nicholas II approved a plan for a tunnel across the Strait (15). Unfortunately, World War I and Lenin stopped the project.

The list goes on. The latest proposal is for a 64-mile road and high speed rail tunnel. Russian's Prime Minister Putin has approved a plan to build a railroad to the Bering Strait that would connect to the proposed TKM-World Link crossing. The projected cost for the Bering Strait tunnel crossing is 65 Billion dollars, over 6 times the cost of the 31 mile English Chunnel, which cost about 10 Billion dollars in 1985 dollars. However, Russia did not promise to supply the funds. (14)

The Bering Strait crossing would use the same approach as the Strait of Gibraltar crossing, prefabricated reinforced polymer concrete tubes anchored to the seabed beneath. The principle difference between the Bering Strait and the Strait of Gibraltar crossing would be the longer length of the Bering Strait crossing, 50 mile miles compared to 14 miles for the Gibraltar crossing, and a shallower ocean – maximum depth of 180 feet, compared to 1500 feet.

The shallow depth enables the floating Maglev tunnel to be anchored only a few feet about the sea bed, reducing the cost of anchoring, and making the assembly process easier and cheaper.

The prefabricated Maglev tubes are 350 feet long, 15 tubes to a mile, with an inner diameter of 20 feet (similar to the English Chunnel) and a wall thickness of 8 inches of reinforced polymer concrete. At 2,000 dollars per cubic yard for reinforced polymer concrete structure, the projected tube cost per one way mile is 17 million dollars. Adding a conservative assembly cost of 1 million dollars per tube for 15 tubes, total construction cost per one way mile would be 32 million dollars.

The total projected construction cost for the 50 mile long 4 tube system – 2 tubes going east across the Strait and 2 tubes going west – would be 6.4 Billion dollars. Amortizing the 6.4 Billion dollars over 30 years, with an annual flow of 10 million TEU freight containers across the Bering Strait connection, corresponds to only 20 dollars per TEU, compared to the 3,000 dollars cost to transport a TEU container by ship.

The Bering Strait Maglev crossing, like the Strait of Gibraltar crossing, can transport very large volumes of freight and passengers. For example, a single one-way tube, with 5 Maglev vehicles each carrying 4 TEU's consists and 1 minute headways could carry 10 million TEU's per year.

The Bering Strait crossing would connect the Russian side of the Strait to the Europe area Trans-Siberia Maglev Line described below. On the Alaska side of the crossing it would connect to the Pan American Maglev Highway, which would run from Alaska's Seward Peninsula down through Canada to the US National Maglev Network described in Chapter 1, and from there through Mexico and Central America, connecting to the South American continent through the Isthmus of Panama. From there, the Pan American Highway runs down the West Coast of South America to its southern tip.

The Europe to Asia connection through Siberia. Figure 2.3 shows a map of the present Trans-Siberian railway that runs from Moscow to Vladivostok (16). The Trans-Manchurian line branches off from the Trans-Siberian Rail at Taranga, 3,898 miles from Moscow, and continues on for another 1,670 miles to Beijing in China. The Trans-Mongolian line branches off from the Trans-Siberian Railway at Ussuriysk, 5,680 miles from Moscow, and continues on for 700 more miles to Pyongyang in North Korea.

Passenger trip time from Moscow to Vladivostok (5,772 miles) is 6 days and 4 hours (16) at an average speed of 39 mph. The fastest freight trip time with good coordination from Beijing to Hamburg, Germany or Europe is about 16 days, for an average speed of approximately 15 mph. Average freight trip-time is about 25 days for an average speed of approximately 10 mph.

Russia plans investing 11 Billion dollars to reduce the cargo trip time across Russia from its Pacific ports to its border with Europe down to 7 days (16). However the Trans-Siberian Railway and other Russian rail lines have a railway gauge problem. Russia's railway gauge is broader than the standard 4 foot 8 inch rail gauge used in China and Europe. As a result, cargo traveling from Europe to Asia and vice versa by rail through Russia has to be unloaded & loaded 2 times – at Russia's Western and Eastern borders. This slows down the average transport speed and increases trip time.

A high speed Maglev route for cargo and passengers across Russia would in general follow the Trans-Siberian railway route. Pre-fabricated Maglev guideway beams and piers would be transported on the existing railway line to assembly sites, there to be erected by cranes on already prepared footings, as planned for the U.S. National Maglev Network.

The cost of construction for the Maglev guideway system would probably be less than the 30 million dollars per 2-way mile in the United States, because of lower labor costs. However, even if it were the same, the large volume of cargo and passengers that it would transport would make it economically attractive.

Transporting 10 million TEU containers per year between Europe and Asia with 20 tons of cargo per TEU, would yield a revenue of 2 million dollars per mile of Maglev guideway at a fare cost of 1 cents per tonne mile (US railway fares are 3 cents per ton mile and highway trucks are 30 cents per ton mile). Add to that 1 million dollars revenue per year from 20 million passengers annually, at a fare cost of 5 cents per passenger mile, yields annual revenue of 3 million dollars per mile of guideway.

Based on the above traffic flow and revenues which will be only a fraction of the total cargo and passenger traffic in 2050, when China's GDP is 50 Trillion dollars annually, the 30 million dollars per 2 way mile guideway construction cost could be paid back in only 10 years. There will be additional costs for Maglev vehicles, personnel, operations and maintenance, but guideway construction cost will be the major cost component.

China to Europe in 1 Day -- not 7 days by rail, or 35 days by ship. And cheaper than by rail or ship. And environmentally benign – no greenhouse gas emissions, no pollutants from dirty bunker fuel emissions from ships that kill many thousands of people annually. Companies that make goods and the companies that buy the goods will love the Trans-Siberian Maglev Line.

The National Maglev Networks in the various countries in Europe and Asia will be connected to the Trans-Siberian Maglev Line, enabling goods and passengers from one country in Europe to go to their destination in an Asian country. For example, from Germany, travelers and cargo would use a Maglev line from Germany to the Maglev line through Poland to connect to the Trans-Siberian Maglev line through Serbia, then leave Russia at the Russia-China Border, traveling from there to their destination in China on a Chinese Maglev line. We describe some potential Maglev routes inside the individual countries that connect to the Trans-Siberian Maglev line later in the Chapter.

The Trans-Siberian Maglev line will have a total 2-way route mileage of approximately 10,000 miles, taking into account the approximately 5,800 main route mileage between Moscow and Vladivostok plus another 4,200 miles for the various Maglev branch routes that connect to the main route. The total construction cost for the 10,000 mile system would be about 300 Billion dollars. World annual GDP is projected to be more than 190 Trillion dollars in 2050 AD (2). Over the next 30 years at an assumed average of about 100 Trillion dollars yearly, World GDP would total approximately 3,000 Trillion dollars, 10,000 times greater than the 300 Billion dollars construction cost. Moreover, the savings in transport costs will far exceed its construction cost.

We turn now to the 4th intercontinental Maglev connection, **The North America to South America connection across the Isthmus of Panama.** Shown in Figure 2.5, the Pan American Maglev Highway would run from the Seward Peninsula in Alaska down to the tip of South America. Total mileage for the Pan American Maglev Highway, including both the official and unofficial routes shown in Figure 2.5 is approximately 30,000 miles, a bit longer than the US National Maglev Network described in Chapter 1. Deducting the approximately 2,000 miles in the US National Maglev Network that would be part of the Pan American Maglev Highway, its total mileage would be about 28,000 miles.

The Pan American Highway directly connects almost all of the countries in North and South America. In North America, it serves (17)

- Canada
- The United States
- Mexico
- Guatemala
- El Salvador
- Honduras
- Nicaragua
- Costa Rica
- Panama

The only North American countries not directly connected to the Pan American Maglev Highway are Belize and British Honduras. In South America the following countries are directly served by the Pan American Maglev Highway (17).

- Suriname
- Guyana
- Brazil
- Venezuela
- Columbia
- Ecuador
- Peru
- Chile
- Argentina

Maglev branches would connect Paraguay and Uruguay to the Pan American Maglev Highway.

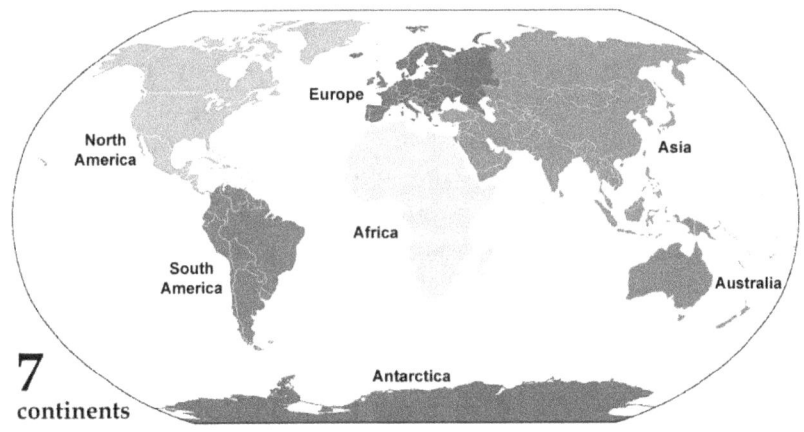

Figure 2.9

Connecting Countries Inside the Continents to Maglev. Figure 2.9 shows a World map of the 7 continents, Asia, Europe, Africa, North America, South America, Australia, and Antarctic. Australia would only connect to global Maglev by air and ship, since its distance from the other continents is too great for an undersea tunnel Australia would have its own National Maglev Network. Antarctica is only populated by visiting research scientists.

So far the Global Maglev Network only 5 of the 7 continents connect together – Asia, Europe, Africa, North America and South America. While anyone at any location of the 5 continents could travel to any other location in any of the 5 other continents, some of the trips would be quite long. To go from Tierra del Fuego at the bottom tip of South America to South Africa at the bottom tip of Africa by Maglev, for example, one would have to take the Pan American Maglev Highway to the Bering Strait crossing in Alaska, connecting to the Trans-Siberian Maglev System across Asia, to connect to the Pan-African Highway System at Egypt, and then travel down the East coast of Africa to South Africa.

Clearly, rather than a person taking Maglev to travel from Tierra del Fuego to South Africa, it would be faster and more convenient to fly. For freight containers, however, Maglev would be much cheaper than air, and much faster than by ship.

Table 2.5 lists the World Continents by population, population density, area, GDP, and GDP per capita, ranked by population (18). Asia and Africa are at the top of the list in population and at the bottom of the list in GDP per capita. Oceana, i.e. Australia and the Island Nations, are at the bottom in population. Europe and North America, with only 20% of the World population, are at the top of the List for GDP per capita. South America with 5% of the World population, is in the middle for GDP per capita, about $9,000 per year.

The GDP's and GDP's per capita in Table 2.5 are priced in standard US dollars. If priced in Purchasing Power Parity, the lower income continents would look substantially better in terms of income per capita. Table 2.6 compares the nominal GDP, GDP (PPP) and GDP (PPP) per capita, for some of the countries in Asia, as estimated in 2013 by the IMF (International Monetary Fund)(19). The GDP (PPP) value corresponds to what people can really buy in their country with their actual GDP, housing, food, services, etc.

Table 2.5

List of Continents Ranked by Population, Included are Land Area, Population Density, GDP, and GDP per Capita

Basis: 2010 Data (18)

Rank	Continent	Population Billions	Area Millions Mi²	Population Density People/Mi²	Annual GDP Trillions $USD	Annual GDP Per Capita $USD
1	Asia	4.2	16.9	246	18.5	2,900
	Africa	1.0	11.7	87	2.6	1,600
	Europe	0.74	3.9	188	24.4	25,000
	North America	0.54	9.5	57	20.3	32,000
	South America	0.39	6.9	57	4.2	$9,000
	Oceana	0.03	3.5	8.3	1.8	39,000
World Total		7.0	57	_____	72	10,000 Avg

Reference (18): List of Continents by GDP (nominal),
en.Wikipedia.org/wiki/list_of_continents_by_GDP_(nominal)

Table 2.6

List of Selected Asian Countries by Nominal GDP, GDP(PPP) and GDP(PPP) per Capita, as Estimated in 2013 by the IMF(19)

Country	Annual National GDP in Trillion USD	Annual GDP(PPP) in Trillion USD	Annual GDP (PPP) per Capita (USD)
China	9.2	13.4	$9,844
India	1.9	5.0	4,077
Japan	4.9	4.7	36,899
South Korea	1.2	1.7	33,189
Taiwan	0.49	0.93	39,797
United Arab Emirates	0.40	0.27	30,122

Reference 19: List of Asian Countries by GDP,
en.wikipedia.org.wiki/list_of_Asian_countries_by_GDP

The ratio of GDP(PPP)/GDP(nominal) varies widely with country. For China, the GDP(PPP) is a factor of 1.5 high than its nominal GDP. For India, GDP(PPP) is a factor of 2.5 higher than its nominal GDP. On the other hand, Japan's GDP(PPP) is slightly lower, 4.7 Trillion US dollars than its nominal GDP of 4.9 Trillion US dollars. And the United Arab Emirates GDP(PPP) is much lower, 0.27 Trillion USD, than its nominal GDP of 0.40 Trillion USD.

However, the times, they are changing. And quickly. In only 35 years, by 2050 AD, the GDP of Asian countries will grow substantially from 20 Trillion dollars annually in 2014 to 94 Trillion dollars in 2050, and will be over ½ of total World GDP. (Table 2.7) That's almost a factor of 5 growth in Asian GDP. Europe and North America GDP will grow significantly, but by a much smaller factor of approximately 2.

Asian countries will need to greatly expand their transport capabilities as they grow. Maglev will be best choice for fast, efficient, low cost transport that does not depend on fossil fuels.

Table 2.7
Present (2014) and Projected (2050) GDP's for the 20 Countries with the Strongest Economies as Ranked by the International Monetary Fund(IMF)
Basis: Nominal Annual GDP in Trillion US Dollars

Continent/Country	GDP(2014)	GDP(2050)
Asia		
China	10.0	50.8
Japan	4.8	7.8
India	2.0	21.0
S.Korea	1.3	3.1
Indonesia	0.86	5.1
Saudi Arabia	0.80	2.6
Turkey	0.77	3.8
Total (% of Total)	20.5(34%)	94.2(51%)
Africa		
Nigeria		3.7
Total (% of Total)	0(0%)	3.7%
Europe		
Germany	3.9	6.2
France	2.9	5.7
UK	2.8	5.8
Italy	2.2	3.9
Russia	2.1	6.2
Spain	1.4	3.2
Netherlands	0.81	--
Switzerland	0.71	--
Total (% of Total)	16.8 (28%)	31.0 (17%)
North America		

US	17.5	38.0
Canada	1.8	3.5
Mexico	1.3	5.8
Total (% of Total)	20.6 (34%)	47.3 (25%)
South America		
Brazil	2.2	7.4
Argentina	0.5	2.0
Total (% of Total)	2.2 (4%)	9.4 (5%)

We now take a look at how the various countries inside the 5 continents are connected to the Global Maglev System. Rather than each continent separately, we look at 3 continental systems:

- The Pan American Maglev System, connecting the countries in North and South America, from Alaska to Tierra del Fuego, would connect to Asia via the Bering Strait crossing.

- The Pan African Maglev System would encircle Africa, directly connecting all of its countries on the East, West and North Coasts of the continent. Island countries would be connected to the Pan African Coastal systems by branch Maglev Systems. Connection to the Pan Europe – Asia Maglev system would be made at 2 points – by the Gibraltar Strait Crossing on Africa's West Coast, and by the Isthmus of Suez crossing on Africa's East Coast.

- The Pan Europe – Asia Maglev System would travel from Britain in the West of the Europe-Asia continent—it really is one continent, with the division between Europe and Asia being cultural, not geographical – to China on the East of the continent. The Pan Europe – Asia Maglev System connects to the Pan American Maglev System at 2 points – the Gibraltar Strait crossing in the West, and the Isthmus of Suez crossing in the East.

To which "continent" do certain countries belong? There is general agreement that France is part of "Europe", for example, and China is part of "Asia". What about Russia? Some put it in Europe, some in Asia. Most definitions split it two –"European Russia", and "Russian Asia". The split seems rather arbitrary and is defined differently by different experts. Most use the Ural Mountains and other physical features to define European and Asian Russia. Generally about 80% of Russia's population lives in "Europe". And Turkey? Some put it in Europe and some in Asia. It makes no real difference. The Pan Europe – Asia Maglev System considers them as an integrated entity, with virtually all of the countries in "Europe-Asia" corrected by Maglev.

The Pan-American Maglev System

The Pan American Maglev Highway, shown in Figure 2.5 with side branches, connects all the countries in North and South America. Table 2.8 lists the countries connected in North America by present population (20), area (21), and nominal GDP (22). Table 2.9 lists the countries connected in South America, also by population (23), area (21), and nominal (22) and PPP (24) GDP. PPP, purchasing power parity, designates what the inhabitants of a country can actually buy with their nominal GDP. For richer nations, nominal and GDP (PPP) are about equal.

The wealthier per capita a country is, the closer the GDP (PPP) is to the nominal GDP. Guatemala has a per capita nominal GDP of about $3,300 and a GDP (PPP) per capita of about $5,300. Mexico has a nominal GDP per capita of approximately $11,000 and a per capita GDP (PPP) of about $15,200. For the US, the nominal per capita GDP and the per capita GDP (PPP) are essentially the same, approximately $52,000 per capita.

Table 2.8 and Table 2.9 show that the continent of North America, principally due to the US economy, is much wealthier than the continent of South America, with total nominal GDP of 20.2 Trillion dollars annually, compared to only 4.1 Trillion dollars for South America, a factor of 5 difference. North America has a somewhat greater population, 514 million than South America's 404 million, but the population difference is much smaller than the GDP difference.

South America's continental economy is probably too small to fully finance its portion of the Pan American Maglev System. Some investment from wealthier North America will probably be necessary.

Total combined annual 2013 GDP is approximately 25 Trillion dollars. By 2050 AD, total annual GDP will be approximately 50 Trillion dollars, measured in today's dollars. The accumulated continental GDP from today to 2050 AD, 35 years from now, will be more than 1,000 Trillion dollars, a thousand times greater than the approximately Trillion dollars investment to create the Pan American Maglev System.

So, on a continental basis, there clearly are ample funds to build the Pan American Maglev System. The economic benefits from it will far exceed the required investment, both in savings from lower cost of transport, and also from the opportunities it will enable the developed countries in North American and South America to grow their economies. Table 2.10 summarizes the parameters for the Pan American Maglev System.

Table 2.8

North American Countries Connected to the Pan America Maglev System: Population, Land Area, and GDP (nominal and PPP)

Country	Population (millions)	Land Area (square miles)	Annual GDP (Nominal) (Trillion USD)	Annual GDP(PPP) (Trillion USD)
Canada	35	3,855,000	1.8	1.5
US	316	3,537,000	16.8	16.8
Mexico	118	758,000	1.3	1.8
Guatemala	15	42,000	0.05	0.08
Belize	0.3	8,400	0.062	0.003
Honduras	8.6	42,800	0.02	0.04
El Salvador	6.6	8,100	0.02	0.05
Nicaragua	6.2	50,300	0.01	0.03
Cost Rica	4.7	19,700	0.05	0.06
Panama	3.6	29,100	0.04	0.06
Total	514		20.2	20.4

Table 2.9
South American Countries connected to the Pan American Maglev System
Population, Land Area, and GDP (Nominal and PPP)

Country	Population (Millions)	Land Area (Square Miles)	Annual GDP (Nominal) (Trillion USD)	Annual GDP(PPP) (Trillion USD)
Columbia	47	441,000	0.38	0.50
Venezuela	30	354,000	0.37	0.40
Ecuador	16	107,000	0.09	0.13
Peru	30	496,000	0.21	0.32
Bolivia	11		0.02	0.05
Chile	16	292,000	0.28	0.32
Argentina	41	1,073,000	0.49	0.76
Uruguay	3.3	70,000	0.06	0.05
Brazil	201	3,287,000	2.2	2.4
Paraguay	6.8	157,000	0.03	0.04
Guyana	0.8		0.003	0.006
Suriname	0.5	63,000	0.005	0.006
Total	404		4.1	5.0

Table 2.10
Summary of Pan American Maglev System

Northern Terminus: Bering Strait, Alaska

Southern Terminus: Tierra del Fuego, Argentina

Countries Directly Served in North America: US, Canada, Mexico, Guatemala, Belize, Honduras, El Salvador, Nicaragua, Costa Rico, Panama

Countries Directly Served in South America: Columbia, Venezuela, Ecuador, Peru, Bolivia, Chile, Argentina, Uruguay, Brazil, Paraguay, Guyana, Suriname

North America Population Directly Served: 514 Million

South America Population Directly Served: 404 Million

33,000 Route Miles of 2-Way 300 mph elevated Maglev Guideway

Maglev System transports Passengers, Freight, Trucks and Autos

Investment for 33,000 miles of Maglev Routes: 1 Trillion Dollars

Present Annual GDP of North & South America: 25 Trillion Dollars

Projected Cumulative GDP, 2015 to 2050: $300 Trillion Dollars

The Pan American Maglev System routes shown in Figure 2.5 use the 2nd Generation Maglev-2000 elevated monorail guideway that carries passengers, trucks and autos at 300 mph. A more detailed description of the 2nd Generation System and its capabilities is given in Chapter 1 on the US National Maglev Network. The projected route mileage of the Pan American Highway route is 30,000 miles (17). This includes approximately 2,000 miles of the US continental National Maglev Network. Deducting this portion results in approximately 28,000 miles for the Pan American Maglev system. However, approximately 5,000 miles of 300 mph Maglev guideway is added to the total mileage, for the extension from Fairbanks to the Bering Strait crossing route, plus a route across Canada to join its West Coast cities to its East Coast cities. Total mileage for the Pan American Maglev System is then approximately 33,000 miles.

The projected construction investment for the 300 mph 2nd Generation Maglev-2000 elevated guideway in the US and Canada is 30 million dollars per 2-way mile. The cost may be less in the countries with lower labor costs and smaller GDPs that connect to the Pan American Maglev System. However, for this study, to be conservative, we take the investment to be constant at 30 million dollars per 2-way mile, for all the countries in the Pan American and Maglev Systems.

At 30 million dollars per 2-way mile, total investment for the 33,000 mile Pan American Maglev system would be about 1 Trillion dollars. Not included are the investments for stations and Maglev vehicles which will be about 100 Billion dollars, making the total investment on the order of 1.1 Trillion dollars.

Not included are the internal investments that would be undertaken by the various countries to provide good access to the Pan American Maglev System that runs through their country. In general, the 300 mph Pan American System would service the principal cities in each country. A country would upgrade its roads and railways to provide quicker and easy access to the Pan American Maglev System at those cities.

In particular, countries with extensive conventional steel wheel on rail systems could upgrade their railways, by adapting the trackage for Maglev travel at a low investment of approximately 4 million dollars per one-way mile, as described in Chapter 1 for US commuter rail and subways.

For example, Mexico, Brazil and Argentina have extensive rail networks (25) but move relatively small numbers of passenger miles and freight ton miles per year.

Parameter	Mexico	Brazil	Argentina
Rail Mileage(miles)	26,704	28,538	39,966
Passenger Miles Per Year (Billions)	<3	<3	5
Freight Ton Miles Per Year(Billions)	41	160	7

In comparison, the US transports 1500 Billion tons of rail freight annually, ten to 100s of times more than Brazil, Mexico and Argentina. This is not surprising given America's greater population and higher GDP. However, as the population and GDP of the other countries connected to the Pan American Maglev System increase in the years ahead, having a substantial internal railway system already in place that can be adapted at low investment to Maglev, enabling fast and easy access to the Pan American Maglev System will be very advantageous.

Brazil's GDP(nominal), for example, is projected to grow in constant dollars, from its present annual value of 2.2 Trillion dollars to 7.4 Trillion by 2050 AD, Mexico, from 1.3 Trillion to 5.8 Trillion by 2050 AD, and Argentina from 0.5 Trillion to 2.0 Trillion, a factor of approximately 4 (2). With the Pan American Maglev System and upgraded internal transport, the growth factor could be even greater, more like India, which is projected to grow by a factor of 10, and China which is projected to grow by a factor of 5(2).

Overall conclusions. The Pan American Maglev System will directly serve all of the countries in North America and South America, providing fast, efficient, low-cost travel for passengers, freight, and motor vehicles. Projected investment for the 33,000-mile Maglev System, which will run from the Bering Strait Maglev crossing that connects North America to Asia, down to the Tierra del Fuego at the bottom of South America, is approximately 1 Trillion dollars.

The present total annual GDP for North and South America is approximately 25 Billion dollars, and is expected to double by 2050 AD, 35 years from now. The total accumulated GDP for North and South America from now to 2050 AD will be on the order of 1300 Trillion dollars, more than 1,000 times greater than the investment for the Pan American Maglev System. The economic earnings and growth enabled by the Pan American Maglev system will be far greater than the investment for it.

The Pan-African Maglev System

Africa's present population (2013) of 1.1 Billion people is projected by the UN (26) to grow to 2.4 Billion in 2050, making it 29% of the World's 9 Billion in 2050. But Africa is a very poor continent, with a present annual GDP of only 2.6 Trillion dollars – 3.6 percent of the World total annual GDP of 72 Trillion dollars (Table 2.6). The average per capita GDP for Africans is only 1600 dollars, compared to the World average of about $10,000 per capita – about 1/6th of the World average.

The total World economy is projected to grow by about a factor of 2.6 by 2050, with a total annual GDP on the order of 200 Trillion dollars (2). The average GDP per capita in 2050, based on 9 Billion people and a 200 Billion dollar annual GDP would be about $22,000 per person,(in constant dollars), more than 2 times the present World average GDP per capita.

Even to maintain its present value of 1/6th of World GDP per capita, Africa would need a per capita value of 1/6 x $22,000 = $3,700 per person. With a growth to 2.4 Billion population, that means Africa would have to have a GDP of 2.4 Billion x $3,700 = 8.9 $Trillion in 2050 – almost 4 times it present GDP.

Whether or not such a GDP growth is possible for Africa is unknown. What is known is that without a modern transport system, it will not be possible. Most of Africa's 58 countries have less than 1,000 miles of railway tracks (25). Only 6 countries have more than 2,000 miles. South Africa (Figure 2.10) has the most mileage (12,000 miles), followed by Egypt (4,000 miles), Sudan (3,300 miles), Algeria (2,500 miles), Democratic Republic of Congo (2,400 miles) and Nigeria (2,000 miles).

African transport is constrained not only by its very limited railway and highway mileage but in the railway sector by the wide variety of different railway gauges in Africa's countries. In South Africa, almost all railways (Figure 2.10) use the Cape narrow gauge (3 feet 6 inches) only a small fraction of its railway is standard gauge (4ft 8.5 in), the widely used standard for most of the World. 35% of South Africa's 12,000 rail mileage carries no activity or very low activity (27).

Table 2.11
List of African Countries and Populations Connected to the Pan-African Maglev System
Northern Coast Route – Morocco to Egypt

5 Countries Directly Connected (Population in Millions)	3 Countries Indirectly Connected (Population in Millions)
Morocco (32.9), Libya (6.3)	Mali (16.7)
Algeria (38.3), Egypt (34.6)	Niger (17.5)
Tunisia (10.9)	Chad (12.9)
Total Population 173 Million	Total Population 47 Million

East Coast Route – Egypt to South Africa

12 Countries Directly Connected (Population in Millions)	16 Countries Individually Connected (Population in Millions)
Sudan (35.2), Tanzania (45.9)	S. Sudan (10.3), Cen. Afr Republic (5.2)
Eritrea (5.0), Mozambique (24.5)	Dem.Rep.Congo (74.6), Uganda (35.4)
Djibouti (0.4), Swaziland (1.1)	Burundi (9.0), Zambia (14.1)
Ethiopia (86.6), Lesotho (1.9)	Zimbabwe (13.1), Rwanda (10.8)
Somalia (9.6), Kenya (43.3),	Malawi (15.3), Botswana (2.1)
South Africa (53.0), Cape Verde (0.31)	Madagascar (21.8), Seychelles (0.09)
	Mayotte (0.22), Comoros (0.74)
	Mauritania (1.3), Reunion (0.86)
Total Population 307 Million	Total Population 215 Million

West Coast Route – South Africa to Morocco

19 Countries Directly Connected (Populations in Millions)	3 Countries Indirectly Connected (Populations in Millions)
Namibia (2.2), Guinea Conakry(11.8)	Burkina Faso (17.3)
Republic of Congo (4.5), Gabon (2.2)	Sao Tome (0.19)
Equatorial Guinea (1.8), Benin (9.7)	Saint Helena (0.004)
Cameroon (20.9), Togo (6.7)	
Nigeria (177.1), Ghana (26.4)	
Liberia (3.9), Western Sahara (0.65)	
Mauritania (3.5), Senegal (13.6)	
Gambia (1.8), Guinea-Bissau (1.7)	
Ivory Coast(23.9), Angola(21.2)	
Sierra Leone (5.8),	
Total Population 389 Million	Total Population 17 Million

Table 2.12
Summary of the Pan African Maglev System

North Coast Maglev Route – Morocco to Egypt

- Strait of Gibraltar West Terminal, Isthmus of Suez East Terminal
- 3,000 Route Miles
- 5 Countries Directly Connected, 173 million population
- 3 Countries Indirectly Connected, 47 million population

East Coast Maglev Route – Egypt to South Africa

- Isthmus of Suez North Terminal; Cape Horn South Terminal
- 7,000 Route Miles
- 12 Countries Directly Connected, 307 million population
- 19 Countries Indirectly Connected, 215 million population

West Coast Maglev Route – South Africa to Morocco

- Cape Verde South Terminal; North Terminal, Strait of Gibraltar
- 8,000 Route Miles
- 19 Countries Directly Connected, 389 million population

- 3 Countries Indirectly Connected, 17 million population
- 75% of Africa's 1.1 Billion population directly connected to Pan-African Maglev System
- 18,000 Route Miles of 2-Way, 300 mph Elevated Maglev Guideway
- Maglev System Transports Passengers, Freight, Trucks, and Autos.
- Investment for 18,000 miles of Maglev Routes: 0.54 Trillion Dollars
- Present Annual GDP of Africa: 2.6 Trillion Dollars
- Projected Cumulative GDP 2015 to 2050: 230 Trillion Dollars

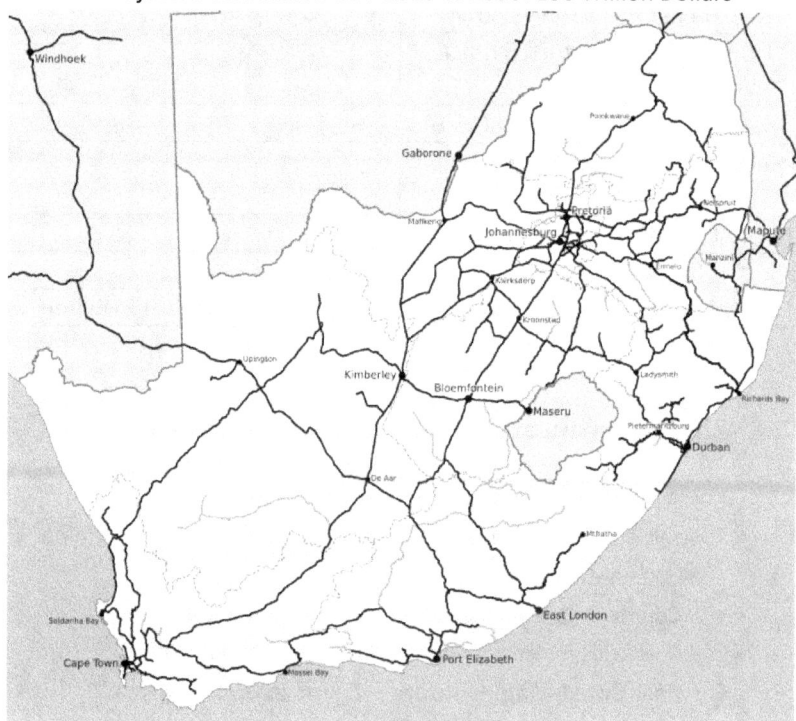

Figure 2.10

Many of Africa's countries operate with Cape Gauge (3 feet 6 inches), including Angola, Botswana, Congo, Ghana, Mozambique, Namibia, Nigeria, Sudan, Zambia, and Zimbabwe (28). Other countries use the narrower metric gauge (1,000 mm, or 3 feet 3 3/8 inches), including Kenya, Uganda, Ethiopia, Cameroon, and Tunisia. There are even narrower gauges, down to as little as 2 feet. Basically, each country operates transport individually, with very little interconnection.

Maglev can connect the countries of Africa into a continental network for high speed, efficient, low cost transport of passengers, freight, trucks, and autos. Figure 2.11 shows a map of the African continent encircled by the Pan-African Maglev system that runs along its North, East, and West Coasts. Table 2.11 lists the countries and their populations that are directly connected to the North, East and West Coast routes that run through their countries. Of Africa's 58 countries (29), 36 are directly connected.

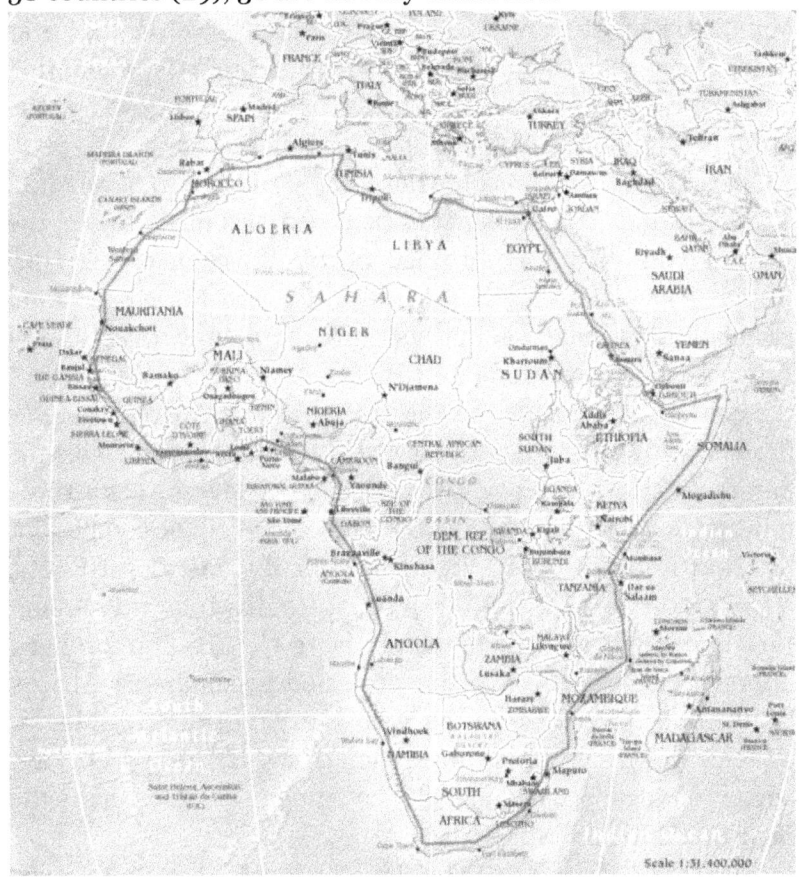

Figure 2.11

Another 14 countries on the African continent (Table 2.11) are indirectly connected to neighboring countries through which the Pan-African Maglev System runs by railways and/or highways. Maglev branches can be built to connect these countries to the main Pan-African Maglev System. An alternative possibility is to adapt existing railway lines for Maglev operation by placing panels of aluminum loops on the crossties of the railroad trackage, as described in Chapter 1. Adaptation to Maglev can be done on any of the various railway gauges, since the rails do not contact the traveling Maglev vehicles. The operating Maglev vehicles on the adapted railway tracks can electronically switch onto and off from the Mainline Pan African Maglev System and travel to whatever destination in Africa they want to. The investment for adapting existing railway tracks for Maglev operation is low, or the order of 4 million dollars per one-way mile.

8 of the 58 African countries, e.g., Madagascar, Seychelles, etc. are in the Indian and Atlantic ocean, too distant from the African continent for a bridge or tunnel connection. They would have to connect by ship. In general, their total population is small, about 25 million, a little over 2% of Africa's total population of 1.1 Billion. Overall, 75% of Africa's population is directly served by the Pan-African Maglev System, with 25% indirectly served. Table 2.12 summarizes the parameters for the Pan-African Maglev Systems. Total route mileage is approximately 18,000 miles, with 3,000 miles for the North Coast Route, 7,000 miles for the East Coast Route, and 8,000 miles for the West Coast Route. Total investment for the 18,000 mile system is 0.54 Trillion dollars, approximately 1/400th of the projected accumulated African GDP from 2015 to 2050. As with the Pan-American Maglev system, the transport savings enabled by the proposed Pan-African Maglev System will be much greater than the investment to build it.

Overall conclusions? The Pan-African Maglev System will directly serve 75% of Africa's population, who live in the 36 countries that the Maglev System passes through. The neighboring countries can connect to the Maglev System using existing railways and highways. With a low investment, existing railway trackage can be adapted for Maglev transport by attaching simple panels of aluminum loops to the railway crossties, enabling Maglev vehicles to switch on to or off from the mainline system to the adapted Maglev trackage.

The investment for the Pan-African Maglev System is approximately ½ Trillion dollars, a small fraction of about 1/400th of the projected cumulative African GDP for the years 2015 to 2050. The economic savings by the Pan-African Maglev System will be much greater than the investment for it.

The Pan Europe-Asia Maglev System

Europe and Asia are not physically separate continents, but rather one giant continent divided by culture, not geography. When historians, economists, marketers, etc. try to divide the one giant continent into 2 separate entities, there's a lot of disagreement. Where is the boundary? Is Turkey in Asia or Europe? Where do you put Russia? Are the Ural Mountains the boundary? And so on.

For our purposes, since we aim to connect all the World's continents into a global transport network, the distinction between Europe and Asia is irrevelant. Our goal is to have the Western terminus of Pan Europe-Asia Maglev System in Britain go through the Chunnel to France, cross "Europe" to Russia and virtually all of the "Asian" countries to the Pacific Ocean and the Bering Strait, where the Pan Europe-Asia Maglev System would connect with the Pan American Maglev System.

And not to leave Africa out, the Pan Europe-Asia Maglev System would connect with the Pan African Maglev System across the Strait of Gibraltar and at the Isthmus of Suez.

Connecting Europe to Asia is nothing new – it has been going on for thousands of years on the 4,000 mile Silk Road (Figure 2.12). Starting in the Han Dynasty around 200 BC, silk and other goods were shipped along the Silk Road, along with technologies, religions, and philosophies (30), plus the

Bubonic Plague. After the Roman conquest of Egypt in 30 BC, trade and travel between China, Southeast Asia, India, the Middle East, Africa, and Europe really took off (30). Roman patrician women really loved silk. The Roman Senate tried repeatedly to forbid wearing silk, but failed. Importing silk caused a huge outflow of gold, and silk clothes were considered immoral. Seneca the younger railed against silk clothes, *"Wretched flocks of maids labour so that the adulterers may be visible through her thin dress, so that her husband has no more acquaintance than any outsider or foreigner with his wife's body."* (30)

Europe finally discovered how silk was made, and spies were sent to China to steal silk worm eggs, with the result that silk production started in the Mediterranean (30). However, the Silk Road continued to flourish after the Roman Empire declined and fell, throughout the Middle Ages. Many countries and groups traveled by caravan along the Silk Road (Figure 2.13). Marco Polo's travels to China in the late 1200's fascinated Europeans, who loved his accounts of his journeys.

The Mongols controlled the Silk Road and benefited greatly from it. And they passed on their control. After Genghis Khan died, his daughters took over (30). But then the Mongol Empire collapsed and by the 1400's, trade largely ceased, because the Ottoman Empire was anti-western (30).

So Europe and China found new ways to trade – by sea. In 1498, Vasco da Gama reached India by sailing from Lisbon around the Cape of Good Hope, establishing the first successful trade route completely by sea. Thousands of sailors had died and dozens of vessels had sunk in the decades before, trying to make the journey (31). After his first journey, trade between Europe and Asia blossomed again, and Portugal reaped the tremendous profits.

Columbus believed that an easier route from Spain to Asia lay across the Atlantic Ocean, because he thought the World was much smaller than it actually was. When he discovered America, he thought it was Asia. It was not, so we have Columbus Day.

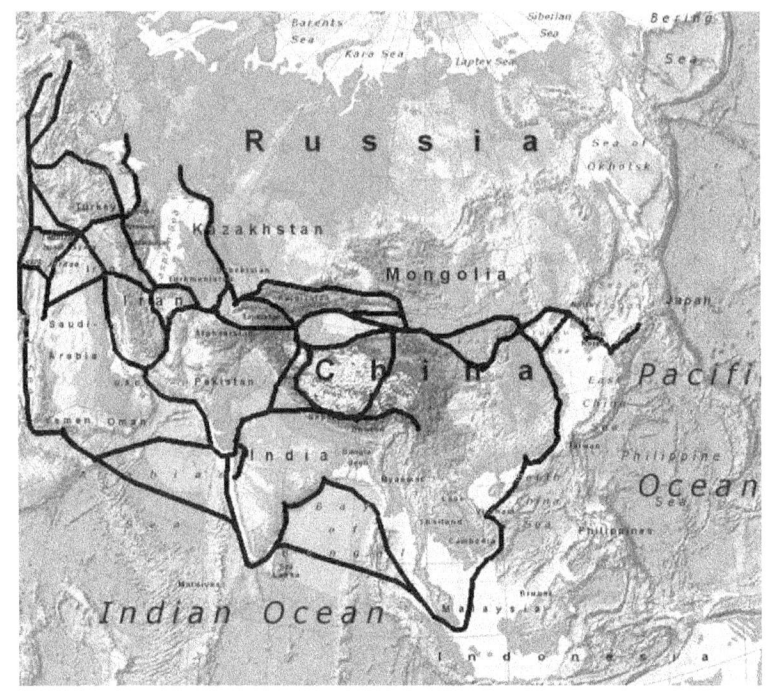

Figure 2.12

Figure 2.13

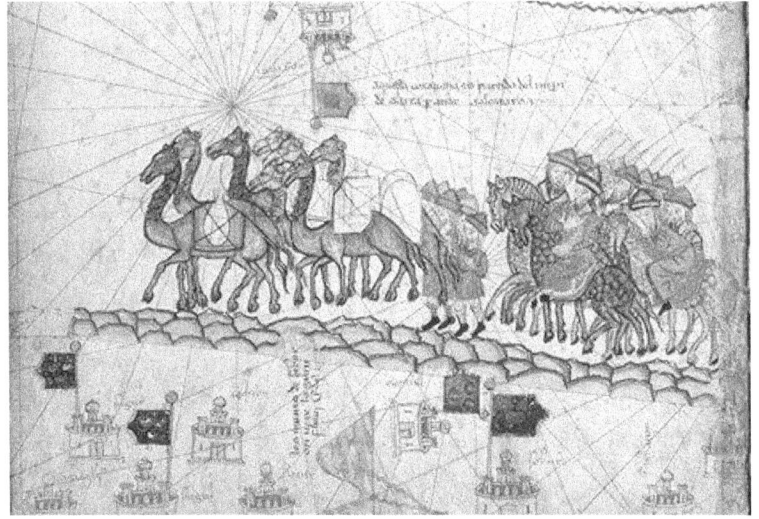

That's been the pattern for travel between Europe and Asia since. Then primarily by ship for passengers and freight until modern air travel. Now most freight travels by ship and most passengers by air. Very little travel by rail: who wants to spend 6 days and 4 hours traveling on the Trans-Siberian Railway from Moscow to Vladivostok? The Silk Road is now a World Heritage Site, honored in memory, but not in practice.

But that will change in the years ahead when the Pan Europe – Asia Maglev System begins operation. Asia GDP is rapidly growing. Today, China's GDP is 10 Trillion dollars annually. In 2050, just 35 years from now, it is projected to be 50 Trillion dollars (2), measured in constant 2014 dollars. India's GDP is 2 Trillion dollars annually, today, and projected to grow to 21 Trillion dollars annually in 2050. The GDP of other Asian countries will also grow by large factors. Europe GDP will grow but its growth factor will be much less (2).

Together, the combined GDP of Europe and Asia will be more than 70% of total World GDP in 2050, with enormous amounts of goods and people moving back and forth between the countries of the "Europe-Asia" continent.

Where will the "New Silk Road" go? The Trans-Asian Railway (TAR) Project carried out by the United Nations Economic and Social Commission for Asia and the Pacific (UNESCAP) has studied this question in detail (32). Figure 2.14A shows a map of the TAR Project.

The TAR Project is based on joining existing railway lines to form a network that would serve 26 countries in Asia. Some short railway links would have to be constructed, but almost all of its 81,000 Kilometer (50,000 mile) railway length is already there.

The Trans-Asian Railway Network Agreement was signed by 17 Asian nations on November 10, 2006 and formally came into force on June 11, 2009. Some progress has been made to implement the TAR Project, but the wide range of railway gauges used by the various countries in the TAR Project complicate things.

As in Africa, there are many different gauges in the TAR network, China, Iran, and Turkey use standard 4 feet 8.5 inch gauge, Russia uses 4 feet 11 and 27/32 inch gauge, India and Pakistan use 5 feet 6 inch gauge, Bangladesh and Vietnam use 3feet 3/8 inch gauge, and Indonesia and Japan use 3 feet 6 inch gauge (32). Joining all the different railway gauges into a seamless system will be very difficult.

The Pan Europe-Asia Maglev System, besides providing 300 mph service for passengers, freight, trucks and autos instead of 30 mph railroad trains for passenger and freight, also solves the railway gauge problem, by adapting at low investment, all the different gauge systems to universally use Maglev.

Figure 2.14B shows the layout of 3 transcontinental routes of 300 mph elevated monorail guideway across Asia, that parallel the 4 principal corridors of the TAR Network and the countries they serve(33).

North-East Asia: China, Democratic People's Republic of Korea, Mongolia, Republic of Korea, Russian Federation. This route would parallel the Trans-Siberian Railway (Figure2.3).

Central Asia and Caucasus: Armenia, Azerbaijan, Georgia, Kazakhstan, Kyrgyzstan, Tajikistan, Turkmenistan, Uzbekistan.

South-East Asia: Cambodia, Indonesia, Malaysia, Miramar, Singapore, Thailand, Vietnam.

South Asia and Islamic Republic of Iran and Turkey: Bangladesh, India, Islamic Republic of Iran, Pakistan, Sri Lanka, Turkey.

The total route length of the TAR Network is 50,000 miles. The total route length of the 3 transcontinental, 300 mph elevated monorail routes is approximately 30,000 miles. The remaining 20,000 railway miles of the TAR Network would connect to the 300 mph Maglev routes using the existing conventional railways. Certain sections of the 20,000 miles of existing railway trackage could be adapted at very low investment to use Maglev, though at lower speeds than 300 mph.

The Pan Europe – ASIA Maglev System will have additional 300 mph elevated Monorail routes to the 30,000 miles in the TAR Network. These additional 300 mph routes would be built in 3 regions.

- The Russian-Far East, from Vladivostok on the Pacific Ocean to the Bering Strait crossing where it would connect to the Pan American Maglev System.

- Europe, where there would be 3 elevated 300 mph Maglev routes: across Northern Europe, Central Europe, and Southern Europe, connecting to the North-East Asia Maglev Route through Russia, and the South Asia, Iran, Turkey Maglev Route (Figure 2.15B).Virtually all of Europe's 50 countries would connect to the Pan Europe-Asia Maglev System.

- The Middle East, where 300 mph Maglev routes would serve Iraq, Saudi Arabia, Yemen, Syria, United Arab Emirates, Israel, Jordan, Palestine, Lebanon, Oman, Kuwait, Qatar, and Bahrain. [Egypt is served by the Pan-African Maglev System, while Turkey and Iran are served by the Pan Europe – Asia Maglev System]. The Middle East Maglev route would connect with the Pan-African Maglev System at the Isthmus of Suez.

The additional 300 mph elevated Maglev guideway route mileage in the 3 regions described above is approximately 10,000 miles. The total mileage for the Pan Europe – Asia Maglev System to 30,000 miles (TAR routes) plus 10,000 miles, equal to 40,000 miles At 30 million dollars per 2-way mile for the Maglev Guideway, this is a total of 1.2 Trillion dollars.

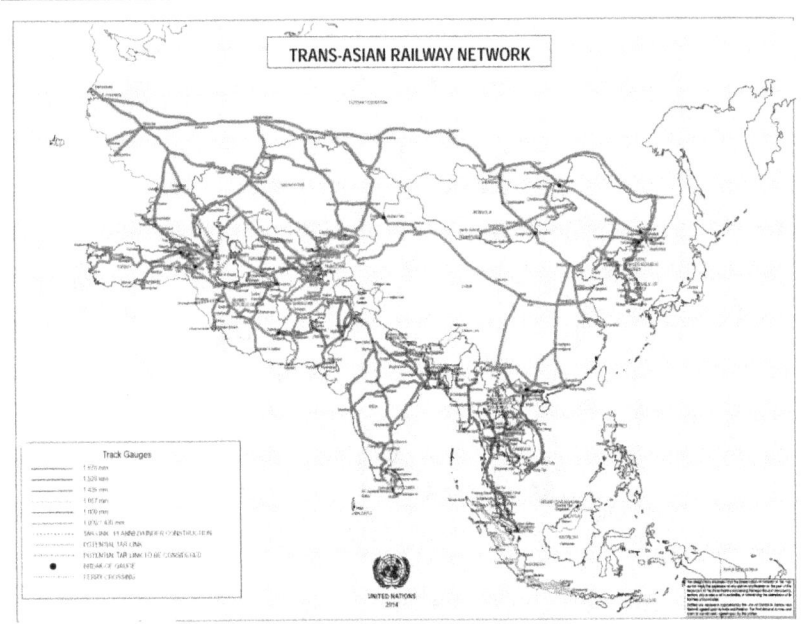

Figure 214A Map of Existing Trans-Asian Railway Network

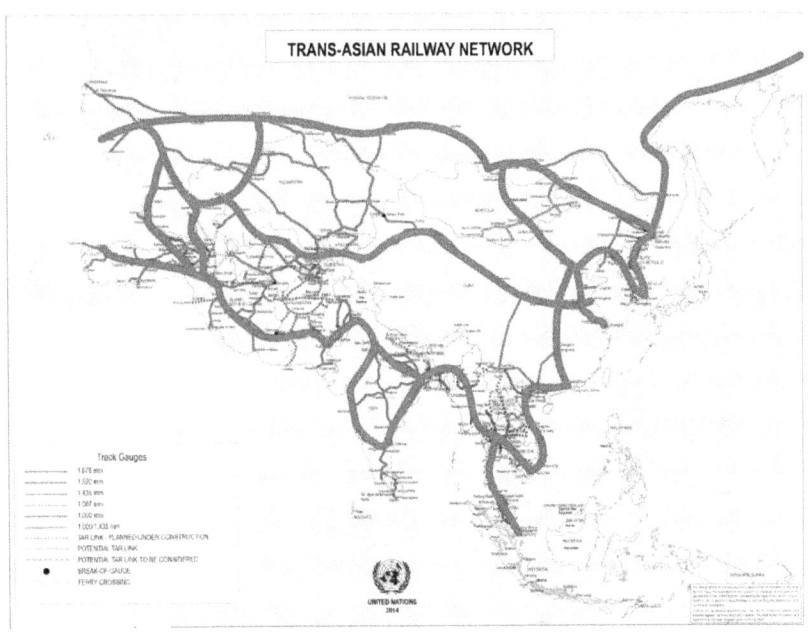

Figure 214B Map pf Pan Europe – Asia Maglev System

133

Table 2.13
Summary of Pan Europe-Asia Maglev System

- Location of 300 mph Elevated Monorail Maglev Routes
    - 3 Routes across Asia – Northern Asia, Middle Asia, and Southern Asia
    - 3 Routes across Europe – Northern Europe, Middle Europe and Southern Europe
    - 1 route in Middle East
- Connections to Maglev Systems in Other Continents
    - Connects to Pan American Maglev System at Bering Strait Crossing
    - Connects to Pan African Maglev System at Strait of Gibraltar Crossing and Isthmus of Suez
- 40,000 Total Route Mileage: 30, 000 miles in Trans-Asia Railway Network (TAR) and 10,000 miles for Routes to Bering Strait, in Europe, and in the Middle East.
- Total Investment for 40,000 Miles of the Pan Europe-Asia Maglev Systems is 1.2 Trillion Dollars
- Cumulative GDP for Europe and Asia from 2015 to 2050 is projected to be more than 2100 Trillion dollars, 1700 times greater than the Pan Europe-Asia Maglev System investment
- 4.9 Billion people would be served by the Pan Europe-Asia Maglev System. 70% of Word's present population of 7 Billion.

To put the investment for the Pan Europe – Asia Maglev System in perspective, consider the population and GDP of the countries it will serve, as shown below:

Populations, Billions		Annual GDP (Trillion USD) (constant $)	
Continent			2050
	2014	2014	(projected)
Europe	0.74	18.5	>31
Asia	4.2	24.4	>94
Total	4.94	43	>125
Total World Pop.	7.0	Total World 73	>188

The annual GDPs for 2050 are taken from the IMF projections of the 20 top economies for that year. Total World GDP will be greater. The projections show that the GDP of Europe and Asia will be an even larger fraction of World GDP in 2050 on the order of 66%, than the already large fraction, 60%, that they are today.

Based on an average annual GDP of 60 Trillion dollars for Europe plus Asia over the 35 year period between 2015 and 2050 which is probably conservative, the accumulated GDP would be 2,100 Trillion dollars, 1,700 times greater than the 1.2 Trillion dollar investment for the 40,000 mile Pan Europe – Asia Maglev System.

Table 2.13 summarizes the principal parameters of the proposed Pan Europe – Asia Maglev system. It will enable a 21st Century Silk Road that will greatly benefit 70% of the World's population, increasing their living standards. The investment required is a very tiny fraction of the GDP of the countries that Pan Europe – Asia Maglev System will serve. It can be built in a relatively short time, following in many regions, the routes of the historic Silk Road, transporting millions of people and billions of tons of freight.

Global Transport of Freight by Maglev Instead of Ships

Today, ships transport an enormous amount of freight. As the Global Maglev Systems begin to operate, this will change. Much of the freight now transported by ship will transition to the Global Maglev Systems.

Before describing how Maglev would transport freight that is presently carried on ships, consider first what kinds of freight ships transport, how big is the shipping industry, how rapidly it moves freight, how much energy it uses, how much it costs, and what are its negative aspects and problems.

World shipping is big business. In 2004, it transported 27.6 Trillion ton miles of freight (34), 10 times greater than the 3 Trillion ton miles of freight transported annually in the US by truck and rail. 6.7 Billion tons are actually shipped, for an average distance of 4,000 miles. Some routes are longer than 4,000 miles, like the 10,000 mile Asia to Europe route.

In 2004, there were 46,200 ships in the World shipping fleet, with a total cargo capacity of 548 million dead weight tons (dwt). Of these, 18,150 were general cargo ships, and 1,733 other ships (34). Ships larger than 80,000 dwt – a major fraction of modern cargo ships – are too big to go through the Panama Canal and must go around Cape Horn.

Ships transport many kinds of cargo, both in bulk form inside the ship – oil, coal, iron ore, etc. – and in closed containers, called TEUs (Twenty Feet Equivalent Unit). TEUs are generally, 20 or 40 feet in length, 8 feet wide, and 8 feet high, with a maximum load capability of 20 short tons (1 short ton = 2,000 pounds). They basically are steel boxes, into which one can load any kind of dry cargo – clothing, electronic equipment, furniture, etc. Large container ships transport thousands of TEUs at a time, as shown in Figure 2.16 for the Hanjin Container ship in San Francisco Bay. The Emma Maersk, one of the World's largest container ships, transports 11,000 TEUs at a time. Total World container ship capacity is 310 million TEUs. Many ports handle 10 million TEUs, or more, per year.

Figure 2.15 Hanjin Container Ship in San Francisco Bay, Approaching Golden Gate Bridge

Suppose country A, in Asia, wants to ship some products from the factory or factories where they are produced – or the farms where they are grown – by truck or rail to a main shipping port, where the cargo is put into a TEU, loaded on board a container ship, which sails 10,000 miles to Europe, taking weeks to do so.

After reaching the appropriate port in Europe, the TEU's are unloaded from the container ship and loaded onto trucks or rail cars for transport to their destination. At 1 to 2 cents per ton mile, depending on the shipping spot price, transporting a 20 ton TEU 10,000 miles would cost in the range of 2,000 to 4,000 dollars.

Using the Pan Europe-Asia Maglev System to transport the 20 Ton TEU 6,000 miles from Asia to Europe at 300 mph would have an operating cost of approximately $1,000 per TEU, including propulsion energy cost, vehicle amortization cost, and system personnel costs.

Added to the operating cost is the amortized investment of the guideway. Based on a 30 year amortization of the 180

Billion dollar investment for the 6,000 mile Northern Asia to Europe route and an annual transport of 10 million TEU's, the amortized guideway cost would be 600 dollars per TEU. Total cost would be 1,600 dollars by the Pan Europe-Asia Maglev System, less than by ship, and much faster, 1 day versus 25 days.

And there are substantial additional economic benefits for shipping on Maglev. No need for long distance expensive transport to and from shipping ports – Maglev stations are much more numerous and much, much more convenient. Much fewer loadings and unloading than for ships. With ships, one needs 1 loading to truck or railcar to get it to the seaport, unloading at the port from the truck or railcar, and 1 loading onto the ship. Then another, 3 unloadings/loadings to get the product to an inland destination on the other side of the Europe-Asia continent, for a total of 6 loadings/unloadings. With Maglev, one loading at a nearby station and unloading near its final destination.

And the much shorter trip time also results in lower inventory costs, economically more efficient match of production to changes in demand, less overproduction, etc. As the old saying goes, "Time is Money". One day trip-times are much better than 25 day trip-times.

On an energy basis, ships are more efficient than trains or high trucks. A typical container ship traveling at 26 knots (30 mph) burns 120 gallons of bunker fuel per mile, corresponding to about 750 ton miles of cargo per gallon of oil fuel.(35) Conventional trains transport freight at about 400 ton miles per gallon of oil fuel, and highway trucks achieve approximately 150 ton miles per gallon.

Consists of multiple Maglev vehicles each carrying 40 tons of cargo, traveling at 300 mph, will achieve approximately 400 ton miles per equivalent gallon, using electric energy, not oil fuel. For routes where ships must travel longer distances than Maglev freight vehicles would, the ton miles per gallon for Maglev can comparable to, or even exceed the ton miles per gallon for ships. For example, for the Asia to Europe route, ships travel 10,000 miles, compared to approximately 6,000 miles for Maglev. The shorter distance for Maglev transport makes its energy consumption per ton for the trip comparable to that for ships. And it only takes 1 day, compared to 25 days.

There are substantial negative aspects for ocean ships. First, ships burn more than 7 Billion gallons of oil per day, almost 10% of the World's oil consumption (36), producing 1.3 Billion metric tonnes of carbon dioxide per year, about 5% of total World emissions. One large container ship emits as much sulfur oxide pollutants as 50 million automobiles do. World ships emit 20 million tons of sulfur dioxide per year, while all of the World's autos emit only 79,000 tons. NOAA (National Oceanic and Atmospheric Administration) estimates that 60,000 people in the World die prematurely from the pollutants and microparticles emitted by the World's ships (37).

The Global Maglev System will provide fast, efficient, low-cost, transport of passengers and freight across continents and interconnect Europe, Asia, North American, South America and Africa into a World Wide Network. Once implemented, Maglev will transport much of the cargo now carried by ocean ships, with major economic and environmental benefits. Ocean shipping will still be necessary, but the amount will be much smaller, greatly reducing the consumption of fossil fuels and their negative environmental effects.

The investment for the Global Maglev Network is projected to be on the order of 2.7 Trillion dollars, less than 1/1000th of the more than 4,000 Trillion dollars of the sum of the annual World GDP's for the period from 2015 to 2050. The transport savings from the Global Maglev Network will far exceed its investment cost, and the environmental and human benefits will be enormous. The Global Maglev Network will enable safer travel with much fewer accidental deaths and injuries.

Figure 3.1

Artist's Depiction Of A StarTram Maglev Launched Cargo Spacecraft Exiting From A Vacuum Launch Tube Constructed On The Side Of A Very High Mountain Peak Like Mt. Saint Elias. Altitude 18.000 Feet, on the Southern Coast Of Alaska. This Is Near Perfect Place For An Over Ocean Launch From US Territory. (Credit Jesse Powell and Mathias Verhasselt)

Chapter Three
StarTram Maglev Launch – The Low-Cost Road to Space

James Powell, George Maise, Charles Pellegrino, Jesse Powell, John Skaritka, and James Jordan

"Why Do You Want to Climb Everest?" – Questioner
Because it's there." – **George Leigh-Mallory**

"Oh, ye'll take the High Road
And I'll take the Low Road,
But I'll be in Scotland afore ye"
Old Scottish Folk Song

StarTram What is It? And Why We Need It.

StarTram (1) is a revolutionary new way to launch payloads and people into space – much less expensive and with much greater volumes than possible using rockets.

Since Neil Armstrong stepped onto the Moon on July 20, 1969, we humans have accomplished amazing things in space – Mars rovers, spacecraft journeys to the outer planets, orbiting satellites for communications, GPS, weather observation, and so on. But we've only gone a little way into space and its been very expensive. The Apollo Program to the Moon cost $170 Billion dollars, in today's dollars (2), and the International Space Station has cost 150 Billion (3). Using rockets, it costs on the order of $2,000 to$5,000 per pound ($4,000 to $10,000 per kilogram) to launch payloads into Low Earth Orbit (LEO) (4). To place payloads in Geosynchronous Earth Orbit (GEO), 22,000 miles from Earth, the cost is 3 to 4 times greater. To the Moon, Mars, and beyond, much, much greater costs.

With its much lower launch costs, StarTram will carry out 4 critically important missions, not possible with rockets, because of their very high launch costs. The 4 missions and their benefits are:

1. Power Earth from space with clean, non-polluting, low cost electric energy, eliminating the need to burn fossil fuels for electricity.
2. Protect Earth from asteroid and comet impacts, and reduce solar energy input, if necessary, to prevent global temperatures from reaching the danger point.
3. Provide a higher quality of life on Earth by more and better communication and observation satellites, and mining scarce resources from the asteroids, moons and planets in the solar system.
4. Permit humanity to really explore the Solar System. Colonize the Moon and Mars, look for evidence of past or present extraterrestrial life on Europa and other bodies in the Solar System that have water/ice, orbit large telescopes to image other star systems in detail, and so on, to better understand humanity's place in the Universe.

Not only will Star Tram provide a safer, more secure, and better life for humanity on Earth, by opening up the vast new World of space, it will serve to unify humanity as it journeys out to Space, helping to eliminate the conflicts and wars that humans have endured for many thousands of years.

Why will StarTram have much lower launch costs than rockets? Because it keeps all of its expensive launch equipment on the ground, to be used many times, where rockets throw most of their equipment away when they launch. Think of a company that wants to deliver a refrigerator to a customer. It has 2 choices. Either deliver it by an expensive helicopter that crashes after delivery, or deliver it by truck, which can be used repeatedly for many deliveries, with its gasoline and driver time the only cost for each delivery. Which choice would the company take? The truck, of course.

Figure 3.2 shows the Saturn V, the rocket that launched Neil Armstrong and his comrades in 1969 to the Moon. 363

feet in height, it was taller than the Statue of Liberty, and weighed 6.5 million pounds. It could deliver 260,000 pounds to LEO. 4 percent of its takeoff weight ended up payload – the rest as trash (5).

Figure 3.2

NASA developed the space shuttle (Figure 3.3) to be a reusable launch system to carry cargo and astronauts to the International Space Station (ISS) with the hope that it would reduce launch costs. The shuttle was successful, but still very expensive. At 450 million dollars operating cost per launch, it cost $8,000 per pound to LEO. Including all costs, the whole space shuttle program (it closed down in 2011) cost 196 Billion dollars when adjusted for inflation (6). For the total of 134 Shuttle missions, the average cost per mission was 1.5 Billion dollars.

Besides the very high launch cost of rockets there is the problem of rocket failures. They average 9% -- that is, about 1 out of 10 launches fails. The space shuttle had a lower failure rate, 2 crashes (Challenger and Columbia) out of 133 launches, with all of their crew killed. That's a fatality rate 100,000 times greater than airplanes.

A detailed description of StarTram and Maglev launch is given in the book, *"StarTram-The New Race to Space"*, by James Powell, George Maise and Charles Pellegrino, available at Amazon.com.

Figure 3.3

Astronauts are willing to accept such a very high risk. Non-astronauts will not. If ordinary humans are to go into space, the accident risk will have to be much lower, closer to airplanes. It is doubtful that rockets will ever be able to achieve such a low fatality rate. StarTram launches should have a much lower accident rate than rockets. The present (2013) World Space Launch rate is approximately 315 tons, annually. (7) There was a total of 81 launches, 48 to LEO, 32 to higher orbits than LEO, and 1 to deep space beyond Earth. There were 3 launch failures for a launch success rate of 96%. Average payload mass per launch was 3.9 tons. The maximum payload launched was 19.7 tons by an Arianne 5-ES rocket launch vehicles. 23 different types of launch vehicles were used for the 81 launches, an average of 3.5 launches per rocket vehicle type.

Despite 4 decades of intense efforts to reduce the high cost of rocket launch and enable much greater launch volume, only marginal improvements have been achieved. The landscape is littered with dead rocket systems – the space shuttle, Titan, Tsyklon, Molniya, START, NASP, Sanger, X-33, X-34, and many others. Rockets appear to have reached their capability limits, with major advances in the future unlikely.

How does StarTram work? Instead of accelerating spacecraft and their payloads with rocket thrust generated by the combustion of propellants, StarTram uses Maglev. The Spacecraft is magnetically levitated and propelled in an evacuated tunnel by the inductive interaction between superconducting loops on the spacecraft and ordinary loops on the tunnel wall. There is no air drag, only the very tiny drag due to electrical resistance losses in the aluminum loops on the tunnel walls.

The cost of the electrical energy required to reach orbital speed in the evacuated tunnel is very small. At a launch velocity of 8 kilometers per second, a kilogram of mass has 32 megajoules (8.9 kWh) of kinetic energy. For a 90% efficient Maglev launch system, the input electric energy would be 10 kWh per kilogram, with an energy cost of $1 per kilogram (45 cents per pound) at 10 cents per kWh. At 10.6 kilometers per second, the launch velocity to reach GEO orbit, the cost per kg is greater, $1.80 per kilogram, but still tiny.

After reaching orbital speed in the evacuated tunnel, the StarTram spacecraft exits into the atmosphere through an MHD (Magneto Hydro Dynamic) window. It prevents the outside atmosphere from flowing into the evacuated tunnel by MHD forces acting on air ionized by an RF discharge. It then coasts upwards to orbit, where an attached small rocket burn finalizes the orbit, preventing the spacecraft from falling back to Earth. For a LEO orbit, the rocket burn is very small, 0.34 km/sec (Figure 3.4). For a GEO orbit the rocket burn is 1.5 km/sec, but still small compared to the 10.6 km/sec required to reach GEO (Figure 3.5).

The StarTram spacecraft are magnetically levitated by the magnetic interaction of the superconducting loops on the spacecraft with the ordinary aluminum loops on the wall of the evacuated launch tunnel. The levitation is inherently strongly stable, with magnetic force keeping the spacecraft centered in the tunnel. In addition to the sequence of aluminum levitation loops on the tunnel walls, there is a second set of aluminum loops carrying AC current. The AC current loops magnetically push on the superconducting loops on the spacecraft, accelerating it. Spacecraft speed is controlled by the frequency of the AC current. As the frequently increases, speed increases.

Figure 3.6 shows the layout of the spacecraft loop geometry for levitation and propulsion and how it keeps the spacecraft centered in the tunnel as it accelerates. The StarTram spacecraft uses the same basic Maglev technology that already operates in Japan's 300 mph 1st Generation Maglev passenger system (Figure 3.7), and the proposed 2nd Generation Maglev System (Figure 3.8), described in Chapter 1 on the National Maglev Network and Chapter 2 on the Global Maglev Network.

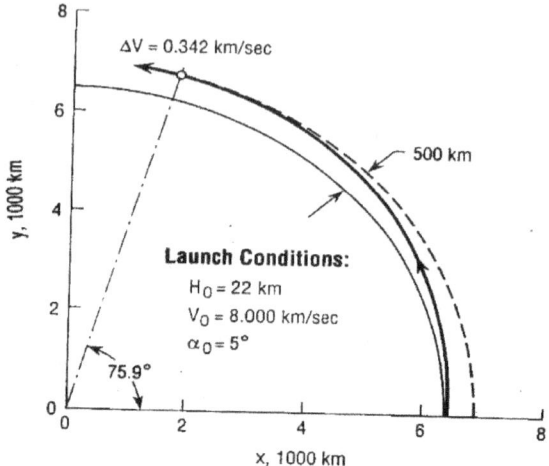

Figure 3.4: Ascent Trajectory to LEO Using StarTram Launch

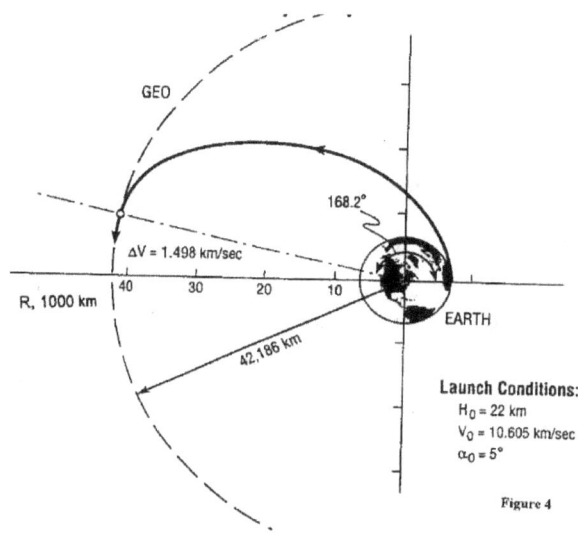

c

Figure 3.5: Ascent Trajectory to GEO Using StarTram Launch

Layout of Cargo Craft Geometry

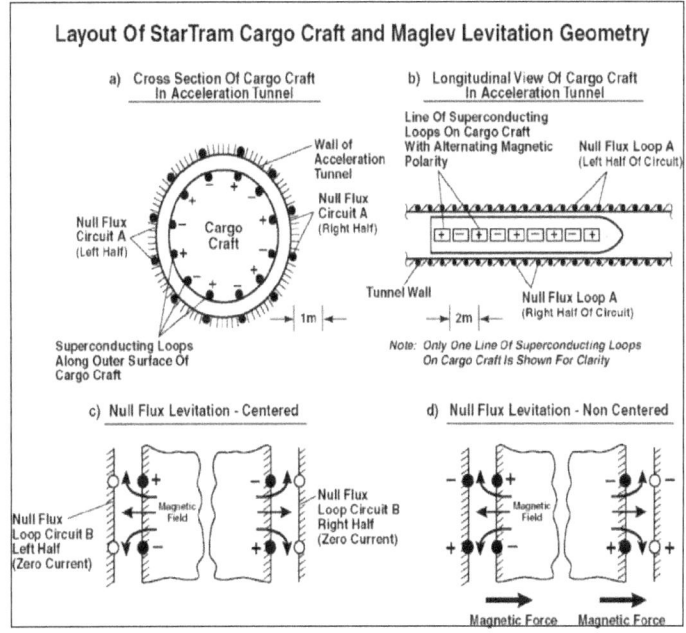

Figure 3.6: Layout of Gen-1 Cargo Craft Geometry.

Figure 3.7

Figure 3.8

Two StarTram Systems have been studied. The near term first generation Gen-1 System would launch spacecraft with cargo but not passengers. After reaching orbital speed in the evacuated tunnel, which would be located in high altitude terrain, e.g. at an elevation of 4,000 meters (13,000 feet) or greater, The Gen-1 cargo craft exits from the tunnel at ground level and climbs upward through the remaining low density atmosphere to space (Figure 3.8).

The aerodynamic deceleration forces and heating of the Gen-1 cargo craft as it ascends through the atmosphere are strong, but manageable – significantly less than the heating and forces experienced by the existing re-entry vehicles for ICBM's (Intercontinental Ballistic Missiles) as they come down from space through the atmosphere on their way to their targets. The re-entry vehicles have to traverse the full atmosphere, including the high density lower portion, while the Gen-1 cargo craft can carry much greater amounts of protective coatings and transpiration coolants than is possible for existing re-entry vehicles, which still manage to survive their downward journey until they hit their targets.

There are many potential high altitude launch sites around the World for the Gen-1 System – in Alaska, Russia, China, the Andes, Africa, Antarctica, Himalayas, and so on. Depending on site location, Gen-1 cargo craft can be launched into polar or equatorial orbits. Some sites are at very high elevation, like Gongga Shan in Szechwan Province in China, where the Gen-1 tunnel exit could be at an altitude of approximately 7,000 meters (23,000 feet).

The atmospheric deceleration forces on the Gen-1 cargo craft are too high, on the order of 8 to 10 g, for passengers. These g forces could be reduced to a tolerable level, e.g. 2 to 3 g by increasing the mass of the cargo craft and incorporating a rocket thruster to counter some of the atmosphere deceleration force. This system, termed Gen-1.5, could transport substantial numbers of persons into orbit.

For the long term, if large numbers of humans are to journey into space, the Gen-2 system may be the way they get there. To avoid high deceleration forces on the StarTram spacecraft as it ascends up through the atmosphere and transitions from the ground level acceleration tunnel into a magnetically levitated evacuation tube (Figure 3.9). The tube has a set of superconducting cables attached to it that magnetically interacts with a set of superconducting cables located on the ground surface beneath the tube. Carrying millions of Amperes of current in opposite directions in the 2 sets of cables, the magnetic repulsive force between them levitates the evacuated tube.

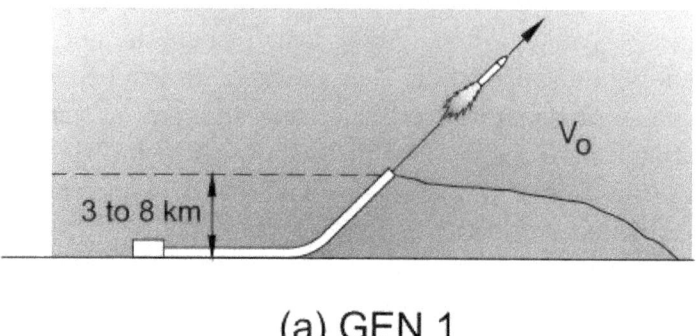

(a) GEN 1

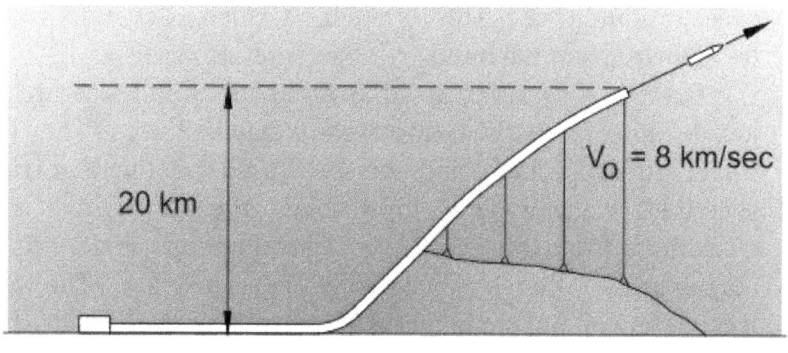

(b) GEN 2

NOTE: NOT TO SCALE

Figure 3.9: Gen-1 and Gen-2 Launch Systems.

The upper end of the evacuated tube is at an altitude of 20 kilometers (65,000 feet). The StarTram Spacecraft exists through the MHD window (Figure 3.10) into the very low density atmosphere, 5% of the air density at ground level, making the aerodynamic heating and deceleration forces very low.

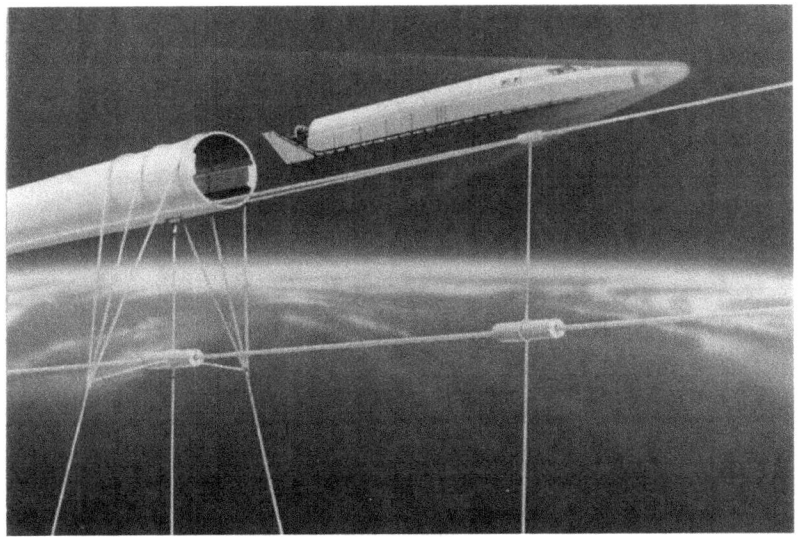

Figure 3.10: View of StarTram Exiting the Launch Tube

The Gen-1 StarTram cargo launch system can be implemented in as little as 10 years with an aggressive, well-funded development program. It should be a cooperative international program that focuses on important peaceful applications. If carried out separately by independent nations, it could become a very dangerous and destructive arms race, as described in more detail later in the chapter.

The longer term Gen-2 StarTram passenger and cargo launch system is much more technically challenging than the Gen-1 system. Gen-1 uses the already existing basic technology for Maglev, MHD, superconducting energy storage, and aerodynamic forces heating on re-entry vehicles. Gen-2 requires developing magnetically levitated high altitude structures, a whole new R&D area. Implementing it will require much greater funding and development.

Table 3.1 gives basic parameters for the Gen-1 system. One Gen-1 launch facility can launch 10 to 20 cargo craft per day, 330 days per year. The 40 metric tonne cargo craft carries a payload of 30 metric tonnes with 10 tonnes for the cargo craft structure and the small AV rocket that establishes the final orbit. Total payload launch rate for one Gen-1 facility can be 100,000 metric tonnes or more, annually. 100s of times greater than the present total launch rate for the whole World. Projected launch cost is 20$ per pound ($43/kg), a factor of approximately 100 times smaller than present launch costs to LEO. Launch cost includes amortization of the construction cost of the StarTram facility, cargo craft cost, and personnel & energy cost.

What are the missions and market for StarTram? The first is Powering Earth, by beaming of electrical power down to Earth from Space Solar power satellites in Geosynchronous orbits. See Figure 3.11 for concept. This application is described in Chapter 4.

Beamed solar power from space is extremely important, because with it we would not need to consume fossil fuels to generate electricity. We already generate non-fossil renewable power from wind and solar farms on Earth, but they have problems – intermittent operation, low average electrical outputs, expensive, negative environmental impacts, and damage from storms and bad weather.

Beamed solar power from space is continuous, 24/7, can readily adjust output to meet peak demands, doesn't harm the environment, cannot be damaged by storms and bad weather, and will be low in cost. Wind power is extremely variable, cannot operate in low or high wind conditions, and its capacity factor is very low, about 30%, delivering power only when wind speed is right. Ground solar farm electrical outputs vary with time of day, cloud cover, etc.

Table 3.1

Gen-1 Projected Capital and Operating Costs			
Capital Costs, $B (Nth Facility)		**Operating Cost Per 30 Ton Cargo, $M**	
Hard rock tunneling (265 km total length, $1500/m3 excavation)	3.4 B	Cargo Craft Structure (5 metric tonnes, $100/kg)	$0.5 M
Aluminum guideway loops ($20/kg)	0.9 B	Superconductors on craft ($2/KA meter)	$0.43 M
SC for energy storage ($2/KA meter)	0.9 B	Personnel (50 man-days @$500/man day)	$0.02 M
Power Conditioning DC to AC for Maglev acceleration ($100/KWe)	1.0 B	Energy cost (10cents/KWH)	$0.04 M
Vacuum & refrigeration systems	1.0 B	Total Operating Cost Per Launch	$0.99 M
Prime power plant [300 MW(e), $3000/KW(e)]	0.9 B	10 year amortized capital cost per launch (3650 launches per Year	$0.52 M
Buildings and facilities	2.0 B	**Total Cost Per Launch =**	**$1.5 M**
Total Capital Cost	9.1 B	**Cost Per kg of Payload =**	**$50/kg**

The proposed SPS-ALPHA DRM-5 Space Solar Power System (8) for delivering 2 Gigawatts (e) [2000 Megawatts (e)] has a mass in GEO orbit of 35,000 metric tonnes, equivalent to 17.5 kg/kW (e). The StarTram launch cost to GEO will be greater than to LEO because of the higher launch velocity required to reach GEO orbit. At $100 per kg, the corresponding launch cost per kW(e) of power delivered to Earth would be $100 x 17.5 or about $1,750 per kW (e).

Amortized over 30 years, the launch cost corresponds to about 0.7 cents/kWh, small compared to the 10 cents/kWh for today's electric energy. There are, of course, substantial additional costs for the space power system hardware and the power receiver station on Earth that the power is beamed to. These additional costs are described in Chapter 4.

Using StarTram, launch cost will not be a significant factor in the economic practicality of space power beaming. Using rockets, with a launch cost of $10,000 per kilogram and a mass of 17 kg/kW(e) in GEO, the capital cost of a kilowatt of power from space would be $170,000 per kilowatt, just for launching it – far too expensive to be practical.

One StarTram facility with an annual launch rate of 175,000 tonnes per year could put 10,000 megawatts of beamed space power capacity annually. Over a 10 year period the accumulated beamed power capacity would be 100,000 megawatts. Over 10 years, 5 StarTram launch facilities could put 500,000 megawatts of beamed power into space – enough to meet all of America's present (2013) annual power needs of 4.26 trillion Kilowatt hours (9).

To meet the present (2013) total annual electrical World production of 23.1 Trillion Kilowatt hours (9), 2,600,000 megawatts, a factor of 5 greater than US electrical generation, of beamed electric power would be required. To do this over a 10 year period, 25 StarTram facilities would be needed, each launching 175,000 tonnes per year. Total launch mass over the 10 year period would be 40 million tonnes in GEO orbit – a factor of 13,000 greater than the payload that rockets would launch over a 10 year period.

And that's at present electrical generation rates. The Energy Information Administration (EIA) forecasts an increase to 39 Trillion hours in 2040 AD.(10) And that doesn't include new markets for electrical consumption like electric cars, electrical heating of homes and buildings instead of using fossil fuels, making synthetic fuels for transport, as described in Chapter 5, and other applications – even greater launch rates will be required.

Clearly, beamed space solar power will not be possible using rocket launch. Not only would the launch costs be much greater than World GDP, but the billions of tons of propellant exhausts would do unacceptable harm to Earth's environment.

In contrast, StarTram's very low launch cost and very high launch volume will enable it to carry out its **1ˢᵗ Major Mission, Powering Earth**. Chapter 4 describes the mission in greater detail.

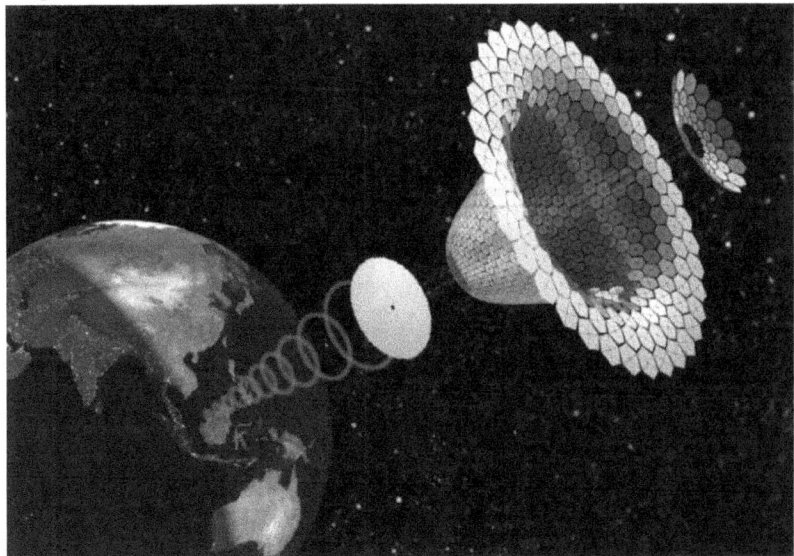

Figure 3.11: SPS-ALPHA concept for space solar power: reflectors concentrate sunlight onto module array. Satellite beams power to receiver from geosynchronous orbit. (Image: John C. Mankins)

We turn now to the 2nd Major StarTram Mission, Protecting Earth. Over its Billion year history, Earth has been hit many times by large objects – asteroids, comets, etc., -- that have had enormous effects. The Moon was ejected from our planet when a very large body collided with the Earth, soon after the birth of the Solar system. A 6 mile diameter asteroid hit the Earth 65 million years ago, wiping out the dinosaurs and letting us mammals take over.

Well, another asteroid, Apophis 99942, is on its way toward us. On Friday, April 13, 2029 it will either pass by us very closely or hit Earth. Figure 3.12 shows a far view of its approach to Earth, with Figure 3.13 a close-up view. The predicted miss distance is 18,300 miles, plus or minus an uncertainty of 8,000 miles. (11)

If Apophis misses us in 2029, it will be back in 2036 for another try. Some scientists predict that perturbation by the gravitation pull of the Earth during its 2029 flyby could alter its path enough to make it hit the Earth in 2026. (12) Figure 3.14 shows one prediction of the 2036 risk path.

Being realistic, nobody knows whether or not Apophis will impact the Earth in 2029 or 2036 or miss us. There are too many unknowns and uncertainties to make precise predictions. One thing we do know. If Apophis hits Earth, the effects will be catastrophic.

The impact energy would be equal to about 500 megatons of TNT – 50,000 times more powerful than the atomic bomb that leveled Hiroshima and twice the energy of the 1883 volcanic eruption of Krakatoa. The Krakatoa explosion was heard 3,000 miles away and killed at least 36,000 people, with some estimates as high as 120,000 (13). Five cubic miles of rock and ash were ejected by the explosion. The shock wave reverberated 7 times around the World. Average global temperatures fell by 1.2 degrees Centigrade and did not return to normal until 5 years later. If Apophis hit in the ocean, the resulting tsunamis would kill millions.

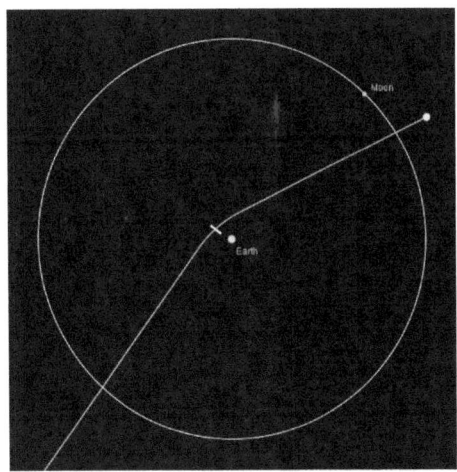

Figure 3.12.
Apophis Pass by of Earth. The Close Approach of 99942 Apophis to the Earth and Moon on Friday, April 13, 2029 Far View

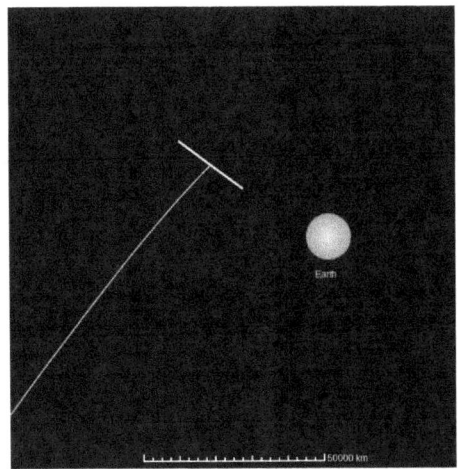

Figure 3.13

The Close Approach of 99942 Apophis on April 13, 2029. The White Bar Indicates Uncertainty in the Range of Possible Positions

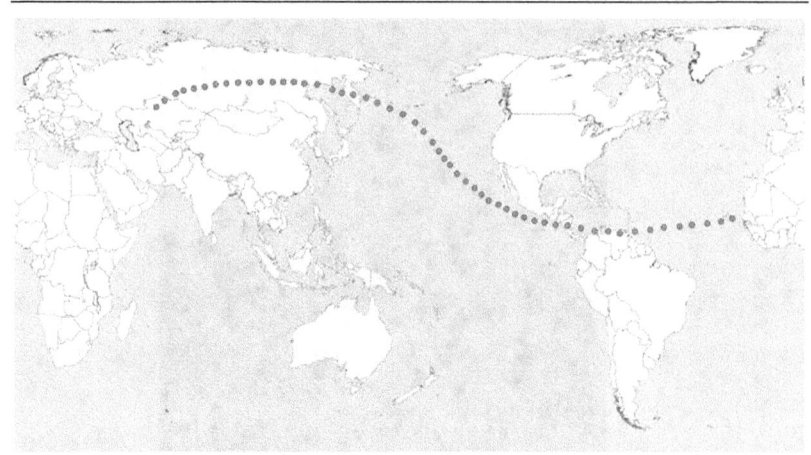

Figure 3.14: Apophis Path of Risk in 2036

Figure 3.15: The Tunguska Meteoroid Knocked
Down Trees Over an Area of Hundreds of Square
Kilometers in Siberia in 1908

The most recent impact was the Tunguska meteoroid
event that hit Siberia in 1908. Small on the scale of Apophis, 10
megatons compared to 500 megatons, it still was devastating.

It flattened 80 million trees (Figure 3.15) over an area of 830 square miles. If it had impacted a major city, instead of remote Siberia with virtually zero people, millions would have died.

In fact, big objects are passing Earth all the time – 2 travel past us every day at a distance from Earth that is less than that distance to the moon. On January 15, 2013, an aircraft sized object (catalog 2012 DA-17) missed Earth by only 17,000 miles. On the same day, a 300 kiloton solid exploded in the atmosphere above Russia, injuring 1,200 people.

Only 1% of the possible asteroid impactors presently in the solar system are catalogued. And there are many more potential unknown impactors that come in at enormous speeds, 50 kilometers per second or more, from beyond the Solar System – the Kuiper Belt, the Oort Cloud, and beyond.

We also need StarTram to protect Earth from asteroid and comet impacts. The protection would be accomplished in 3 ways.

First, by sending up telescopes that can fully catalogue all potential hazardous objects in the solar system.

Second, by sending spacecraft to the potential hazardous objects that would sufficiently shift their orbits so that they would never come close to the Earth. A safe distance for passing Earth would be on the order of a million miles.

Third, by stationing a large fleet of high velocity interceptors in Earth orbit that could rapidly respond to hazardous asteroid and comets coming in from beyond the Solar System, intercepting them and destroying them at a far distance from Earth.

With StarTram, the chances of a disastrous asteroid or comet impact on Earth should be virtually zero, because of the very large amounts of interceptors that it can launch into space. Rockets alone will not be able to completely protect the Earth from asteroids and comet impacts.

However, asteroids and comets are not the only danger that threatens Earth. Global warming from greenhouse gas emission is heating the Earth more rapidly than at anytime in its geologic history. At some point, as discussed in the

Prologue, the warming will become irreversible and impossible to stop, even if the consumption of fossil fuels stops, due to CO_2 and methane releases from the warming permafrost and methane hydrate deposits in the ocean beds.

What recourse will humanity have if global warming is reaching the tipping point for runaway warming? There are 2 possible ways to avoid the resulting environmental catastrophe. The first is to extract carbon dioxide from the atmosphere and bury it underground. This would require enormous amounts of electric power, much greater than the amounts the World now generates. In principle, StarTram beamed power from space could be used, but the launch rate would have to be much greater than that described earlier for powering the Earth.

The second possible way to counter global warming would be to reduce the amount of sunlight that strikes the Earth. Angel (14) has investigated the feasibility of placing a cloud of very thin discs at Lagrange Point, 1 between the Earth and the Sun, that would block out a portion of the sunlight that normally would strike the Earth.

The LaGrange Point 1 is at a distance of about 1 million miles from Earth. Angel estimates that a total reflector disc area of 4.7 million square kilometers at the LaGrange Point 1 would reduce sunlight falling on the Earth by 1.8 percent, enough to have a substantial cooling effect. The estimate launch weight using electromagnetic coil guns is 20 million tons.

StarTram would provide a more practical way to launch reflector discs to the LaGrange Point. Or it could launch an alternative proposed approach to reduce the sunlight falling on the Earth, a set of thin Fresnel lens that would spread the sunlight beam instead of reflecting it (Figure 3.16)

Rather than using Angel's concept of launching a vast number of 1 meter diameter reflecting discs to block the sunlight, magnetically inflatable reflecting discs could be used. Launched by StarTram as small compressed packages into space, upon energizing the current in a network of attached superconducting cables, the disc would expand into its full size of 1 square kilometer disc of 1 micron thick reflecting film. With its cables and refrigeration equipment plus reinforcement structure, the total weight of the 1 square kilometer reflecting disc is 10 tonnes.

For a 4.7 million square kilometer sunshield, the total launch cost would be 2.4 Trillion dollars at a StarTram Launch Cost of $50 per kilogram. Adding in the manufacturing costs for the disks, the total cost for the sun shield would be 5 Trillion dollars, less than 1% of the 10 year World GDP. Protecting Earth against environmental catastrophe for 1% of its economic output – seems worth it.

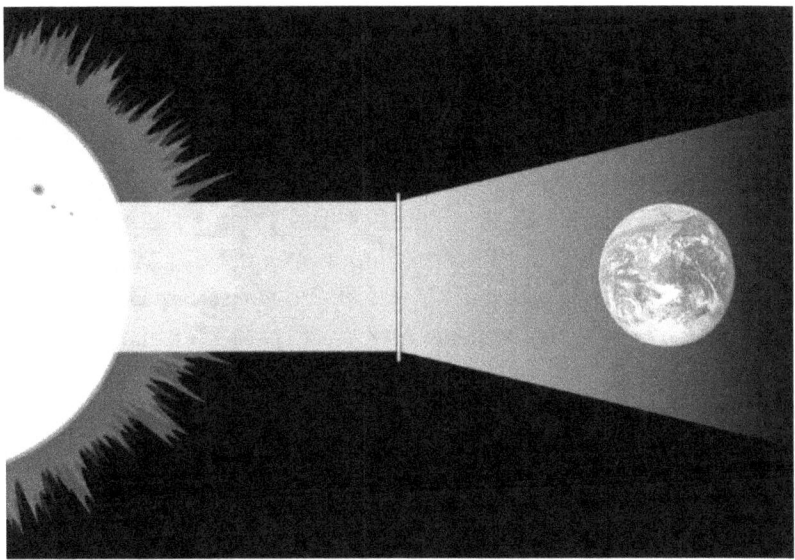

Figure 3.16: How a Space Lens Would Mitigate Global Warming

The 3rd mission, **Providing a Higher Quality of Life on Earth**, through being able to launch many more satellites with much greater capabilities than presently possible, is clearly very important. The present launch rate is very low, 81 per year, and the average weight only a little over 3 tons of payload per launch.

StarTram's much lower launch costs will enable placing much greater numbers of satellites in orbit. It will also enable much cheaper satellites since their launch mass budget can be much greater, allowing lower, heavier, more rugged materials, mass production economy of scale, and simpler, lower cost designs.

Instead of 81 satellite launches annually, there could be thousands. Satellite mass could easily be increased by a factor of 10, enabling much greater capabilities. Low cost, world-wide communications, no need to lay cables or build cell towers – full broadband phones, TV, internet everywhere, even in remote and poverty areas anywhere in the World.

Much greater satellite observation capabilities to monitor weather, track storms and storm damage, drought effects in much greater detail, with much greater predictive and assessment abilities. With very detailed measurement of ground movements from space, it may be possible to predict earthquakes. Highway traffic flows could be determined much more accurately. Pollution sources and their effects could be detected much better and remedied more quickly. And so on.

Turning now to the 4th Mission, **Permitting Humanity to Explore the Solar System**; StarTram will at last allow humanity to move out into the Solar System and beyond. In the last 50 years, humans have done wondrous things in space – astronauts on the Moon, the Mars Rovers, the Voyager Spacecraft, journeys to Jupiter, Saturn, Venus, Mercury, asteroids, comets, and other bodies, the International Space Station (ISS), the Hubble Telescope, and much else.

One may ask, how will exploring the solar system, help humanity on Earth? In many ways – mining asteroids for rare materials, bringing Helium-3 back to Earth for future fusion reactors, and so on. The big benefit will be to help unify humanity with a common purpose – to find out what and who is out there. The Universe is much bigger than Earth. If humanity is to have a greater destiny than conflicts in wars on Earth, it must unify.

However, we have only scratched the surface. With StarTram, humanity will be able to establish bases on the Moon and Mars for people to live, work, and explore. Figure 3.17 shows a NASA concept drawing of the Lunar Base.

Figure 3.17: Lunar Base Concept Drawing from NASA

The much lower launch cost and much greater launch volume using StarTram will enable practical transport of supplies to maintain large permanent bases on the Moon at lower cost than the Apollo Program, in which only 2 astronauts per mission stayed on the Moon for a few days before they ran out of supplies and had to return to Earth. 170 Billion dollars for landings on the Moon, each lasting only a few days. With StarTram, at $150 per kilogram, 3 times the cost to LEO, for launch to the Moon, 15 Billion dollars would put 100,000 tons of supplies and equipment on the Moon's surface – enough to build and supply some real bases.

Similarly, StarTram will enable permanent colonies on Mars. Direct launches to Mars using StarTram alone are not practical. The Delta V requirements are too great. Instead, StarTram would launch cargo into High Earth Orbit (HEO), to be loaded into spacecraft for the journeys to Mars. The cargo would be transported by unmanned or manned spacecraft, depending on mission requirements. For fast, efficient journeys, nuclear rockets would be the propulsion choice, with hydrogen propellant supplies and equipment launched from Earth.

The North Polar Cap of Mars, viewed by the Hubble Telescope (Figure 3.18) is the ideal location for a permanent base/colony on Mars. It has the raw materials, water from the ice cap and carbon dioxide and nitrogen from the atmosphere to manufacture large amounts of breathable air, propellants, grow food, plastic construction materials, and other products, greatly reducing the supplies required from Earth – primarily small amounts of medicines, electronics, and other hi tech equipment.

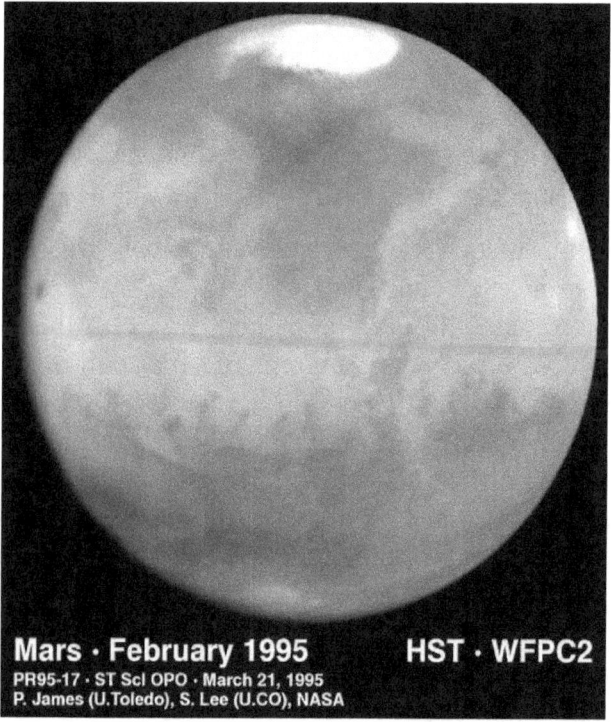

Mars · February 1995   **HST · WFPC2**
PR95-17 · ST ScI OPO · March 21, 1995
P. James (U.Toledo), S. Lee (U.CO), NASA

Figure 3.18: Hubble Space Telescope View of Mars, Showing Polar Ice Cap

Moreover, the breathable air, propellants, food, etc., can be manufactured robotically by equipment landed on the Mars site before humans take off for Mars, and already there waiting for them when they land. The robotic ALPH factory would be powered by a compact nuclear reactor, generating on the order of 1 megawatt (e) electrical power and 4 megawatts of heat. The reactor would utilize existing nuclear technology. Using heat from the reactor, ALPH would melt large cavities inside the Polar Cap (Figure 3.19), for storage of the oxygen, propellants, food, and other supplies it would manufacture. ALPH would also create thermally insulated cavities inside the Polar Cap as habitats for the humans living on the Mars Base. The ice above the living areas would shield the inhabitants from cosmic radiation – a serious problem for astronauts that traveled to Mars and were located on its surface.

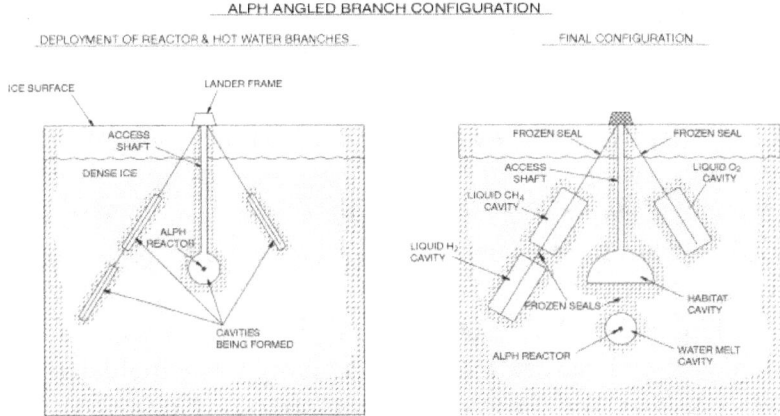

Figure 3.19: ALPH Construction of Manned Base Inside North Polar Ice Cap

All of the above supplies and habitats would be available to the first humans that landed on Mars – they only would have to step out of their spacecraft, and begin operations. Using the North Polar Cap as their base they could explore the entire planet in detail, using robotic and human operated flyers and land rovers, manufactured at the base. Using robotic probes they could explore the kilometer Polar Cap in great detail, to determine Mars past geologic history from the deposits within it from volcanic eruptions, wind storms, and asteroid/comet impacts. They could also search for evidence of past life on Mars – bacterial, fossils, etc.

As time went on, the number of humans at the Polar Cap would increase from tens of persons to hundreds, becoming a real colony. Other bases would be established on Mars. Inhabitants would not have to permanently stay on Mars, but could return to Earth after a few years, if they desired.

An important mission enabled by StarTram will be to search for life forms in the Solar System that did not originate on Earth. We had our first life form, LUCA, the Last Universal Common Ancestor (15) spring into being about 4 Billion years ago, probably at hot hydrothermal vents in the ocean floor. From our LUCA ancestor, all present life forms evolved – in the sea and on land. Today, the hydrothermal vents are populated by a dazzling variety of life. Archaea bacteria ingest dissolved minerals from the vents, converting them to food, using the energy from chemical reactions, not sunlight. They form the basis for extended food chains, supporting the Pompeii worms, vent crabs, lobsters, barnacles, octopi, limpets, scale worms, clams, eelpout fish, and many other weird types of organisms that are only found at hydrothermal vents.

Liquid oceans exist beneath the surface of some of Jupiter's and Saturn's moons. With the strong tidal forces from Jupiter and Saturn, hydrothermal vents probably exist. Europa, Jupiter's Moon, is a prime candidate for searching deep oceans extending all around Europa. The strong tidal forces from Jupiter heat the rock under Europa's ocean, probably producing hydrothermal vents.

Launched by StarTram, the NEMO (Nuclear Europa Mobile Ocean) spacecraft, named after Captain Nemo in Jules Verne's "**20,000 Leagues Under the Sea**", would journey to Europa. Landing on Europe's icy surface it would deploy a compact nuclear powered probe that would melt down through the multi-kilometer thick ice sheet to the ocean below (Figure 3.20), where it would deploy a compact nuclear powered submersible that would investigate Europa's oceans for evidence of extraterrestrial life forms. Samples would be collected and loaded on to the NEMO Spacecraft for return to Earth, using on-board nuclear power to produce propellant for the return journey by electrolyzing water from the ice sheet.

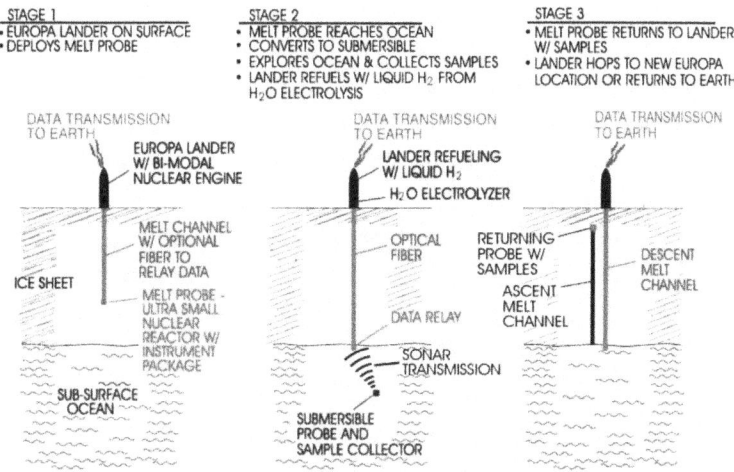

Figure 3.20. Exploration of Subsurface Europa Ocean by the NEMO Probe

StarTram's ability to launch large volumes of payload at very low cost will generate a New Era in the exploration of all of the bodies in the Solar System in great detail, something not possible with rockets.

With StarTram, humanity will be able to explore all of the Solar System's planets and moons, fly nuclear powered aircraft in the atmospheres of the giant planets, bring Helium-3 extracted from the atmosphere of Uranus, to be fuel for future DHe-3 fusion reactors on Earth, travel to the gravitational lensing point beyond Pluto, where the Sun's gravitational pull has focused light from distant stars so that very detailed images of their planets can be obtained, and many other amazing journeys beyond Earth.

StarTram can enable a wonderful future with enormous benefits for humanity. However, it also has the potential to enable a dark and dangerous future, depending on the choice that humanity makes. Life altering technologies can lead to dark futures as well as bright futures – ICBMs and nuclear bombs or peace space rockets and nuclear power, for example.

The potential dark side of StarTram is the capability to weaponize space. The very low cost of launch would enable a nation to place many thousands of Kinetic Energy Weapons into orbit. Termed *"Rods from God"*, they would be like simple poles of a dense material like tungsten, with the ability to deorbit on command using a small attached rocket motor. They could strike any point on Earth within a few minutes after the command.

A 10 ton "Rod from God" striking its target on Earth would be equivalent to 50 tons of TNT. The Rod would strike as a single impactor, or MIRV into a number of smaller objects just prior to impact, so as to destroy a larger area. All kinds of targets could be destroyed using GPS to guide the Rod – airports, seaports, factories, military bases, highways, bridges, power plants, water supplies, railroads, etc.

A country using StarTram to launch Rods into space could completely dominate all other countries on Earth. The launch cost would be very low -- $500,000 per 10 ton Rod. A fleet of 20,000 rods would only cost 10 Billion dollars to place into orbit – less than 1 week of the DOD's annual budget. No need for large standing armies and fighting offshore wars – just a small group of domestic controllers to target and dispatch the Rods. Do what we say or else! We are the dominant country!

Development of StarTram by individual countries has the potential to become a dangerous and disastrous Arms Race to Space. Instead, a cooperative, peaceful International StarTram Program is needed, and needed soon, before an Arms Race starts.

The most likely outcome of the Arms Race is not, however, having the winning country emplace *"Rods from God"* in orbit, and dominating everybody else. Instead, the countries losing the Race would launch enormous amounts of debris into orbit, basically small metal ball bearings, using rockets. With the rocket systems now available, sufficient debris could be launched into orbit to create a space debris cloud (Figure 3.21) that would destroy any orbiting Rod, plus everything else in orbit, in a few weeks. Humanity would then be cut off from space for many thousands of years, Marooned on Earth.

Let us avoid this Dark Future, and develop StarTram as a peaceful, cooperative International Program.

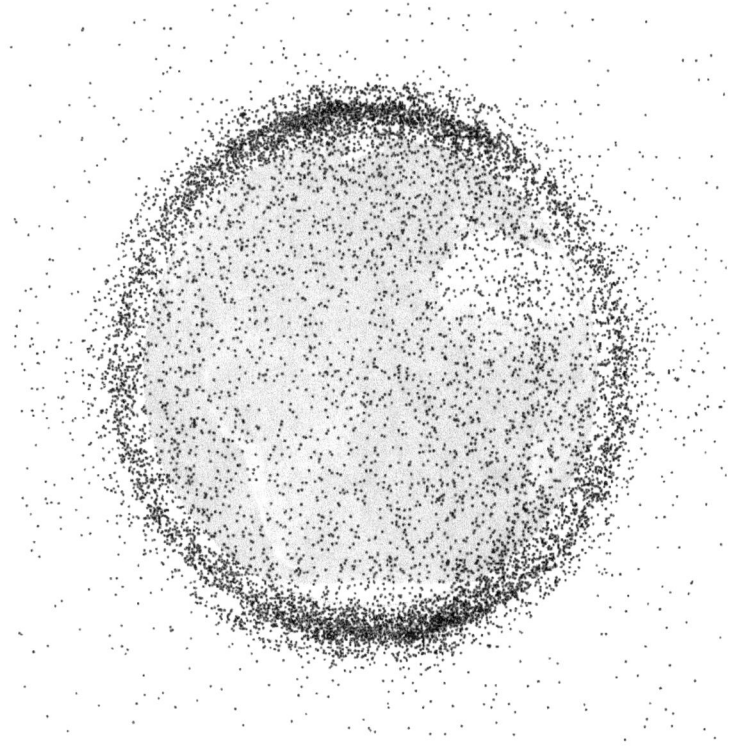

Figure 3.21: Space Debris Cloud in Low Earth Orbit (LEO)

Figure: 4.1

1904 image of Wardenclyffe Tower located in Shoreham, Long Island, New York. The 94 by 94 ft (29 m) brick building was designed by architect Stanford White. Wardenclyffe Tower, also known as the Tesla Tower, which began construction in 1901, was an early wireless transmission station designed by Nikola Tesla in Shoreham, New York and intended for commercial trans-Atlantic wireless telephony, broadcasting, and proof-of-concept demonstrations of wireless power transmission. A grass roots campaign to save the site succeeded in purchasing the property in 2013 with plans to build a future museum dedicated to Nikola Tesla. (From Wikipedia)

Chapter Four

Powering Earth from Space with Clean, Non-Polluting Energy

James Powell, George Maise, Charles Pellegrino, Jesse Powell,
John Skaritka and James Jordan

"The chief discovery which satisfied me thoroughly as to the practicality of my plan, was made in 1899 in Colorado Springs, where I carried on tests with a generator of 15 hundred kilowatt capacity and ascertained that under certain conditions, the current was capable of passing across the entire globe and returning from the antipodes with undiminished strength... I saw in a flash that power in virtually unlimited amounts could be conveyed through the Earth at any distance, limited only by the physical dimensions of the globe, with an efficiency as high as ninety-nine and one-half percent."

— Nikola Tesla, "World System of Wireless
Transmission of Energy," Telegraph and Telegraph Age,
October 16, 1927

Nicolai Tesla was an incredible genius. Among his many amazing inventions, he gave us the Alternating Current (AC) power generator and transmission systems that now power the whole World. Without electrical power we would still be in the early 1800's.

Tesla's AC power system was much more practical than Edison's DC power system and finally because the World standard after numerous efforts by Edison to kill AC, including electrocution of dogs, cats, and elephants with AC to show how dangerous it was. Edison even managed to have New York State adopt the electric chair using AC current to electrocute prisoners. New technologies often have to travel a very rocky road to implementation.

Tesla's big dream was to develop worldwide wireless transmission of electric power, without the need to construct hundreds of thousands of miles of transmission lines (1). In his Wardenclyffe Laboratory at Shoreham, Long Island, 1/2 mile from Powell's home, he experimented with beamed wireless power, with funding from J.P. Morgan. However, when J.P. Morgan realized that if Tesla did succeed, power would be free to everyone – no profit there – and withdrew his funding.

To be practical, wireless transmission of net electric power must be beamed from its source in a narrow beam to a receiver that is equipped to efficiently capture the beam. This can be done on Earth, and has been demonstrated in experiments in Hawaii by Dr. John Mankins (2) and his fellow scientists from Kobe University in Japan, the Jet Propulsion Laboratory, and Texas A&M University, put together as a team by Managed Energy Technologies, Inc.

They demonstrated (2) a first of a kind long distance wireless system that transmitted power from the crest of Haleakala on Maui to the slopes of Mauna Loa, a distance of almost 100 miles (149 kilometers).

While Earth to Earth beamed power transmission could be potentially useful in special situation and remote locations, its major application will be to beam very large amounts of clean, non-polluting, power from solar power satellites down to receivers on Earth.

In 1941, the science fiction writer Isaac Asimov published a short story, "Reason", (3) depicting a space station transmitting energy it collected from the Sun by microwave beams to planets.

How did the idea of beaming power from space get started?

In 1968, Dr. Peter Glaser, published in Science magazine (4) his groundbreaking paper on beaming power from Solar Powered Satellites in orbit down to the Earth and filed a patent for it. It took the Patent Office 5 years to finally grant the patent in 1973, probably because it went far outside of the usual Patent Office box.

NASA and ERDA (Energy and Research Agency) started the first serious effort on Glaser's concept in the late 1970's but stopped in 80 (2). However, work on the SPS concept (Space Solar Power Satellite) continued in other countries. Then in 1995, NASA got back into the SPS game. In the 20 years since, interest in, and work on, SPS has continued to grow, with studies and experiments in Japan, China, Europe, the U.S. and other countries. (2)

Glaser understood not only how power could be beamed down to Earth from SPS satellites in orbit, but why we would need it. He recognized that Earth's fossil fuel resources were limited and would inevitably run out. He also realized that the options solar and nuclear energy were the only real options to supply the energy that modern civilization depends on. As discussed in the *Prologue*, he was correct and we will need it soon, not only because we will run out of the known reserves in a few decades, but as become evident in the last 20 years, the global warming resulting from fossil fuel emissions of carbon dioxide, will cause environmental catastrophe before the end of the 21st Century.

The basic principles of beamed solar power from space are illustrated in Figure 4.1. The satellite receives sunlight from the Sun at the same rate as the Earth, about 1300 megawatts (th) per square kilometer. However, if the satellite is in orbit high about the Earth, say in Geosynchronous Orbit (GEO) that is 27,286 miles (35,786 Kilometers) from the Earth at sea level, it sees the Sun virtually continuously.

There is no night in GEO, no sunrise and no sunset. There are no clouds and storms that block out the Sun. As a result, the power generated by a square kilometer of solar collector in GEO will be about 4 times greater than the average power generated by a square kilometer of solar collector located on Earth. Moreover, space solar power beamed down to Earth is continuous. There is no need to store electrical energy to meet demand when the Sun is not shining where one is located. Space solar power beams will continue to provide you with lights at night, and operate your TV, computer, refrigerator, air conditioning, and so on. No need to pay for expensive energy storage in batteries, pumped hydro lakes, etc.

GEO orbits are special. In lower orbits closer to the Earth, spacecraft in their orbits more rapidly than the Earth rotates, so they don't hover above a particular point on the planet underneath them. On higher orbits above GEO, spacecraft move more slowly in their orbit, than the Earth rotates beneath them. Again, they do not hover above a particular point on the planet. Only in GEO orbit can a spacecraft stay hovered above a particular point on Earth. The ability to stay hovered is extremely important for beamed solar power, since it enables 24/7 power beaming to a given ground receiver – the space solar power satellite SPS satellite that beams the power is always in view.

There are limitations.

With multiple SPS satellites distributed around the GPS orbit, all of Earth is in view from the SPS satellites, 24/7. However, there are limitations to what locations on Earth are able to receive power from space. Sites with latitudes greater than 60 degrees North of the Equator and 60 degrees South have too great an angle for efficient transmission from the SPS satellite on the equatorial GEO orbit. Fortunately, only a very small fraction of the 7 Billion, World population live above 60 degrees South, as shown in the World map (Figure 4.2).

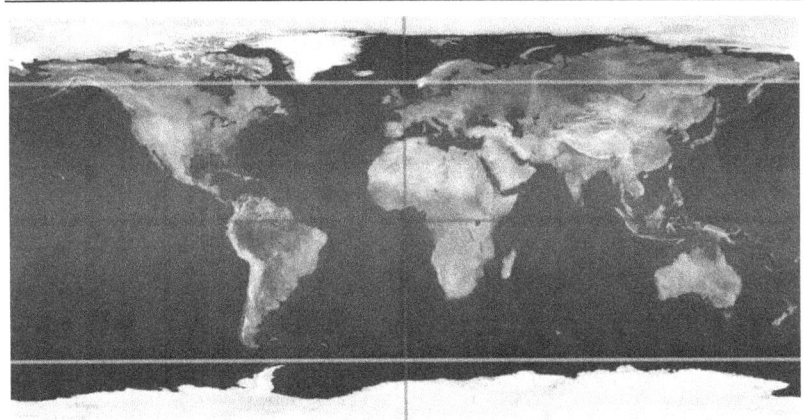

Figure 4.2

For the Northern continents, North American, Europe, and Asia, only a tiny part of their countries' population live above 60 degrees North – Alaska in the US and the Eskimos in Northern Canada, in North America. In Europe, only Finland, Sweden, Norway, and Iceland.

In Asia, only Russia, and that in Northern Siberia. All together, total population North of 60 degrees latitude less than 50 million people, compared to the slightly less than 7 Billion that live where SPS could supply their power.

Now to the technology of SPS satellites, described as responses to the questions that people frequently ask:

1. How do the SPS satellites generate electrical power?
2. How do they beam it down to Earth?
3. How do the Earth stations receive the beam and deliver the beamed energy to the electrical grid?
4. How large is the SPS satellite and its Photovoltaic Array transmitter?
5. How big is the microwave receiver on the Earth?
6. Is the microwave beam safe? Will it harm humans, animals and birds that move through it?
7. How are the SPS Satellites launched and assembled in orbit?
8. How much will beamed space power cost? Can it provide most of Earth's electric power?

9. How soon can SPS units be implemented? What are the benefits?
10. How does the public get World Leaders to implement Space Solar Power?

First question. How do the SPS satellites generate electric power? Very likely SPS will use photovoltaic cells (PV) similar to those already on Earth to generate electric power in large solar arrays that supply the electric grid, and on the roof tops of houses that generate power for the home owners.

An alternate approach would be to have reflecting mirrors focus sunlight on a high temperature receiver that used a conventional Rankine or Brayton cycle to generate power. This approach is used by some solar power facilities on Earth. Disposing of the waste heat to the atmosphere or water sink from the power cycle for Earth based power plants is relatively simple on Earth. However, it is not simple in space, and the waste heat radiators are heavy and complex. Thus, it is very likely the SPS satellites will use PV cells to generate their electric power.

SPS PV arrays will have a number of important advantages over solar power PV arrays located on Earth.

- Much greater electric output per unit area, compared to arrays on Earth,
- Not subject to damage by storms and bad weather
- Able to serve multiple areas widely distributed across the planet without needing transmission lines
- Very low environmental impact on an array of PVs
- Continuous power, not intermittent

At 30% efficiency, sunlight to DC electricity, an array of one square kilometer of PV cells will generate approximately 400 megawatts(e) of electric power. The same array on Earth would generate only 1/4th as much power, i.e. 100 megawatts (e).

Besides having much more power per unit area of the Solar PV Array, the other advantages of SPS are also very important. Space based solar PV arrays cannot be damaged by storms, can transmit continuous power and meet peak demands at many different locations, without needing transmission links – things Earth based arrays are not capable of.

Second question – how do the SPS satellites beam power down to Earth? The preferred approach is to convert the DC electric power from the PV arrays to microwave power that is focused by a large transmitter and beamed down to receivers on Earth. Optimum microwave frequency is in the range from 5 to 10 gigahertz. In this frequency range, atmospheric attenuation of the microwave beam is very small, less than 1% (2). Including focusing lenses, the overall beaming efficiency is in the range of 92% to 94% of the microwave power put into the beam ends up as microwave power at the receiver on Earth.

Based on a conservative conversion efficiency of 70% from DC to microwave power (2), 1,000 megawatts of DC electric power generated by the PV array, the microwave power level reaching receivers on Earth is on the order of 0.70 x 0.93 x 1,000, equaling 650 megawatts. With increased DC to microwave efficiency, which appears possible, the microwave power level at receiving stations would be greater.

Third question – How do the Earth Stations receive the beam and deliver the beamed energy to the electrical grid? The ground based rectenna (Figure 4.3) receives the microwave beam, and converts it to ordinary 60 Hertz AC current that is fed to the electrical grid that it is connected to. The conversion of efficiency, microwave to AC power is on the order of 85% (2).

Putting the individual efficiencies together, the overall end-to-end efficiency, DC power in space to AC power in electrical grids on Earth, is then:

End-to-end efficiency = 0.70 x 0.93 x 0.85 = 0.55. That is, for every 1,000 MW€ generated by PV arrays in space, about 550 MW(e) ends up power in electricity grids.

Figure 4.3
Space to Ground Microwave, Laser Pilot Beam

Fourth and Fifth questions – how big is the satellite PV array and its transmitter and the microwave receiver on Earth? The two questions are tied together by the physics of focusing microwaves. The diameters of the receiver and transmitter are related, plus the wavelength of the microwave beam and the distance between transmitter and receiver are related by the equation below (2).

$D_{TRANS} \times D_{RCVR} = 2.44 \; \lambda_{BEAM} \times R_{Distance\;Trans\;to\;RCVR}$

For an SPS Satellite in GEO orbit, R = 3.58 x 10^7 meters

For a 10 gigahertz microwave beam

$\lambda_{BEAM} = 3x \; 10^{-2}$ meters

Incorporating R & λ, equation (1) is then

$D_{TRANS} \times D_{RCVR} = 2.44 \times 3 \times 10^{-2} \times 3.58 \times 10^7 = 2.62 \times 10^6 m^2$

Expressed in kilometers

$D_{TRANS} \times D_{RCVR} = 2.62 \; km^2$

If we make D_{TRANS} the same size as D_{RCVR}, then

$D_{TRANS} \times D_{RCVR} = 2.62 = 1.62$ km or about 1 mile in diameter.

If we D_{TRANS} make bigger. E.g., with $D_{TRANS} = 2$ kilometers, D_{RCVR} is smaller= 1.3 kilometers

The sizes of the transmitter and receiver are independent of the amount of power that is transmitted, so it pays to transmit a lot of power per transmitter/receiver combinations. For an AC power of output of 2,000 megawatts(e) at a receiving station on Earth, one would have to generate 2,000/0.55 = 3,600 megawatts(e) of DC power in space, based on an end-to-end efficiency of 55%. The total area of the solar cell arrays to provide 3,600 megawatts(e) would be 9 km^2, based on 400 megawatts(e) per km^2 of area. Multiple PV arrays, each with a smaller area, could be connected to the transmitter.

Sixth Question – is the microwave beam safe? Will it harm humans, animals, and birds that move through it? Based on studies of flora and fauna exposed to microwave, there appear to be no harmful effect. To assure safety the International Academy of Astronautics (IAA) has recommended in its 2008-2011 study of space solar power that the maximum intensity of wireless power transmission should be less than full summer sunlight at the equator, i.e., 1,000 watts/m^2 (2). A 2 km diameter receiver receiving 2,350 megawatts of microwave power and delivering 2,350 x 0.85 = 2,000 megawatts of electric power to the grid, would have a microwave beam intensity at the receiver of:

2,350 x 10^6/(π/4) (2,000)^2 = 748 watts/m^2, well under the IAA recommended limit. Microwave intensity outside of the receiver, which would be fenced off, would be much smaller.

Seventh question – how is the SPS system launched and assembled in orbit? Many different designs and assembly approaches have been proposed Figures 4.4 and 4.5 show 2 of the proposed designs. The Sun Tower approach (Figure 4.4) is a long column of modular units with the microwave transmitter at the bottom of the column, pointed towards Earth. The column is gravity stabilized, so the transmitter always remains at the bottom, facing the receiver on the surface of the Earth, to which it radiates power.

The Modular Sandwich approach is shown in Figure 4.5. This design reduces the area of the PV array that generates DC power by using a set of mirrors to concentrate sunlight on the PV cells at higher intensity than the ambient 1,300 watts per m².

Figure 4.4

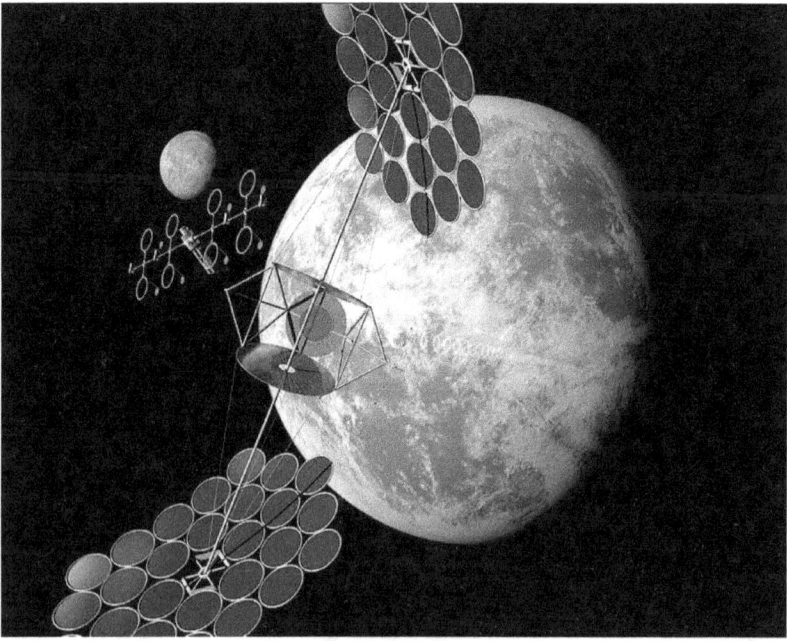

Figure 4.5

Mankins (2) describes in detail the wide variety of SPS designs that have been proposed. In general, the designs require human or robotic assembly of a large number of modules that have been launched into space by rockets.

An alternate approach is based on magnetically inflated cable (MIC) structures as shown in Figure 4.5. Launched as a compact collapsed package of superconducting cables, tethers, and thin PV films, when the superconducting cables are energized with current, the resultant magnetic forces automatically expand the collapsed package into its final full size structure. (5)

Two versions of a MIC Solar PV array are shown in Figures 4.6 and 4.7. The first (Figure 4.6) is a simple loop of superconducting cable, with tethers of Kevlar or other high strength fibers that restrain the outwards magnetic forces acting on the cable loop.

The second version (Figure 4.7) uses superconducting cables to form a large thin-film reflecting mirror that concentrates sunlight onto a small solar cell PV array, reducing its mass and cost. A concentration ratio in the range of 5 to 10 can be achieved. For a 2 mil (50 microns) thick reflecting film of aluminized plastic, the mass of the solar collector would be less than 0.2 kg/kw(e), very small compared to other SPS designs.

A third and even simpler version of a MIC space power satellite is a small flat rotating MIC loop structure with extended fins that hold the PV cells. The extended fins would extend beyond the disc to a distance of 3 to 4 times than the radius of the MIC loop, with the outwards centrifugal force from the rotation keeping them as flat sheets.

For an effective fin thickness of 100 microns, structure plus PV cells, the unit weight of the fins would be only 0.3 kg/kw(e) at a solar cell efficiency of 30%. The weight mass of the rotating MIC loop would be much less than the mass of the fins.

THIN FILM SOLAR CELL SYSTEM

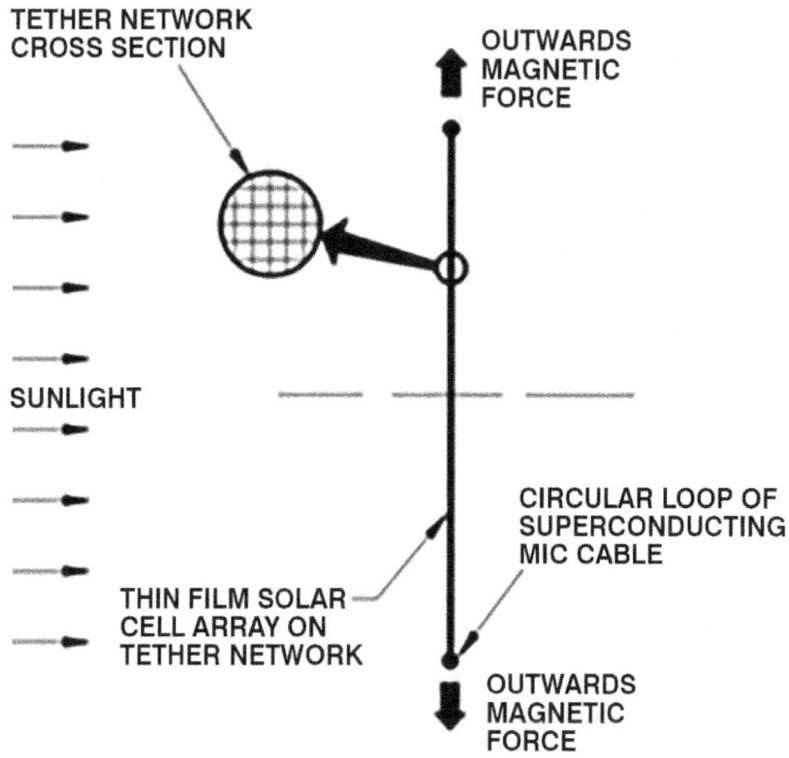

TETHER NETWORK
CROSS SECTION

OUTWARDS
MAGNETIC
FORCE

SUNLIGHT

CIRCULAR LOOP OF
SUPERCONDUCTING
MIC CABLE

THIN FILM SOLAR
CELL ARRAY ON
TETHER NETWORK

OUTWARDS
MAGNETIC
FORCE

Figure 4.6

SOLAR CONCENTRATOR SYSTEM

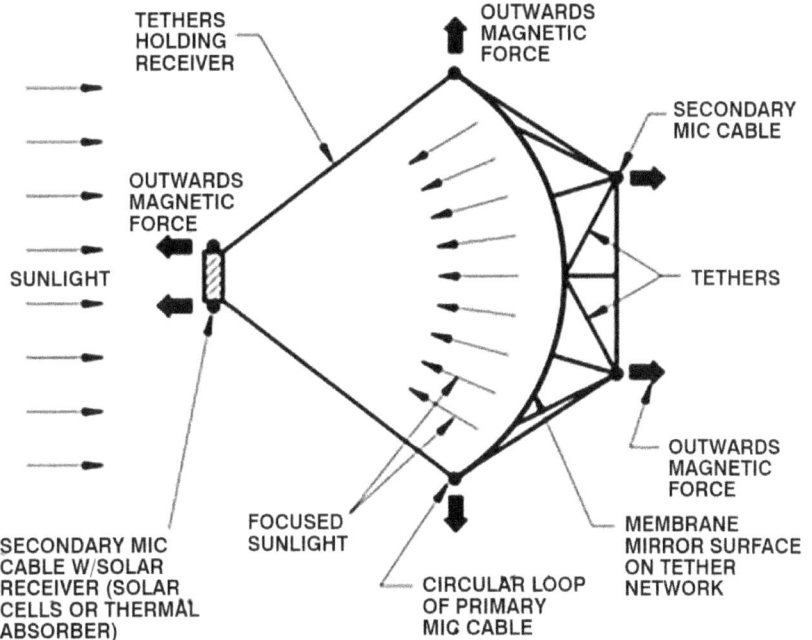

Figure 4.7

185

The solar PV array is just part of the total SPS mass in orbit. To assess total SPS mass and its cost, including launch cost, we use Mankins Design Reference Mission (DRM) #5 for the SPS-ALPHA concept described in his book, The Case for Space Solar Power (2). Mankins describes the SPS-ALPHA concept in detail – its various components, their mass and cost, how they are assembled in orbit, and so on. Five DRM versions of SPS-ALPHA are described, with increasing levels of beamed power to Earth.

DRM#1 and DRM#2 are small scale demonstration SPS systems in LEO orbit. DRM #3 is a subscale integrated demonstration in GEO orbit. Two beamed power levels for DRM#3 are analyzed – one at 2 megawatts (e) and the other at 18 megawatts (e). DRM#4 is a commercial SPS system delivering 500 megawatts (e) to markets on Earth.

DRM#5, the design we have selected, would deliver 2,000 megawatts(e) [2 gigawatts(e)] of power to markets on Earth. Table 4.1 summarizes the principal parameter for DRM#5.

Table 4.1
Parameters for 2000 MW SPS-ALPHA Space Power Systems
Basis – DRM Case #5

Parameter	Value
Beamed Power Input Electrical Grid on Earth Per Unit	2,000 megawatts(e)
Location in Space	GEO orbit
SPS Platform Hardware Mass in Orbit per Unit	34,800 tonnes
DRM#5 Hardware Cost Per Unit	$5.7 Billion
Earth Receiver Cost Per Unit	$0.7 Billion
Total Hardware Cost (Launch Cost Not Included)	$6.4 Billion

Eighth Question – how much will beamed space power cost? Can it provide most of Earth's electric power? Based on Mankin's cost projections (2) for the SPS ALPHA-DRM#5 design and a StarTram launch cost of $100 per kilogram to GEO orbit, the parameters for the SPS units launched by StarTram are shown in Table 4.2.

Table 4.2
Parameters for StarTram Launch Facility to Place 2,000 MW(e) SPS ALPHA-DRM# Units in GEO

- 40 metric tonne payload per launch

- 12 launches per day

- 175,000 metric tonnes launched per year

- 5 SPS ALPHA – DRM#5 Units emplaced in GEO per year (2,000 MW(e) per unit, 34,800 tonnes per unit.

- 30 Billion dollars Capital cost of StarTram facility

- $100/kg launch cost per kg to GEO (amortized)

- 3.5 Billion dollars cost to launch 1 SPS ALPHA-DRM #5 Unit [$1250/kw(e)]

- First StarTram launches begin in 2025

- 2 potential rates of StarTram Facility Construction Studied:

  o **Case 1**: 2 facilities constructed per year in years 2025, 2026, 2027, 2028, and 2029 10 total Star Tram facilities constructed. Total Construction Cost of $300 Billion Dollars.

  o **Case 2**: 4 facilities constructed per year in years 2025, 2026, 2027, 2028, 2029. 20 total StarTram Facilities constructed. Total construction cost of $600 Billion dollars.

- 10 Billion dollars total capital cost for one 2000 MW(e) SPS Unit Space and Receiver Hardware Plus Launch Cost.

- 2 cents per KWH cost of SPS power beamed to Earth (Amortized Capital plus O&M costs)

The StarTram launch cost of $100 per kilogram to GEO is 2 times the $50 per kilogram to LEO, reflecting the high launch speed for a direct launch to GEO, higher delta velocity (1.5 km/sec), and heavier rocket required to establish the final orbit in GEO. An alternative approach is to launch to a lower orbit, e.g. LEO, and use a small high Isp electric thruster powered by the solar PV array to slowly ascend to GEO orbit. Detailed analysis will be needed to determine which approach is best.

Using StarTram, launch cost is approximately 3.5 Billion dollars for the 2,000 MW(e) SPS systems, about 1/2 of the systems hardware cost of 6.4 Billion dollars, resulting in a total cost of approximately 10 Billion dollars, equivalent to $5,000/kW(e). If present rocket launch costs to GEO, on the order of $10,000 per kilogram were to apply, the total capital cost would be 360 Billion dollars for the 2,000 MW(e) SPS System, equivalent to $180,000/kW(e), 36 times greater than launching by StarTram. A completely impractical cost.

Using StarTram, the 2000 MW(e) SPS system can deliver 1.75×10^{10} KWH/year of electric power at a cost of 2 cents per KWH. The US consumed 4.3 Trillion KWH of electric power annually in 2013.(6) 245 SPS units in space could supply all of America's 2013 electric power generation. At 34,000 tonnes per unit, 1 StarTram launch facility with a launch capability of 175,000 tonnes per year could launch 5 each 2,000 MW (e) SPS units annually and 75 unites over a 15 year period. 5 StarTram facilities could launch the 245 units in slightly less than 10 years. The World consumed approximately 20 Trillion KWH of electrical power in 2011, about 5 times the annual US power consumed in 2011 (7). A correspondingly greater number of 2,000 MW(e) SPS units would be required, about 1140.

But, World electric power needs will grow substantially in the decades ahead. The EIA forecasts that World electric power generation will increase from 20 Trillion KWH per year in 2011 to 39 Trillion KWH/year in 2040, a factor of 2 increase to supply all of the World's electrical power needs. Extrapolating the 2040 EIA projection to 2050, World electric generation will further increase at approximately 0.6 Trillion KWH/year for a total of 45 Trillion KWH/year. To supply all of the World's electrical power in 2050 AD would require 2,400 SPS units in orbit (Table 4.3). And, as discussed in Chapter 5, even more electric power will be required if synthetic gasoline and diesel fuel for motor vehicles and jet fuel for airplanes are manufactured from carbon dioxide extracted from the atmosphere.

Yes, beamed solar power could supply virtually all of Earth's future electric power needs, and supply it indefinitely with no environmental damage and no additions of greenhouse gases to the atmosphere.

And, it can accomplish the above, while at the same time, saving enormous amounts of money for electrical consumers, because of its much lower cost per KWH. SPS power to the grid will cost about 2 cents per KWH – even less as SPS designs evolve and become cheaper and lighter – while commercial power from conventional fossil fuel plants now costs about 10 cents per KWH(e), and will become even more expensive in the years ahead (8).

The capital cost to put 2400 SPS units in orbit to supply all of the World's power in 2050 AD, is 24.6 Trillion dollars including the cost of constructing 20 StarTram launch facilities.(Table 4.4) The **Net World Savings** from SPS power at 2 cents per KWH, as compared to purchasing conventional power at 10 cents per KWH is 40.6 Trillion dollars over the period from 2025 to 2050AD (Table 4.4). The savings in the cost of power is almost 2 times the cost of putting the SPS units in orbit. And, as noted above, it is likely that SPS units will become even cheaper as the technology evolves, further increasing the savings.

World GDP in 2016, amounted to about 75.4 trillion U.S. dollars, annually and probably will increase substantially in the coming decades. Projections vary as to what the increases will be for the various nations of the World. Using the predictions of Price Waterhouse Coopers (9) for the 20 largest economies in the World the following are the top 10 ranked countries in 2014 and 2050AD.

2014 Annual GDP			2050 Annual GDP	
Country	GDP (in 2014 Billions USD		Country	GDP (in 2014 Billions USD
1. U.S.	17.5	1.	China	50.9
2. China	10.0	2.	U.S.	38.0
3. Japan	4.8	3.	India	21.0
4. Germany	3.9	4.	Japan	7.8
5. France	2.9	5.	Brazil	7.4
6. United Kingdom	2.8	6.	Russia	6.2
7. Brazil	2.2	7.	Germany	6.2
8. Italy	2.2	8.	United Kingdom	5.9
9. Russia	2.1	9.	Mexico	5.8
10. India	2.0	10.	France	5.7

China's GDP grew by a factor of 5 from 10.0 in 2014 to 50.9 in 2050. India grew by a factor of 10.5 from 2.0 in 2014 to 21.0 in 2050. The 2 most populous countries on Earth, with approximately 1/3 of the total World population in 2050. European countries – France, Germany, United Kingdom – grew only by a factor of about 2. Italy dropped out of the top 10, also growing only be a factor of 2. Japan grew by only a factor of 1.6 and the U.S. grew by a little more than a factor of 2. Russian and Brazil grew by a factor of 3.

China, the US, and India will be the dominant economic powers in 2050, with their combined GDPs being almost 60 percent of total annual World GDP, which well be on the order of 200 Trillion dollars, 3 times greater than today's (2014) annual World GDP.

And many other developing nations in Africa, Asia, and South America will want to catch up to the big GDP nations, so 2050 is not a final state of affairs. Given low cost, environmentally clean SPS power, they will seek to use it to develop their economies and a higher standard of living, so that electrical power demand will still continue substantially over the remainder of the 21st Century.

The 10 year Phase 1 program would develop and test the components of the StarTram launch system and the SPS system to be ready for commercial deployment in 2025. Here we examine only the development and testing of the StarTram System – planning the program for the development and testing of the SPS system will be carried out by other scientists and engineers.

Based on our studies of StarTram technology and its development requirements, described in more detail in Chapter 3, we believe that StarTram launch facilities can begin operation in 2025. We assume that the SPS technology will also be ready for implementation in 2025. However, this will have to be validated by R&D on the SPS system. If SPS technology is not ready in 2025, the implementation schedule will slip to a later date. We believe that 2025 target date for the SPS system should be feasible, but if it is delayed to 2030 or 2035, SPS power can still meet a major portion of the World's power needs in the 21st Century.

Based on a 2025 start date for SPS launches, implementation schedules have been analyzed (Table 4.2) for Phase 2.

Case 1: 10 Total StarTram launch facilities constructed in years 2025 through 2029, 2 per year.

Case 2: 20 total StarTram launch facilities constructed in years 2025 through 2029, 4 per year.

As each StarTram launch facility is completed and begins operation, each year it launches 5 SPS units into GEO orbit. Table 4.3 gives the total SPS units in orbit at the end of year 2029. For Case 1, 10 total StarTram launch facilities, there are 150 SPS units in orbit. For Case 2, 20 total StarTram launch facilities, there are 300 SPS units in orbit.

Already in 2029, the SPS units supply a significant fraction of the total World electric power generation of 32.1 Trillion KWH. For Case 1, SPS supplies 2.7 Trillion KWH (Table 4.3), 8.4 percent of the World Total for Case 2, SPS supplies 5.40 Trillion KWH (16.8 percent).

Going forward from 2030 in Phase 3 the already operating StarTram launch facilities continue to launch more SPS units, increasing the amount of SPS power beamed to Earth. (Table 4.3). In 2050 AD, Case 1 has 1260 SPS units in orbit, supply

48% of total World electric generation while Case 2 has 2400 SPS units in orbit supplying 76% of total World generation.

The corresponding annual investments for the SPS units, together with the annual savings in World electric power cost, based on 2 cents /KWH average cost for power from conventional sources (8) – coal, gas, nuclear, wind, and ground solar PV – are shown in Table 4.4.

By 2035 the net annual savings offset the net annual investment in SPS units. From then on the net annual savings are greater than the annual investment. By 2050, for Case 1 the total SPS investment over the period 2025 is 12.30 Trillion dollars, while the total savings in power cost is 20.28 Trillion dollars, resulting in a net savings for the period of 20.28-12.30 = 7.98 Trillion dollars. By 2050, for Case 2, total investment is 24.60 Trillion Dollars, total savings are 40.56 Trillion dollars, and net total savings are 15.96 Trillion dollars.

Total annual GDP in 2050 for the top 20 largest economies in the World is projected to be 188 Trillion dollars, expressed in 2014 US dollars (9). China's annual GDP is 50.8 Trillion USD, America is 38.0 Trillion USD, India, 20.8 Trillion USD, accounting for more than ½ of the 190 Trillion US dollars total. The other 18 economies, Japan, Russia, Brazil, Germany, etc., down to $20, Switzerland at 2.0 Trillion dollars annually, make up the balance. It is noteworthy that China will have a bigger GDP than America, and that India will have a DDP that is more than 1/2 that of the US.

In the years following 2050 Case 2 will have a net annual savings in power cost of about 45 Trillion KWH x $0.08/KWH=3.6 Trillion dollars, about 2% of World Annual GDP. Not included in the savings enabled by the SPS-StarTram programs are the much greater external costs of generating electric power from fossil fuels – the costs of global warming from greenhouse gas emissions, strip mining, oil spills, pollutants that damage health, contamination of aquifers from fracking for gas and oil, and so on. The benefits from eliminating the external costs are even greater than the economic benefits.

Tenth Question. How does the Public get World leaders to implement Space Solar Power? This is the most difficult question of all. It's easy to make a very strong case for beaming Space Solar Power to Earth as a replacement for electric power generated from fossil fuels. There are large economic and environmental benefits, humanity can avoid the real possibility of the collapse of modern civilization when we run out of fossil fuels and/or environmental catastrophe occurs.

In the absence of opposition from vested powerful economic interests, there would be strong political support from World leaders, who would initiate an aggressive program in SPS because of its many major benefits to their constituents.

Unfortunately, political leaders also listen carefully to powerful existing economic interests that would be negatively impacted by implementing SPS. These include the fossil fuel industries that produce oil, gas, and coal, electric utility companies, industries that manufacture conventional equipment – turbines, boilers, etc. – for generating electric power, and so on.

To counter this opposition, it will be necessary for the public and environmental groups to recognize the benefits of SPS and strongly push for its implementation.

In summary, the StarTram SPS project is extremely important. It offers the opportunity to meet World electrical power needs without using polluting, environmentally damaging fossil fuels, at lower cost than conventional power. It can provide sustainable electric power for humanity indefinitely. Future generations will not have to struggle for existence in an increasingly environmentally devastated World as fossil fuels run out.

Table 4.3

Implementation Schedule for SPS ALPHA DRM#5 Units in Orbit and SPS Power Beamed to Earth.

Basis: 5 SPS Launches Per Year from 1 Star Tram Facility

2,000 MW(e) per SPS Unit

EIA Projections for World Power Demand ( )

Year	Number of StarTram Facilities		SPS Launches per year		Number of SPS Units In Orbit		SPS Power Supply Trillion KWH/year		Total World Power Demand
	Case 1	Case 2	Case 1	Case 2	Case 1	Case 2	Case 1	Case 2	Trillion KWH/YR
2025	2	4	10	20	10	20	0.18	0.36	29.5
2026	4	8	20	40	30	60	0.54	1.08	30.1
2027	6	12	30	60	60	120	1.08	2.16	30.8
2028	8	16	40	80	100	200	1.80	3.60	31.4
2029	10	20	50	100	150	300	2.70	5.40	32.1
2030	10	20	50	100	200	400	3.60	7.20	32.7
2035	10	20	50	100	450	900	8.1	16.2	35.8
2040	10	20	50	100	700	1400	12.6	25.2	39.0
2045	10	20	50	100	950	1900	17.1	34.2	42.0
2050	10	20	50	100	1200	2400	21.6	43.2	45.0

Case 1: 10 StarTram Launch Facilities – SPS supplies 48% of Total World Power Demand

Case 2: 20 StarTram Launch Facilities – SPS Supplies 96% of Total World Power Demand

Table 4.4

Investment In and Savings in Cost of Beamed World Power from SPS ALPHA – DRM# Solar Satellites in GEO Orbit

Basis: 10 Billion$ Capital Cost Per 2,000 MW(e) Launch Unit (5,000/KW(e))

- Includes Hardware & Launch Costs
- 30 Billion $ Capital Cost of StarTram Launch Facility
- 2 cents/KWH from SPS Units (Amortized Capital + Operating Cost)
- 10 cents/KWH from conventional power plants
- 8 cents/KWH net savings for SPS power
- Case 1: 10 Star Tram Launch Facilities
- Case 2: 20 StarTram Launch Facilities

Year/ Period	Number of SPS Units in Orbit			SPS Investment During Year/Period In Trillion Dollars			SPS Power Savings During Year/Period in Trillion Dollars	
	Case 1	Case 2		Case 1	Case 2		Case 1	Case 2
2025	10	20		0.16	0.32		0.01	0.02
2026	30	60		0.26	0.52		0.04	0.08
2027	60	120		0.36	0.72		0.09	0.18
2028	90	180		0.46	0.92		0.14	0.28
2029	150	300		0.56	1.12		0.22	0.44
2030-2034	400	800		2.50	5.00		2.16	4.32
2035-2039	650	1350		2.50	5.00		3.96	7.92
2040-2044	900	1800		2.50	5.00		5.76	11.52
2045-2050	1200	2400		3.00	6.00		7.90	19.80
			Total SPS Investment in Trillions $	12.30	24.60	Total SPS Power Savings in Trillions $	20.28	40.56

Net Savings From SPS System=Total SPS Power Savings – Total SPS Investment

Case 1: Net Savings = 20.38 T$ - 12.30 T$ = 7.98 Trillion $

Case 2: Net Savings = 40.56 T$ - 24.60T$ = 15.96 Trillion $

U.S. Energy-Related Carbon Dioxide Emissions by Sector, 2011

million metric tons carbon dioxide

2011 total = 5,480.63

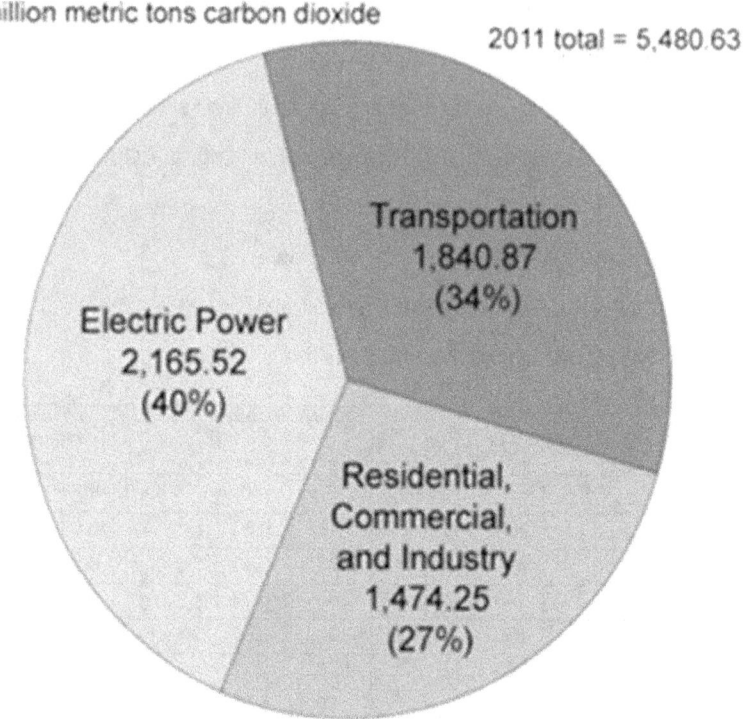

Transportation
1,840.87
(34%)

Electric Power
2,165.52
(40%)

Residential,
Commercial,
and Industry
1,474.25
(27%)

Note: Totals may not equal sum of components due to independent rounding.

Source: U.S. Energy Information Administration, *Monthly Energy Review*, Tables 12.2-12.5 (May 2012).

Figure 5.1

Chapter Five

Gasoline and Diesel Fuel from Air and Water Will Never Run Out
James Powell, James Jordan, and Jesse Powell

"Wither goest thou, America, in thy shiny car in the night."
Jack Kerouac

Question to the reader. What do you and your automobile have in common? Flesh vs metal and plastic? No. Wheels vs legs? 100 mph Speed vs 4 mph? No. Intelligence and the ability to communicate? No, though Google's self-driving autos are on the way there.

Both us and cars have one thing in common. We both take in fuel for energy and excrete waste products when we burn the fuel. Sure, the fuel is different – beef and chicken, milk and donuts, french fries, etc., for humans, vs gasoline or diesel for cars – but leaving aside certain human waste products, both cars and humans expel carbon dioxide and water vapor generated by burning fuel or food.

Water vapor is harmless. It's around us all the time. It is sometimes annoying when it condenses to rain, especially in severe storms, or alternatively when it doesn't rain and droughts occur. But carbon dioxide? It can be dangerous. Earth's temperature depends on its concentration in the atmosphere, as evidenced by millions of years of geologic data. Increasing CO_2 concentration in the atmosphere makes the Earth warmer; decreasing it makes the Earth cooler. The Permian-Triassic mass extinction event which occurred 250 million years ago, wiped out 90 percent of Earth's species. It appears to have been caused by very massive volcanic eruptions that released vast amounts of lava and carbon dioxide, which warms the planet and acidifies the oceans, killing off much of the marine life.

So, how much carbon dioxide (CO2) do humans and cars emit? On average, each human breathes out 2.3 pounds (1 kilogram) of CO2 per day (1) equal to 0.365 metric tonnes, annually, from the digestion of the food they eat. A metric tonne is1,000 kilograms. Cars emit much more. Burning 1 gallon of gasoline emits 19.65 pounds(8.9 kilograms)of CO2: burning 1 gallon of diesel fuel emits 22.3 pounds (10.1 kilograms) of CO2(2).

The average American drives his/her car about 10,000 miles per year, averaging 27 miles per day. Assuming that it is a fuel efficient car 25 mpg, and not some gas guzzler, that's about 1.1 gallons of gas per day. The car's CO2 emission is approximately 10 kilograms daily, 10 times more than its driver.

Not only is the car's CO2 emission much greater than the driver's, there is an extremely important difference in how the two kinds of CO2 emission affect the Earth's environment.

Human CO2 emissions into the atmosphere are taken up by the plants that make the food that humans eat. There is no net addition of CO2 into the atmosphere. The environment is in balance – animals, including humans, eat plants, releasing CO2, which is taken up by the plants that grow the food for the animals. It's beautiful, a Garden of Eden situation, sustainable forever, as long as some does not disturb it.

Well, humans are disturbing the Garden of Eden. The CO2 emitted by cars does not come from today's Nature, but from immense amounts of fossil fuels deposits formed in the Earth millions of years ago. And it's being consumed at rates that are too rapid for today's Nature to handle.

As a result, the concentration of CO2 in the atmosphere does not stay constant, as it did before we humans started burning fossil fuels a couple of hundred years ago, but is rapidly increasing with time. Since the beginning of the Industrial Revolution, atmospheric CO2 concentration has increased from 280 ppm (parts per million) to today's 400 ppm (1), and is increasing by approximately 2 parts per million per year. By 2100, with an increase in World population from today's 7 Billion people to 9 Billion, and the large increase in economic activity of much of the population, CO_2 atmospheric concentrations are likely to be on the order of 600 ppm or more. As described in the Prologue, if we keep burning fossil fuels that would be an environmental catastrophe.

Current World CO_2 emissions from fossil fuels are about 32 Billion metric tonnes per year (3). That's hard to visualize. Think of the gaseous CO_2 turned to liquid CO_2, with a density of 1.18, compared to the density of water at 1.0. Then visualize a container 1 square mile in area that extended 6.5 miles up into the sky. That's how big a container one would need to contain 32 billion tonnes, one year's worth of liquefied CO_2.

Transportation is a fundamental pillar of modern society. We need cars, trucks, airplanes trains, ships, pipelines, tractors, etc., to bring us food and goods, and let us travel for work and pleasure. Without modern transport we would be back to horses, oxen drawn wagons, sailing ships, and walking – life as it was a few hundred years ago.

And transport consumes a great deal of fossil fuel. Figure 5.1 (4) shows United States CO_2 emissions from fossil fuels by Energy sector in 2011. Electric power was the biggest (40%), followed closely by transport (34%), with the other sectors residential, commercial, and industry totaling only 27%. Most of the US transport CO_2 emissions come from cars and trucks. Of the total 1.84 Billion tonnes of CO_2 from transport, 1.095 Billion tonnes came from gasoline and 0.427 Billion tonnes from diesel fuel. The remaining 0.3 Billion tonnes came from airplanes, ships, trains, etc.

Petroleum is by far the biggest fuel for transport. Figure 5.2 (5) shows where energy for the US economy comes from and where it goes. Petroleum is the biggest source of energy, 37% of total energy input. Natural gas trails at 25%, followed by coal at 21%, with renewables (wind, solar, hydropower and biomass) far behind at 8%, essentially tied with nuclear (9%).72% of petroleum energy goes to transport, with it constituting 94% of the energy consumed by transport.

Renewables, i.e., ethanol and a tiny amount of biodiesel, account for only 3% of transport energy input, tied with natural gas, also at 3%.

Figure 5.2

US Energy Flows: 2009

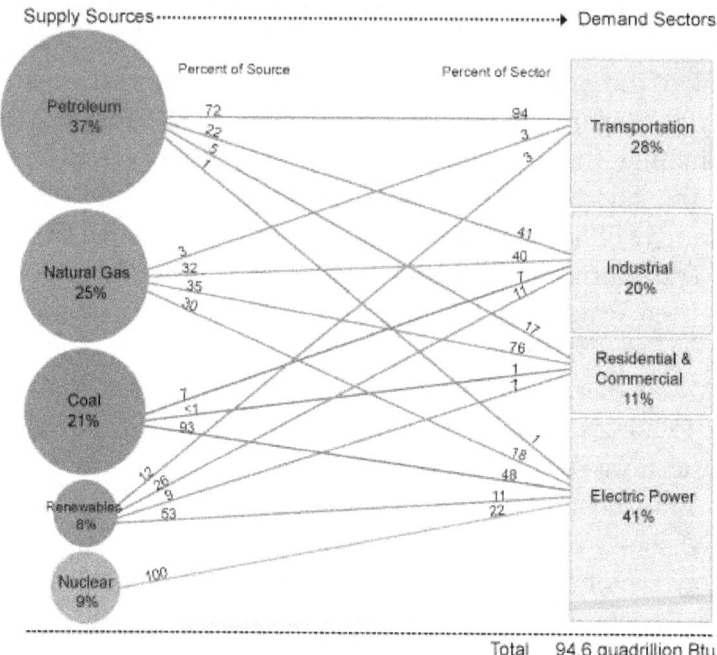

Source: Energy Information Administration, *Annual Energy Review 2009*

The global transport picture is pretty much the same as the US, though less intense. The US with 310 million people consumes 18.5 million barrels of oil daily (6), 72% of which goes for transport. The other 6,700 million people in the World consume (92 - 18.5 = 73.5) million barrels of oil daily [92 million barrels per day total World consumption (6)], 61.5% of which goes for transportation (7).

So the US uses 18.5 x 0.72 =13.3 million barrels of oil per day for transport. The rest of the World uses 73.5 x 0.615 = 45.2 million barrels per day for transport. Put in per capita terms, the average American consumes 0.043 barrels per day, equal to 42 x 0.043 = 1.8 gallons per day of oil for transport.

The rest of the World average is equal to 42 x 0.0067 = 0.28 gallons per capita per day.

Conclusion? The per capita rate of US oil consumption for transport to the rest of the World's per capita consumption is 1.8/0.28 or more than 6/1. Further conclusion? As the less developed countries like China, India, and others continue on their paths to bigger GDPs and greater industrialization, their transport needs and energy consumption will grow.

As discussed in Chapter 2, there are currently approximately 1 Billion cars in the World. By 2050 AD, it is projected that there will be 2.5 Billion automobiles, with a World GDP almost a factor of 3 greater than today's 70 Billion dollars GDP. Annual passenger miles and freight ton miles will be double today's values.

Figure 5.3, taken from Chapter 2 shows the number of automobiles per 1,000 population in a country as a function of the GDP per capita for the country. The relationship is essentially linear. At $50,000 GDP per capita, the linear curve projects approximately 800 automobiles per 1,000 persons. At $10,000 GDP per capita, 5 times lower, the curve projects 160 automobiles per 1,000 population. 800 cars per 1,000 population approaches the limit, although some Americans do own 2 or more personal cars.

As World GDP grows, so will the number of World cars. Average GDP World GDP is propelled to grow from about 70 Trillion dollars today to over 180 Billion dollars in 2050 AD(8). Combined with the population increase from 7 Billion today to 9 Billion in 2050 AD, the per capita World average GDP will grow from $10,000 today to $20,000 in 2050, a factor of 2 increase. The average number of cars per 1,000 population also increases by a factor of 2 from 140 to 280.

Combining today's CO_2 transport emissions from the US, 1.8 Billion tonnes per year, and 4 x 1.8 = 7.2 Billion tonnes from the rest of the World, we get a total of 9 Billion tonnes per year from transport, 28% of total global CO_2 emissions – almost 1/3 rd.

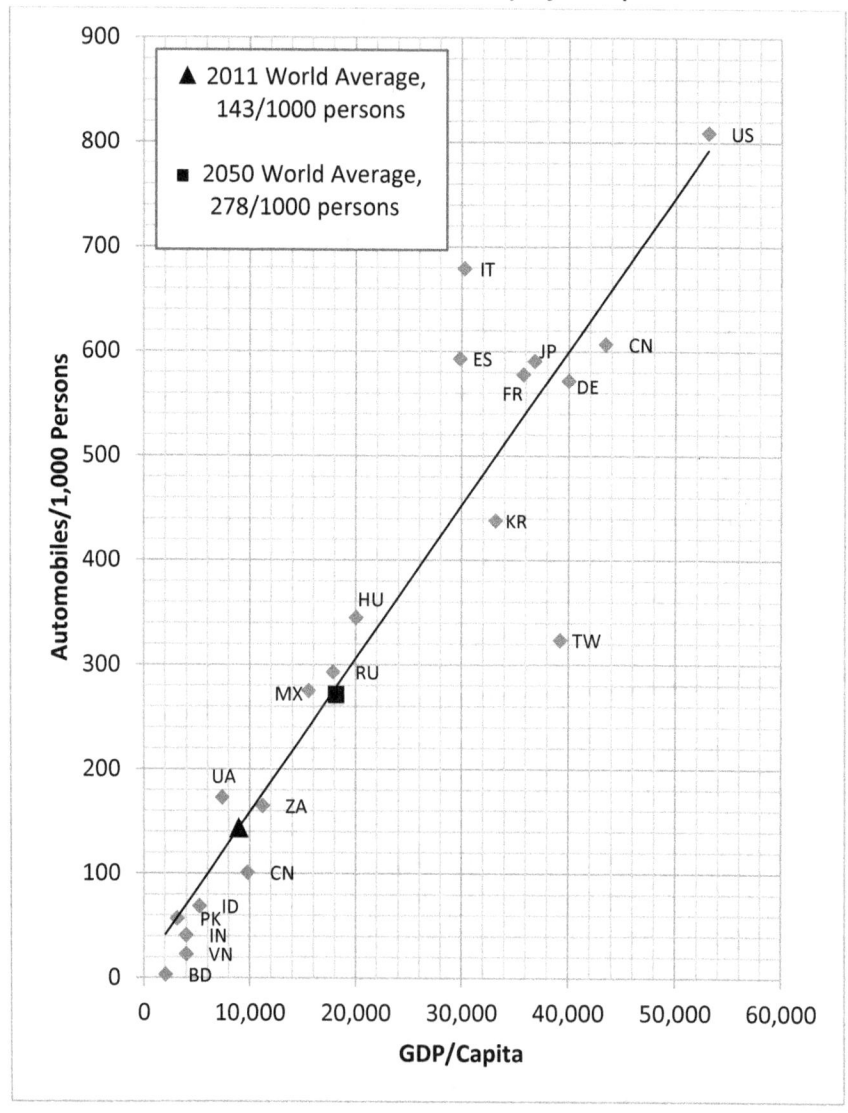

Figure 5.3

Automobiles Per 1,000 Persons as a Function of GDP Per Capita for Different Countries

Basis: 2010 – 2013 Data PPP (Adjusted)

If transport requirements essentially double over the next 35 years to 2050 AD, World CO_2 emissions will also double, to about 18 Billion tonnes per year, assuming transport continues to depend on oil fuel. There would be no hope of maintaining atmospheric CO_2 concentration at 400 ppm, the goal if we are to avoid environmental catastrophe.

Transportation is between a rock and a hard place. On one hand, to maintain and hopefully grow World living standards, we must continue and grow our transport systems. Without them, modern society would collapse. On the other hand, if we continue to use fossil fuels for transport, we will increase atmospheric CO_2 concentration, causing environmental catastrophe and the collapse of modern society.

Conclusions. Humanity now has a real choice to make. If we want to avoid environmental catastrophic and social collapse, we must begin to transition to non-fossil fuel transport as soon as possible. We cannot continue to have climate deniers keep saying that there's no problem, or procrastinators saying that we can postpone action to sometime in the future.

We must acknowledge reality, not deny it. It is ironic. If some species of Earth were emitting 32 Billion tonnes of CO_2 per year, increasing its atmospheric concentration to dangerous levels, humanity would not deny it was dangerous. And they would take action, hunting down and destroying the species that was emitting the CO_2. But since humans are doing it, we deny it and take no real action. Ah well, that's human nature.

What are the alternatives to fossil fuels for transport? Are they feasible? Practical? Too expensive? Limited capability? Limited applicability? Environmentally harmful? The alternatives to fossil fuels for transport are:

- Biofuels
- Hydrogen fuel
- Batteries, i.e. "electric" cars
- Electrically powered Maglev
- **Synfuels from air and water**

Biofuels are feasible, but not practical. Growing Biofuels compete with growing food. Already, on the order of 1 Billion people in the Word are hungry and malnourished. The US consumes 135 Billion gallons of gasoline, annually, from oil, plus 13 Billion gallons of ethanol mixed with the gasoline (9), derived from 40% of the US corn crop. On an energy basis, 1.5 gallons of ethanol equals 1 gallon of gasoline. On a net energy basis, after one deducts the energy to grow the corn and process it to ethanol, 3 gallons of ethanol in an automobile equals only 1 gallon of gasoline.

The total US consumption of liquid fuel, including gasoline, diesel fuel, and jet fuel is about 200 Billion gallons annually. To produce the equivalent amount of liquid fuel energy using ethanol at 1.5 gallons equivalent to 1 gallon of gasoline, forgetting about the extra energy required to produce it, would take 870 million acres of US farmland, 3 times greater than the 300 million acres we grow our food on.

Conclusion? If we used all our farmland to make ethanol for transport and didn't eat, we could travel 1/3rd as much as we do today. But hey! We wouldn't be using oil. Even if we manufactured ethanol from cellulosic feed material and not corn, we still would need vast amounts of arable land to grow the material that otherwise could be used to grow food for a malnourished World with an increasing population. It is silly to expect that biofuel will be able to meet more than a very small fraction of future World transport needs.

As discussed in Chapter 1, hydrogen fueled cars are not practical. There has been lots of enthusiasm for hydrogen cars – clean energy, only water out of the tailpipes, etc. – but the realities do not support using hydrogen as a fuel in cars. Yes, we can make hydrogen by electrolyzing water using low cost electricity, and we proposed to do that for synthesizing gasoline and diesel fuel from carbon dioxide and hydrogen.

But, using hydrogen as a fuel stored in automobiles and trucks is extremely frightening. Imagine 230 million cars and trucks driving 70 mph on US highways, each with a tank of very cold liquid hydrogen at close to absolute zero temperature, or a tank of compressed hydrogen gas at 5,000 psi. If the car is seriously damaged in an accident, and there are many thousands of such accidents every year on US highways, it is very possible that hydrogen gas would leak from the tank, mix with air, and explode. The explosive force would be equivalent to hundreds of pounds of TNT. The explosion could then cause other nearby cars to explode.

Or, you're filling your car with hydrogen at a filling station, and experience a hydrogen leak, from your car, another car, the pump, or the station storage tank. And there is a hydrogen explosion. It doesn't matter who is responsible – you're dead. Or, you park your hydrogen car in an underground garage along with dozens of other hydrogen cars. One car leaks and explodes, causing all the other cars to explode and bringing down the building with its occupants.

Hydrogen cars would be very dangerous if used to fuel the World's 2.5 billion cars in 2050 AD.

In contrast to hydrogen cars, electric cars are practical. In fact, electric cars have been driving around American for well over 100 years. Figure 5.4 shows Parker electric car, produced from 1899 to 1915. It had a top speed of 14 mph and a range of 50 miles (10). Figure 5.5 shows Thomas Edison with an electric car in 1913.

Figure 5.4 Thomas Parker Electric Car

Figure 5.5 Photo of Thomas Edison with Electric Car 1913

Electric cars were popular at the beginning of the 1900's. 40 percent of US automobiles were powered by steam, 38 percent by electricity, and 22 percent by gasoline. There were 33,842 electric cars registered in the United States (10). There even was a battery swapping service provided by a subsidiary of General Electric. Well-to-do upper class families loved their electric cars. They were comfortable, luxurious, noiseless, didn't smell like gasoline cars, didn't require long start-up times – 45 minutes in cold morning for Stanley Steamers – and didn't require a hand crank for startup, as did gasoline cars.

James Jordan's father started his working life as an 11 year old chauffeur in a Baker electric car driving an upper class family around in Winston Salem, North Carolina. There were many electric cars in the town. Everybody loved electric cars.

However, things started to change, and gasoline cars took over. Charles Kettering invented the electric starter in 1912 (10) which eliminated the hand crank, a really big deal, and gasoline cars started to use mufflers. Gasoline became much more plentiful, and Henry Ford invented the assembly line, which made gasoline cars much cheaper. By 1912 an electric car cost twice as much as a gasoline car (10).

The electric car industry withered and died. From time to time, there were efforts to revive it, but they did not succeed. Figure 5.6 shows the 1960 Henny Kilowatt, based on the European Renault Dauphine. Top speed was 60 mph, with the capability to travel for almost an hour on a single charge (10). A good car, but it did not last long. Production stopped in 1961.

Electric cars began their revival in the 1990's. Today, there are lots of makes and models, both all-electric and electric/gasoline hybrids. Manufacturers include BMW, Mercedes Benz, Chevrolet, Honda, Nissan, Fiat, Mitsubishi, Kia, Ford, Toyota, and Tesla.

Tesla electric cars receive the most media attention and are top rated. Figure 5.7 shows a photo of the Tesla roadster. However, the other electric car makes and models are also excellent automobiles.

Figure 5.6 1960 Henney Kilowatt Electric Car

Figure 5.7 Tesla Roadster

Top sellers in the global market from 2008 through June 2014 – total sales for the 6 year period – are

Car	Total Sales (2008-2014)
Nissan Leaf	124,000

Tesla Model S	39,163
Mitsubishi-MiEV	32,000
Renault Kangaroo ZE	14,542
Renault 200	12,631

All electric vehicles still constitute a very small percentage of new car sales Worldwide. The top 10 countries that had the largest market share of all-electric cars in 2013 are listed below, along with the percentage of new car sales that were all electric vehicles (11).

Rank	Country	All-Electric Car Market Share (2013)
1	Norway	5.75%
2	Netherlands	0.83%
3	France	0.79%
4	Estonia	0.73%
5	Iceland	0.69%
6	Japan	0.51%
7	Switzerland	0.39%
8	Sweden	0.30%
9	Denmark	0.28%
10	United States	0.28%

All of the above 10 countries have high per capita incomes, yet electric car sales are very small – growing, yes, but not rapidly.

In the US, 250,000 plug-in electric cars have been sold since 2008, about 40% of them all-electric cars and 60% plug-in hybrids. That's equivalent to 1/1000th of the approximately 250 million cars on US highways. (11)

Worldwide, about 300,000 all-electric cars and light utility vehicles have been sold since 2008. That is 1/3000th of the Billion cars now operating Worldwide.

This very small fraction of the car market for electric cars, and the very small fraction of the World's total number of cars, raises questions.

- What is holding back the sales of electric cars?
- What are their limits and constraints?
- Can electric cars become a major part of long-term, sustainable transport for the World?

There are a number of limits and constraints that hold back electric car sales, as listed below:

- Range limits
- Long charging times
- Availability of Charging Stations
- High cost
- Heating and cooling vehicle
- Safety / operating reliability
- Battery cycle lifetime
- Resource limits for batteries.

A very important constraint is the limited range of all-electric cars, those that use only battery packs as their energy source. Plug-in hybrids like the Chevrolet Volt, that have a gasoline fuel engine in addition to the battery pack, which can power the car when the battery energy runs out, have much longer ranges, limited only by the capacity of the gasoline tank.

The top selling all-electric car, the Nissan Leaf, has an EPA rate range of 73 miles with an EPA combined fuel/electric, economy of 34 KWH per 100 miles of combined city and highway driving (11).

At 10,000 miles per year the typical annual mileage for US cars, the average daily mileage is about 27 miles per day, well within the range capability of the Nissan Leaf. However, while most trips are short, Americans do take long drives of a hundred miles or more for business, visits to friends and relatives, and vacations.

Such trips would exceed the range capability of an all-electric vehicle with the possible exception of a Tesla car, which has a range of over 200 miles (11). However, battery packs are expensive. At the present cost, on the order of $300 per kilowatt hour of storage (11) and an EPA economy of 34 KWH per hundred miles, the cost of the battery pack alone for a 200 mile range would be 34 x 2 x 300 = $20,000, more than the entire cost of a typical gasoline car. The US Department of Energy has set research targets for $300 per KWH in 2015, and $125 per KWH in 2022.

Prosperous people can afford all-electric long range cars just as they did in the early 1900's. Most Americans, however, if they want an electric car will have to settle for a less expensive version, with a maximum range on the order of 75 miles. For longer trips, they will have to recharge on the way.

If one could charge electric cars as rapidly as gasoline cars can refill their tanks, and there were many thousands of recharging stations around the country, then drivers might be more amenable to owning an electric car. However, there are problems. One can fill up with gasoline in 3 minutes or less, and there are 250,000 filling stations around the US. No long lines except in emergencies.

If you plug in your electric car overnight at home with a normal 1.5 KW(e) outlet, you can charge it for 10 hours to 15 KWH, sufficient for about 43 miles range, more than the daily driving average of 27 miles. With rewiring in the house, to deliver more power to the electric car's battery, additional energy storage can be accomplished. The Nissan Leaf has an on-board 3.3 Kw charger (11) and could be fully charged to its 73 mile range in a few hours.

For long trips at greater distances than the range of the electric car, stopping one or more times during the trip to recharge it would be unacceptable, if recharging it took a few hours. Traveling long distances on a 300 mph Maglev vehicle with your electric car on-board would be practical and attractive, however. Not only would you get to your destination much faster and at lower cost than by highway, but the electric car could be fully charged enroute on the Maglev vehicle, ready to go when you drove the car off the Maglev vehicle.

The other options for long trips at distances greater than the electric car range are:

- Fast charging stations
- Battery swapping stations

For national electric car usage, one would require a lot of stations, whether they were fast charging facilities, or battery swapping facilities. The US has on the order of 250,000 gasoline stations, about 1 per 10 square miles. Assume that there are 30,000 charging stations for long trips, one station for every 100 miles of US roads. The total number of US vehicles miles traveled in 2012 was 2.9 Trillion miles (12).

Assuming that 10% of the total US vehicle miles traveled was on long trips, i.e., 10% of 2.9 Trillion vehicle miles, and that the average electric energy consumption per vehicle was 0.35 KWH per mile on average, electric charging station would deliver 3.4 million kilowatt hours per year, about 10,000 KWH per day, with an average power demand of 400 KW.

That's a lot of power. To meet a peak to average demand factor of 3/1, the average station would need a power capacity of 3 x 400 KW or 1.2 megawatts. The power capacity for 30,000 stations would be 30,000 x 1.2 = 36,000 megawatts.

With 1.2 megawatt and a 200 mile range auto charging in 30 minutes at 0.35 KWH per mile, the fast charging station could handle a peak demand of 8 cars being charged at the same time.

So, visualize taking a long trip, greater than 100 miles in an all-electric car. Partway through the trip you need to charge it. Maybe there's been congestion, speed is slow, its been very hot and the air conditioner has drained your battery pack or it has been very cold and the car's heater has drained the battery. This is one of the constraints on all-electric cars that doesn't get mentioned very often. If you have to heat or cool the car, you have to use battery power, which can be a big drain on a 100 degree or zero degree day.

Anyway, you desperately look for one of those 30,000 charging stations, one every 100 miles on America's 3 million miles of roads. You are lucky. You find one before the battery dies. You pull into the station, but there already is a line of cars waiting to be charged. After 1 hour of waiting, you are next in line for a charge – you connect, and then charge for another 30 minutes. As you wait, you think, if I had a gasoline car, or a plug-in hybrid with a gasoline engine that takes over if the battery pack dies, I could have found a gasoline station much faster and refueled in only 3 minutes! Maybe my next car won't be an all-electric one, but a plug-in hybrid!

A further problem with fast charging stations is that the charging system will have to be the same at all US stations, and capable of handling different battery packs on different model cars. Today, there are a dozen different kinds of all-electric cars.

As an alternative to "fast", i.e., 30 minute recharging stations, battery swapping stations have been proposed. You pull your car into the station, and attendants remove your 500 pound depleted battery pack and insert a fresh one in a few minutes. Sounds great, but there are questions:

- Do they stock battery packs for every electric car model? Are they out of batteries?
- Are you getting a good battery or a defective one? You think that it well last for another 100 or 200 miles, but will it?
- Is it safe? Lithium ion batteries do catch fire.
- How much does it cost? At $300 per KWH the capital cost of a 50 KWH battery back would be $15,000. Can a battery swap station afford an inventory of hundreds of battery packs so that they can service many different models of cars, and not run out? Ten different types of battery packs, with 20 batteries per type, equals an inventory of 200 batteries. At $20,000 per battery that's 4 million dollars inventory value.
- If your battery has degraded by charging to rapidly or at the wrong temperatures is the battery swapping station obligated to take it and suffer a cash loss? Not likely. "Sorry, sir, we can't take your battery".

Or, as has been proposed, one could rent a trailer with extra battery power for long trips, towing it behind you as your drive (11). BMW is offering to provide a conventional gasoline powered BMW for long trips for a given number of days per year (11).

All of this does not give one a lot of confidence about taking a long trip with an all-electric auto. If your battery dies on the highway before you can reach the next recharging station, what can you do? It's not like a gasoline car, where your can call a friend or the AAA to bring a can of gasoline. Instead you would have to have load them on a truck and taken to the nearest recharge station. Not cheap, and it would cause congestion and delays for other drivers, who would not be sympathetic.

So range limits and charging times & availability are very important issues in determining whether or not all-electric cars will play a major role in future world transport. As stated above, heating or cooling all-electric cars draws power from the battery pack and reduces range. This can be important for drivers in hot or cold climates like Arizona or Minnesota. The possibility of battery fires, either spontaneous, or in accidents is also a serious consideration, as is the possibility of running out of batter power on a busy highway.

The cost of electric cars is very important. Advocates concentrate on comparing the relative costs of gasoline and electric power per vehicle mile. They point out that an all-electric car with an energy consumption of 34 KWH per 100 miles and electricity cost of 12 cents per KWH, costs only 4 cents per mile for electricity. Then they say that a 30 mpg gasoline car, with regular gasoline at $3.00 per gallon costs 12 cents per mile.

Cost savings for a 100,000 mile car? ($0.12 - $0.04) x 100,000 = $8,000. Great. What they fail to point out is that a 200 mile range electric car with a battery pack that costs $300/KWH has an expense of 34 x 2 x 300 = $20,000 – over twice the savings in energy cost. This does not point out, the battery pack degrades over 10 years and 100,000 miles, the owner will have to buy one or more new battery packs over the life of the car. Batteries do degrade as users of cell phones, computer, etc., know.

The last issue we consider here is the amount of resources needed to manufacture the batter packs used in the electric vehicles. The favorite battery type is the Lithium ion battery. World lithium resources are estimated at 9.9 million tonnes (13) found in brine pools (Figure 5.8) and other sources.

Estimates of the amount of lithium required per KWH for Lithium-ion batteries vary. Using a nominal value of 10 KWH storage capacity per kilogram of lithium (14), World Lithium resources would correspond to 1 Billion KWH of storage capacity, enough to power 4 Billion Nissan Leaf all-electric cars (14).

Of course, there are many other uses for lithium that will compete with electric cars. Also, over the long term for a sustainable, non-fossil fuel World, lithium and other types of battery packs will have to be recycled. Recovering 100% of the Lithium in all of the recycled batteries for many, many recycles does not appear likely, so that even if world lithium resources can initially sustain billions of electric cars, there could be a problem over many decades.

Conclusions. Electric cars, as well as Maglev, will play a major role in a sustainable World that does not consume fossil fuels. Both all-electric and plug-in hybrids will operate. Many consumers will favor plug-in hybrids that can switch to gasoline if their electric battery pack runs out of energy, giving them longer range, easier and more convenient refueling, and less worry about Moreover, there are important transport systems that cannot be running out of energy on the highway. Gasoline will be needed for the plug-in hybrids.

Electrically powered, i.e., airplanes, long-haul heavy duty trucks, large farm equipment, etc. So, even with electric cars and Maglev, gasoline, jet and diesel fuel will still be needed.

Figure 5.8 Lithium in Brine Pools on the Pacific Coast of South America
Photograph from US Geologic Survey

Currently, the US uses 13.3 million barrels of oil per day for transport, with the rest of the World using 45 million barrels per day for transport. Total oil consumption for transport is 13.3 + 45 = 58 million barrels per day, 63 percent of the total daily consumption of 92 million barrels daily. Of the 33 Billion barrels of oil consumed annually by the World, 21 Billion go for transport.

As described earlier and in Chapter 2, the number of cars in the World is projected to considerably increase from the present value of 1 Billion cars in the decades ahead, reaching a level of about 2.5 Billion cars in 2050 AD. World passenger miles and freight ton miles are expected to be twice today's values in 2050 AD.

If transport were to continue to depend on oil based fuel, World oil consumption for transport would double from todays 21 Billion barrels per year, to more than 40 Billion barrels annually – an impossibility, given World oil reserves and the resulting carbon dioxide emissions.

For the World economy to grow and living standards to increase without environmental disaster electric transport, both Maglev and electric cars must become dominant modes of transport in the decades ahead. However, liquid fuels will still be needed for airplanes, long distance trucks, plug in hybrids and other transport applications. What fraction? Probably in the range of 5 to 10 billion barrels of gasoline, diesel, and jet fuel, annually, up to a quarter of transport energy demand in 2050 AD.

Is there an alternative to obtaining these liquid hydrocarbon fuels from fossil fuels, either from crude oil or by synthesis from natural gas or coal? Yes, there is – by reacting carbon dioxide extracted from the atmosphere with hydrogen obtained by electrolyzing water. Consumption of the resulting synthetic fuels will not cause an increase in carbon dioxide concentration in the atmosphere, since CO_2 from burning the synthetic fuel is offset by CO_2 extraction from the atmosphere to make the synthetic fuel.

We now turn to describing how synthetic gasoline and diesel fuel can be manufactured at prices competitive with present fuels derived from crude oil using the carbon dioxide extracted from the Earth's atmosphere, plus hydrogen from the electrolysis of water, with low-cost electric power from space solar power satellites launched by StarTram, as described in Chapters 3 and 4.

There is nothing new about making synthetic fuel, both gasoline and diesel, without using oil. The Fischer-Tropsch process, invented in 1925 (15) reacts carbon monoxide (CO) and hydrogen (H_2) to make a range of hydrocarbons that can be processed into gasoline and diesel fuel. Alkanes, the straight line hydrocarbons, are produced by the chemical reaction shown below.

$$(2n+1) H_2 + nCO \rightarrow C_nH_2n + 2 + nH_2O$$

Octane, a major component of gasoline, has 8 carbon atoms, with the formula C_8H_{18}. Diesel fuel contains higher molecular weight hydrocarbons, e.g. decane C, $C_{10}H_{22}$ and above. Gasoline and diesel fuel are complex mixtures of many different hydrocarbons, both straight line molecules, branched molecules, and aromatic molecules. Process parameters and product refining are controlled so as to obtain the desired type of synthetic fuel.

Because of fuel shortages for its army and air force in World War II, Nazi Germany produced large amount of synthetic fuels using the Fischer-Tropsch process. In early 1944 German synthetic fuel production reached more than 124,000 barrels per day (16).

Worldwide, currently 260,000 barrels per day of synthetic liquid fuels are produced using the Fischer-Tropsch process and other processes that react syngas, i.e. a mixture of carbon monoxide (CO) and hydrogen (H2) to manufacture the synthetic fuel (16) 260,000 barrels per day. That's 100 million barrels per year, 1% of our goal of 10 Billion barrels per year of manufacturing synthetic gasoline and diesel fuel from air and water. Still, it shows that the process is practical and affordable if the cost of the raw materials, CO2 and H2 is acceptable. South Africa, which has limited oil reserves, produces 150,000 barrels per day of synthetic liquid fuel using the Fischer-Tropsch process (16).

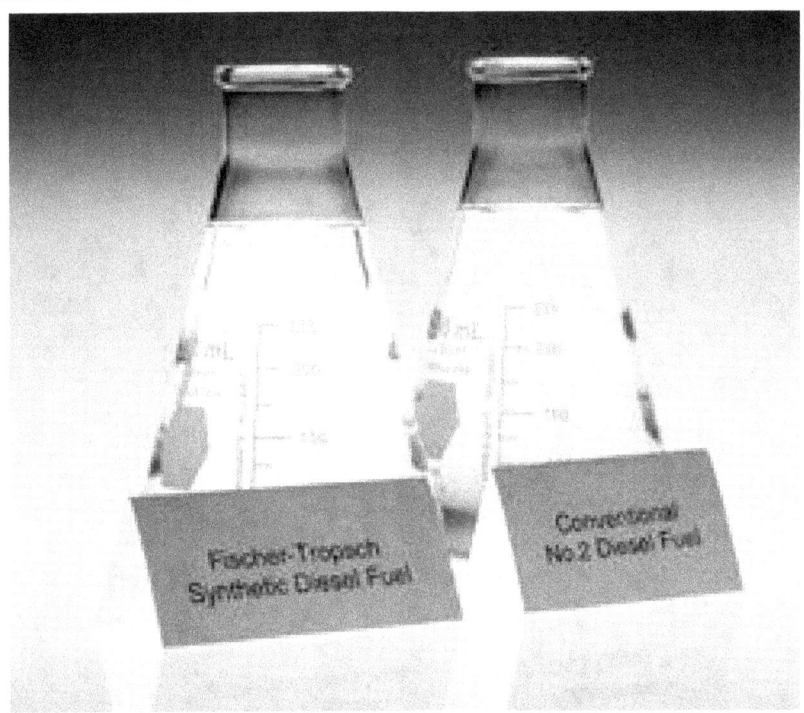

Figure 5.9 NREL FT diesel vs conventional diesel photo

An important benefit of synfuels production is that the product is less polluting than fuel obtained by refining crude oil. Figure 5.9 compares a container of conventional diesel fuel refined from crude oil with a container of diesel produced by the Fischer – Tropsch (FT) Process. The FT diesel is much clearer, with much lower concentrations of sulfur and aromatics (16).

Figure 5.10 shows the reduction in diesel exhaust emissions achieved using FT diesel fuel relative to conventional diesel fuel from crude oil. The reductions in hydrocarbon (HC) emissions, carbon monoxide (CO), carbon dioxide (CO_2), nitrogen oxides (NOx) and particulate matter (PM) are substantial.

Figure 5.10 OSD Clean Fuel Initiative FT Diesel Emissions Presentation

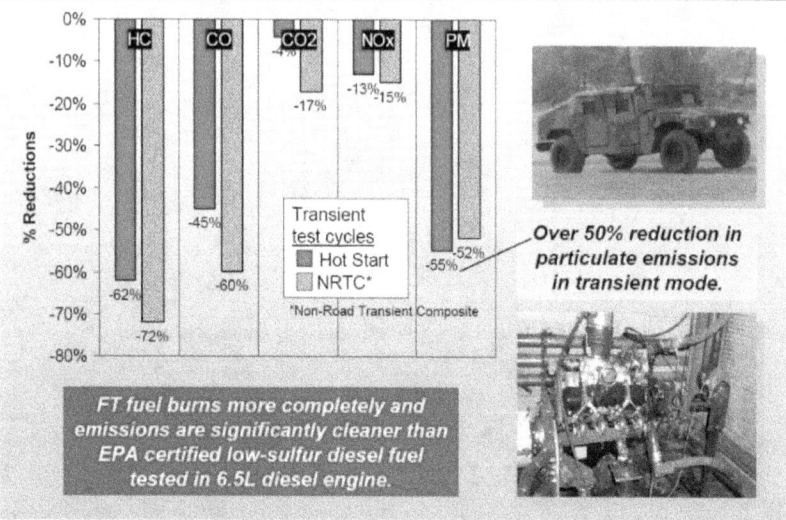

The FT process and other synfuel production process starts with Syngas, a mixture of carbon monoxide (CO) and hydrogen (H2) to produce synthetic gasoline and diesel fuel, as described above.

Today, the syngas for the Fischer-Tropsch process can come from a variety of sources, natural gas, coal, biomass, and the source that does not involve possible fuels or biomass – namely, carbon dioxide from atmosphere air and hydrogen from water.

After extracting carbon dioxide (CO2) from the atmosphere, it would be converted to carbon monoxide (CO) by the reverse water-gas-shift reaction using hydrogen as shown below:

$$CO_2 + H_2 \rightarrow CO + H_2O$$

The reverse water gas shift reaction is simply the old water gas shift reaction in reverse, e.g., the water gas shift reaction is:

$$CO + H2O \rightarrow CO2 + H2$$

Depending on reaction condition, one can drive the reaction to make CO from CO2 by supplying hydrogen, or make hydrogen by supplying CO with high temperature steam. The water gas shift reaction was discovered by the Italian physicist Felice Fontana in 1789 and been in commercial use for a hundred years (17).

Combining the Fischer-Tropsch reaction and the reverse water gas shift reaction we have

$$n\ CO_2 \quad + \quad (3\,n + 1)\,H2 \quad \rightarrow \quad C_nH_{(2n+2)} + n\ H_2O$$
$$\text{(atmospheric air)} \ \text{(hydrogen fm electrolysis)} \quad \text{(gasoline)} \quad \text{(water)}$$

Burning a gallon of gasoline releases 19.64 pounds (8.9kg) of CO2 into the atmosphere (2). To synthesize gasoline from air and water we extract the same amount of CO2 from the atmosphere 19.64 pounds, as raw material for the reverse Water Gas Shift/Fischer-Tropsch process.

The amount of hydrogen feed required per gallon is given by

$$3x\ \frac{\text{molecular weight of H2 x19.64 pounds of H2 per gallon}}{\text{molecular weight of CO2}}$$
$$= 3 \times 2/44 \times 19.64 = 2.7 \text{ pounds (1.2 kg) of H}_2$$

The above mass requirement assumes no loss of material, and do not include the 1 term in the expression $3\,n + 1$. However, for octane, $n = 8$, so $3\,n + 1 = 25$ is essentially the same as $3n = 24$ for preliminary analysis.

The World's air and water inventories are enormous. The effect of withdrawing CO2 from the atmosphere and water from the environment to make synthetic gasoline and diesel will be completely negligible and undetectable.

Figure 5.11 Denali Mt McKinley

The World's atmosphere, a portion of it shown in the beautiful photo of Mount McKinley – now called Denali (Figure 5.11) – has a total mass of 5×10^{15} metric tonnes. In more familiar terms that's 5 million Billion metric tonnes. To obtain the CO_2 for 10 Billion barrels of synthetic gasoline and diesel flow, annually, one would extract 10^7 (10 million tonnes) of CO_2 from the atmosphere daily, about 12 percent of what we currently emit to the atmosphere each day by burning fossil fuels. To extract the CO_2 at its concentration of 400 parts per million, the CO_2 extraction plants would take it from about one millionth of the World's atmosphere.

Similarly, electrolyzing water to provide the hydrogen for manufacturing the synthetic gasoline and diesel fuel would have a negligible impact or the World's water resources. Figure 5.12 shows a view of the Mississippi River. Electrolyzing a cubic meter of water produces 110 kg of hydrogen, sufficient to manufacture 92 gallons of synthetic gasoline. For 10 Billion barrels of synthetic gasoline and diesel fuel per year, one would require a water flow of 150 cubic meters per second.

Figure 5.12 View of Mississippi River from Fire Point in effigy Mounds National Monument, Iowa, USA

The average water flow rate of the Mississippi River is 16,792 cubic meters per second (18), more than 100 times greater than required for generating 10 Billion barrels per year. The Amazon is much greater still, with a discharge rate 8 times that of the Mississippi. There are lots of other big rivers around the World from which one would take a very tiny bit of their discharge into the ocean, to generate the synthetic gasoline and diesel. Also, it is likely the synthetic fuel plant will recycle the H2O produced by the Fischer-Tropsch process and electrolyze it to generate hydrogen. This would reduce the amount of input water for the synfuels plant by 30%.

A variety of processes have been investigated for extracting CO2 from flue gases and the atmosphere. Extraction from the atmosphere is more difficult than from flue gases emitted by fossil fuel fired power plants and other sites, because of the much lower concentration of CO_2 in the atmosphere, only 400 parts per million.

CO_2 can be scrubbed from the atmosphere by chemical reactions with materials like CaO (quicklime), or an aqueous solution of calcium hydroxide [CA(OH)2], or sodium hydroxide (NaOH) or other materials (19). The CO_2 is released from the resultant carbonate and the absorbing material by heating to high temperature. Georgia Tech is working on capturing CO_2 from the atmosphere using a ceramic honeycomb structure coated with an absorbent, i.e., a dry amino-modified silica material. The CO_2 is removed and the absorbent regenerated by passing steam through the structure. The Georgia Tech researchers estimate that a structure the size of a shipping container – 8x8x20 feet long – could remove 1,000 tons of CO_2 per year. To produce 3.7 Billion tons per year for manufacturing 10 Billion barrels of synthetic gasoline/diesel would require 3.7 million shipping containers, a fraction of the present world inventory of 17 million shipping containers. Georgia Tech projected cost for CO_2 extracted from the atmosphere is $100 per tonne (20).

A promising approach for CO_2 extraction is the "artificial tree" under development by Klaus Lackner of Columbia University and associates (21, 22). The artificial tree consists of the multiple plastic sheet surfaces coated with a resin that absorbs CO_2 from the air. To recover the CO_2, which is then compressed to be used or sequestered, the resin dries in low humidity air to be regenerated.

Lackner estimates that approximately 1.1 megajoules of electrical energy per kilogram of CO_2 per tonne of CO_2 is required for pumping and compressing the captured CO_2 (21). Per tonne of CO_2 fed to the Fischer-Tropsch process that corresponds to 1100 megajoules or 305 KWH. At 2 cents per KWH using beamed solar power from space, that's only $6 per tonne energy cost, a small amount. The main cost elements for the artificial tree process will be the capital cost of the structure and absorbent sheets, plus O&M costs. Some estimates project the cost of CO_2 removal will be much higher than $100 per tonne as much as $600 per tonne (23).

As we show later, even at $300 per tonne, synthetic and diesel fuel can be manufactured from air and water at less than $5 per gallon which is less than the cost of today's gasoline and diesel in most countries. CO_2 extraction technology is evolving, so that costs will drop substantially. There is general agreement that the cost of extraction of CO_2 from flue gases emitted by power plants and industrial processes will be on the order of $100 per tonne. In the initial phases of manufacturing synthetic gasoline and diesel fuel, CO_2 from such sources could be used, reducing the emissions of CO_2 from fossil fuels.

How much intake area of atmospheric air into CO_2 extraction units will be required to produce 10 Billion barrels per year of synthetic gasoline/diesel? One's first impression is that it must be enormous. However, that's not correct. In fact, the inflow intake are is much smaller than the air flow area through the windmills now operating around the World.

Visualize a 1 m^2 (10.8 square feet) intake area through which air flows at 10 meters per second (23 mph) to be processed for CO_2 extraction. How much CO_2 would be retrieved per year? 245 metric tonnes. Per acre of intake flow area, that's, 1 million tonnes of CO_2 extracted per year. To manufacture 10 Billion barrels of synthetic gasoline/diesel per year, total intake flow required is only 3700 acres.

To put the intake flow are in a different perspective let's compare it with the swept flow area of windmills (Figure 5.13).

Figure 5.13 Windmill Farm

Shown below are the operating parameter for a typical windmill, the GE 1.5 sle (24).

- 1.5 MW(e) power capacity
- 38.5 meter (126 ft) blade length
- 80 meter (262 ft) hub height
- 118.5 meter (389 ft) total height
- 4,657 m² (1.15 acre) area swept by blades

Wind mills are BIG structures. The GE 1.5 sle windmill has a total height of 389 feet, 84 feet higher than the torch of the Statue of Liberty (Figure 5.14). It is dwarfed by the 2.75 MW Vestas V 100 windmill (24), which has a total height of 492 feet (150 meters), almost 200 feet higher than the torch on the Statute of Liberty (25).

Total air flow area swept by the blade on a GE 1.5 sle windmill is 4657 m² (1.15 acre). A CO_2 extraction plant with the same intake air flow at 10 meters/sec would produce 1.15 million tonnes of CO_2 per year. 3200 such sized air intakes would produce sufficient CO_2 for 10 Billion barrels of synthetic gasoline/diesel per year.

Figure 5.14 Statue Of Liberty

In practice, however, the CO_2 production rate of 245 tonnes per m^2 per year at 10 meters per second is the maximum possible, assuming that recovering the absorbed CO_2 and regenerating the absorbent material can be accomplished in a time much shorter than the time that the absorbent is exposed to intake air flow. Also, the average air flow velocity may be lower than 10 meters per second. An interesting note is that CO_2 removal rate scales as air flow velocity to the 1st power i.e. $(V)^1$, while windmill output power scales as the 3rd power, i.e. $(V)^3$, so that windmills are much more sensitive to wind speed than CO_2 extraction units will be. Also, wind mills shut down if wind speed drops much below the rated value, while CO_2 extraction units will operate over the full range of speeds from zero to maximum rated.

To be conservative we take the capacity factor of the CO_2 extraction units to be 1/3 (33%) of the 245 tonnes/m^2 maximum. This allows for down time to recover the absorbed CO_2 and regenerate the absorbent and an average wind speed less than 10 meters per second. At 33% capacity factor, 1 m² of airflow intake area would produce 0.33 x 245 = 80 metric tonnes of CO_2 per year.

The total area of windmill airflow intake to produce 1.0 Billion barrels per year would then be 3 x 3200 = 9600 equivalent 1.5 MW GE sle windmills. As of the end of 2012, there were over 200,000 windmills operating in the World, with a nameplate capacity of 282,452 megawatts (26). World total wind power generation was 534 terawatt hours, about 4% of the World's total electric power production, with a capacity factor on the order of 40 percent of nameplate rating.

So, in 2012, the swept area of 282,482 windmills was 282, 482/9600 = 30 times greater than the air intake area to produce 10 Billion barrels of synthetic gasoline/diesel per year. And the number of windmills in the World is rapidly growing. In a few years there will be 50 times as much windmill swept area as needed for CO_2 extraction, in a decade or so, 100 times as much swept area.

Having the intake area for CO_2 extraction be such a small fraction of the swept area of already operating windmills is very impressive. Also very impressive is the value of the CO_2 produced. For the swept area of the GE 1.5 MW sle windmill, 4657 m², at 100 tonnes per m² per year and $100 per tonne value, operating for 1 year would yield a CO_2 value of 46 million dollars. In comparison, the 1.5 MW(e) turbine, operating at a capacity factor of 40% and an electric value of 4 cents per KWH(e), would produce $210,000 of electric energy annually, a ratio of 220 in value of product.

Table 5.1 summarizes the principal parameters for production of 10 Billion barrels of synthetic gasoline/diesel from air and water using the Fischer-Tropsch process.

Table 5.1

Parameters for Production of Synthetic Gasoline and Diesel Fuel From

Air and Water

Parameter	Value
Barrels of Synthetic Fuel Produced per Year	10 Billion (420 Billion Gallons)
Amount of CO_2 Per Gallon extracted from the atmosphere per year	3.7×10^9 Billion Tonnes,
Kilograms of CO_2 per gallon	8.9 Kilograms
Amount of H_2 Produced by Water electrolysis per year	500 million tonnes
Kilograms of H2 per gallon	1.2 Kilograms
Electrical Energy Per Kilogram of H_2	50 KwH (80% efficient)
CO_2 Extraction from Atmosphere Tonnes/year per M^2 of intake area at 33% capacity factor	80 tonnes/year
Total CO2 Extraction Air Intake Area for 10 Billion Barrels per Year	11,000 Acres
Swept Air Flow Area for 1.5 MW(e) GE Windmill	1.15 acre
Amount of CO2 Extracted from Atmosphere with Intake Area Equal to Windmill Swept Area at 33% Capacity Factor	380,000 tonnes per year
Total Cost Per Gallon of Fuel for $100 per tonne of CO_2	$2.81 per gallon
Total Cost Per Gallon of Fuel for $300 per tonne of CO_2	$4.67 per gallon

What will be an acceptable price for synthetic gasoline manufactured from air and water in a non-fossil fuel World economy. It's difficult to predict the future. Absolutely, all we can be certain of is Yogi Berra's prediction – "The future lies ahead."

However, we can get a reasonable idea by look at what people pay today for a gallon of premium gasoline in countries around the World. Table 5.2 based on data given by Bloomberg (27). Cost per gallon ranges from $9.69 in Norway to $8.56 in Germany to $7.58 in Japan, to $6.41 in Brazil, to $6.06 in India, to $5.39 in Chia, tol $4.19 in the US, to $3.71 in Russia, to 1.79 in Egypt, to $0.61 in Saudi Arabia, to $0.29 in Venezuela with a crude oil price of $100 per barrel, the crude oil alone costs $2.50 per gallon, plus the cost of refining and shipping. Clearly at a price of less than $4 per gallon, the gasoline is being subsidized.

For synthetic gasoline and diesel from air and water, a target cost of about $5 per gallon appears very reasonable. A large portion of the World pays even more than $5 per gallon today, even in countries with large populations, like India and China.

So, what is the projected cost for synthetic gasoline and diesel from air and water? For simplicity, we project the cost for gasoline, since it is a considerable bigger market than the market for diesel fuel. The cost for diesel will be somewhat higher, but close to the cost for gasoline, perhaps 10 to 20% greater.

Table 5.2
Price per Gallon of Premium Gasoline Around the World (27)

Price Range per Gallon ($)	Countries in Price Range (Highest Price Countries First)
$10 to $9	Norway, Denmark, Italy, Netherlands, Greece
$9 to $8	Sweden, Hong Kong, Portugal, U.K., Belgium, France, Finland, Germany, Ireland
$8 to $7	Switzerland, Slovakia, Hungary, Czechoslovakia, Japan, S.Korea, Spain, Slovenia, Austria, Malta, Latvia, Luxembourg, Lithuania, Estonia
$7 to $6	Cyprus, Bulgaria, Australia, Singapore, Romania, Chile, Brazil, India
$6 to $5	Canada, South Africa, Seychelles, Argentina, China
$5 to $4	Thailand, US, Indonesia
$4 to $3	Russia, Malaysia, Mexico
$3 to $2	Iran, Nigeria
$2 to $1	UAE, Egypt
$0 to $1	Kuwait, Saudi Arabia, Venezula

Table 5.3 gives the cost of gasoline based on the 4 principal cost elements using the Fischer-Tropsch process with the Reverse Water Gas Shift Reaction.

1. Cost of CO_2 extracted from the atmosphere
2. Cost of H_2 produced by electrolysis of water
3. Capital cost of the Fischer-Tropsch plant
4. Operation and Maintenance costs for the FT plant.

Since there is no large-scale production plant experience for CO_2 extraction from the atmosphere, we consider a range of costs from $100 to $300 per metric tonne of CO_2. $100 per tonne is projected by a number of researchers. But on the other hand, some project as high as $600 per tonne based on present experience. It appears likely that as CO_2 extraction technology evolves over the next 10 to 20 years, the cost per tonne should be in the range of $100 to $300 per tonne. For 8.9 kg of CO2 per gallon of gasoline, this corresponds to $0.89 per gallon at $100/tonne, $1.78 per gallon at $200 per tonne, and $2.67 at $300 per tonne.

Per gallon of gasoline, the Fischer-Tropsch process requires 1.2 kilograms of hydrogen. A 100 percent efficient electrolysis requires 40 kilowatt hours of electric energy per kilogram of hydrogen produced. At 80% efficiency, a practical value of 50 kilowatt hours per kilogram of H2 is required. Using beamed power from space solar power satellites at 2 cents per KWH, the cost of 1.2 kilograms of hydrogen per gallon of gasoline would be $1.20.

The 3rd cost component of synthetic gasoline/diesel is the capital cost of the Fischer-Tropsch plant that processes the CO2 and H2 feed materials into gasoline and diesel. Sasol estimates the capital investment for a 96,000 barrels per day FT plant in Westlake, Louisiana to be about 11 Billion dollars (15).

Over a 30 year amortization period, the Sasol plant would produce 96,000 x 365 x 30 = 1.05 Billion barrels. Per barrel the 11 Billion dollar investment corresponds to a cost of $10.5 per barrel. Per gallon of gasoline that is $0.25 per gallon for the capital cost of the FT plant.

The 4th cost component is the operating and maintenance costs for the FT plant. A reasonable projection for O&M costs is 10% of the capital investment cost per year, i.e., 1.1 Billion dollars per year. At 96,000 barrels per day, the FT plant would produce 3.5 x 10^7 barrels per year. The corresponding O&M cost would then be 1.1 x 10^9/3.5 x 10^7 = $31.4 per barrel or $0.75 per gallon.

Table 5.3 shows the total estimated cost per gallon of gasoline, including the above 4 principal cost elements. For $100 per tonne of CO2, total cost per gallon is $3.09, well under the average price paid worldwide. At $200 per tonne, cost is $3.98 per gallon, and at $300 per tonne, cost is $4.87

per gallon, less than most countries already pay for gasoline (Table 5.2).

Table 5.3

Projected Cost of Synthetic Gasoline per Gallon from Air and Water

as a Function of Cost per Tonne of CO2 Extracted From Air

Cost per Gallon of Gasoline, in Dollars

Cost Component	$100/tonne of CO_2	$200/tonne of CO_2	$300/tonne of CO_2
CO_2 From Atmosphere	0.89	1.78	2.67
H_2 From Electrolysis of Water	1.20	1.20	1.20
Amortized Capital Cost of FT Process Plant (30 year Amortization)	0.25	0.25	0.25
O&M Costs of FT Plant	0.75	0.75	0.75
Total Cost Per Gallon	$3.09	$3.98	$4.87

Conclusion? Manufacturing synthetic gasoline and diesel fuel from CO2 and H2 is nothing new. The Fischer-Tropsch and other processes have been operating for many decades at large scale. Currently, 260,000 barrels of synthetic fuel are manufactured daily at acceptable prices. Manufacturing large amounts of hydrogen at acceptable cost by electrolysis of water is not new either, and in fact, is the basis for the hydrogen fuel cell automobile program.

The new element for the proposed program for synthetic fuel generated from air and water is the extraction of CO2 from the atmosphere as feed material for the synthesis. Various processes for CO2 extraction have been tested and found feasible but as yet, no large scale plants have operated.

Estimates of the cost of extracting CO_2 range from $100 per tonne to much higher values, on the order of $600 per tonne.

Depending on the cost of the extracted CO_2, the estimated cost of synthetic gasoline from air and water on the order of $3 per gallon at $100 per tonne of CO_2 and $5 per gallon at $300 per tonne of CO_2. The current price of gasoline derived from oil is more than $5 per gallon in most countries around the World, so that $5 per gallon for synthetic gasoline from air and water would be acceptable for most countries.

As a first step, CO_2 emitted from fossil fuel fired power and industrial plants, for example, the cement industry, which accounts for about 5% of global CO_2 emissions, could be used as feed material for the synfuels process as the technology for CO_2 extraction develops and cost figures become more precise. Because of the higher CO_2 concentrations in emissions from fossil fuel fired power plants and industries, the cost per tonne of CO_2 is relatively low, about $100 per tonne at maximum. As CO_2 extraction from air technology evolves, and fossil fuels are phased out, the synthetic gasoline and diesel fuel plants would transition to CO_2 extraction from the atmosphere.

By 2050 AD, it appears possible to have a mix of Maglev, electric cars, and synthetic gasoline/diesel fuel for the World's transport needs, with no need for fossil fuels, and no net CO_2 emissions into the atmosphere.

Introducing LISA
James Powell

"I attach the greatest importance to the prompt examination of these ideas," he wrote. "The advantages of a floating island or islands, even if only used as refueling depots for aircraft, are so dazzling that they do not need at the moment to be discussed. There would be no difficulty finding a place to put such a 'stepping-stone' in any of the plans of war now under consideration" [P, SEP]

Quote by Winston Churchill in 1942, upon learning of Geoffry Pyke's proposal to build a very large, unsinkable aircraft carrier code-named "Habbakuk" from ice. Habbakuk would attack the German U-boats in the Atlantic that were sinking ships carrying vital cargo from America to Britain in World War II.

Reference: <u>Ice</u>, by Mariana Gosnell, University of Chicago Press, 2007.

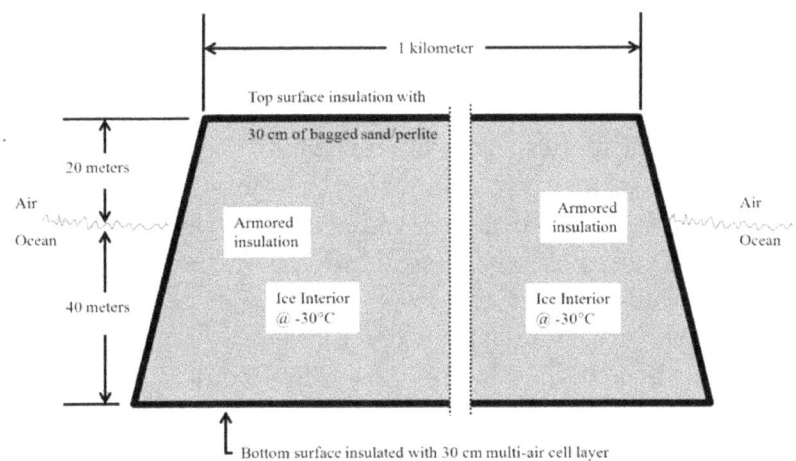

Figure 1

Most people hate ice and snow -- it's cold and uncomfortable, scary for walking on and flying in icy conditions. Pipes freeze, it costs lots to heat the house, sidewalks need shoveling, roofs collapse, trees break, power lines go down, etc., etc.

While ice and snow are definitely not friends to most people, they are to some. Skiing, ice skating, ice fishing – there are wonderful things one can do on ice and snow. When I was young, I loved ice climbing in the mountains and ice skating. I remember skating on Walden Pond when the ice was perfectly transparent and I could see the fish swimming under my skates. I can remember climbing the ice and snow slopes on Mt. Assiniboine and Mt Athabasca and the Columbia Ice Field in the Canadian Rockies, the Matterhorn in Switzerland and other mountains. It was great fun until after a few near misses from death, I realized I could die and stopped climbing. Ice and snow sort of lost their appeal after that.

But, a new Nice Age of Ice is about to dawn in which virtually everyone will view ice as a great friend and protector. How can this be? What will make us love ice?

Very simply, by thermally insulating the ice and refrigerating it, large very low cost ice structures can be built and used for a wide range of important applications that will greatly benefit society.

We call these large ice structures LISAs (Large Ice Structure Assemblies). They are frozen into their desired shape and size using existing commercial refrigeration technology, and are thermally insulated with existing, readily available materials and methods – sand, perlite (expanded volcanic rock), air cells, asphalt, etc.

The refrigeration power per unit area to maintain the thermally insulated ice interior at a temperature of -30 degrees Centigrade (-22 degrees Fahrenheit) is very small. For a LISA ice island floating in the ocean with an ambient surface temperature of +30 degrees Centigrade (+86 degrees Fahrenheit) and an ice interior at -30 degrees Centigrade, the electric refrigeration power for 2,000 square feet (200 square meters), about the floor area of a typical house, is only 500 watts (e). The thermal insulation is simple – a 1 foot (30 centimeters) thick layer of plastic bags containing a 50/50 mixture of dry sand and perlite.

Think of it – a typical house uses hundreds of watts for lights, refrigerator, air conditioning, TV, computers, etc. – comparable to that needed to live on a floating LISA island in the ocean.

Figure 1 illustrates the features of a LISA floating ice island in the ocean. Put in another way, the 30 year cost for refrigerating 1 square foot of LISA's surface at 5 cents per kilowatt hour is only $3 – a small fraction of the $100 to $200 per square foot that constructing a typical house costs. Depending on application, it can have various shapes – square, circular, rectangular, annular, etc. – and range in size from a few hundred feet to miles across.

In the example shown, the LISA island is square in shape and 1 kilometer (0.6 miles) across. Its top surface is 20 meters (66 feet) above the ocean surface, high enough to prevent waves from washing over it. Its bottom surface is 40 meters (132 feet) below the surface of the ocean. There are air cavities inside the -30 degrees Centigrade ice interior to provide positive buoyancy, enabling it to ride much higher above the surface than a natural iceberg does, where the depth below the surface is 10 times the height above the surface.

LISA's top surface is thermally insulated with a 1 foot thick sandbag layer, with the bags containing a 50/50 mixture of dry sand and perlite, an expanded volcanic rock. LISA's bottom surface is thermally insulated by a foot thick layer of multiple air cells, similar to multi-cell air mattresses. The side surfaces of the LISA island are subject to wave action, so the thermal insulation is armored with steel plate cover and an underlayer of asphalt paving.

Total refrigeration power for the 1 kilometer LISA ice island is 4.5 megawatts. This power can be supplied by on-board wind power or solar cells, or by OTEC (Ocean Thermal Energy Conversion) power cycle that operates using the temperature difference between warm surface water at 25 to 30 degree centigrade and deep cold water at +5 degree

Centigrade. At 5 cents per kilowatt hour, the annual cost for refrigeration of the 1 square kilometer LISA ice island is only 2 million dollars.

For a 1,000 families living on the LISA island with nice big houses or apartments, 2 thousand square feet, that's only a cost of 170 dollars per month. Compare that with Manhattan, where a 400 square foot apartment costs 3,000 dollars per month and the population density is 10 times greater. Chapter 7 describes the benefits of low cost housing on pleasant, low population density LISA ice islands that are adjacent to existing coastal cities with high density, high cost housing.

Chapter 7 also describes how LISA ice barriers (Figure 2) can completely protect vulnerable coastal cities from devastating storm surges that accompanied Hurricane Sandy and Hurricane Katrina. Sandy caused 85 Billion dollars of damage. Katrina flooded 80 percent of New Orleans. LISA ice barriers would have completely protected New York and New Orleans from the devastating storm surges. Figure 3 shows the flooding of New Orleans during Hurricane Katrina. Detailed descriptions of the LISA coastal barriers and other LISA applications are given in *The Ice World Cometh*, by James Powell, Jesse Powell and John Powell, available on Amazon.

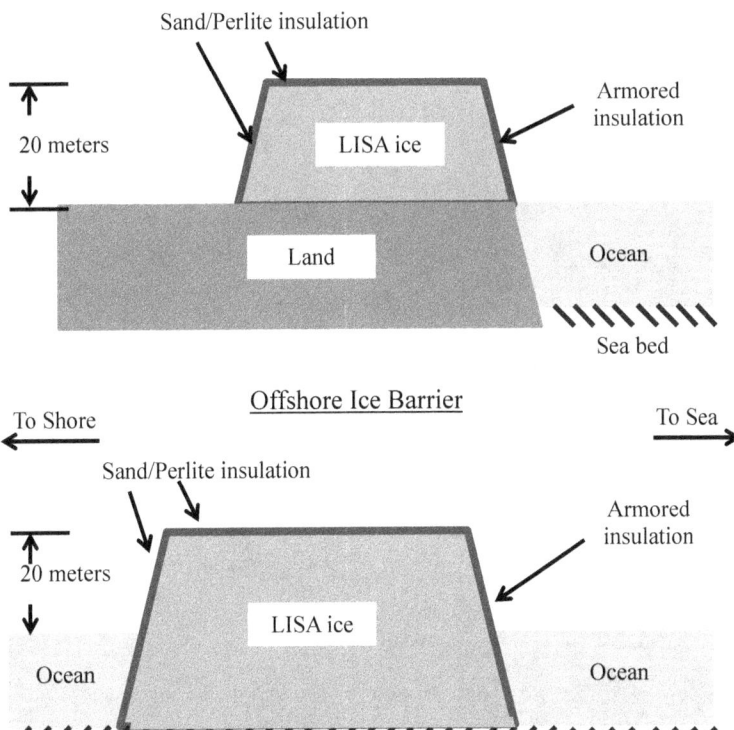

Figure 2. LISA Ice Barrier

Figure 3. Katrina Storm Surge Flooding New Orleans

And LISA can completely protect coasts from tsunamis, which can be even more deadly. Strong Earthquakes can generate enormous waves, as illustrated by Hokusai's famous wood cut, the Great Wave off Kanagawa (Figure 4). The recent 2012 Fukushima tsunami killed 20,000 people and destroyed 4 nuclear reactors, with subsequent radioactive releases that contaminated thousands of square miles. Total economic loss was in the range of 250 to 500 Billion dollars.

Figure 4. Great Wave off Kanagawa by Hokusai.

Then there's the 2004 Indonesian tsunami that killed 280,000 people. LISA ice barriers would have completely protected Japan and Indonesia from these tsunamis – no deaths and no economic losses. There is a very large tsunami waiting to happen on the West Coast of the United States. Every 300 years or so, the Cascadian fault, which is on the boundary between the Juan de Fuca Plate as it slides under the North American continental plate, ruptures, causing a magnitude 9 earthquake and massive tsunamis.

The last 2 ruptures were in 1500 AD and 1700 AD. When the Cascadian Fault ruptures, the sea bed can bounce as much as 20 feet setting off tsunamis as much as 100 feet high! The July 26, 1700 earthquake and tsunami on the Cascadian fault was a monster. When the resulting tsunami crossed the Pacific at 500 mph and hit Japan's Honshu Island, thousands of miles away, it was 16 feet high. Along the West Coast of North

America? There are no historical records, but it would have been much greater.

Scientists believe there is a 50% probability of a magnitude 8.0 or more on the fault in the next 50 years, and a 15% chance it will be a magnitude 9.0 or greater. (1) When it happens – not if, a massive rupture is inevitable – the loss of life and economic damage to the US will be far greater than any previous natural disaster in America's history, if California, Oregon, and Washington are not protected. We must prepare for it. Ice coastal barriers offer the best protection and are affordable.

Ocean levels are rising, thanks to global warming. Already coastal cities are experiencing flooding as the World's ice caps and glaciers melt. In a paper published in the National Academy of Science (2) a team of researchers concluded that a billion people are at risk from rising sea levels. Depending on the projected level of sea rise, 400 million people will be flooded annually in 2100 AD, with an expected annual loss of as much as 9.6 percent of global GDP – trillions of dollars per year.

Netherlands understands the threat of rising sea levels. Without their 2,100 mile of dikes, half the country would be under water (Figure 5). The US is also vulnerable along with many other countries and islands to rising sea levels. Studies (3) project that a 3.3 foot rise (1 meter) in sea level by 2100. A real possibility, such a rise would put 3.7 million Americans underwater, more than 1% of our population. Another study (4) projects that a sea level rise of only 2 feet would put 10% of Florida's land area underwater, erasing the homes of 1.5 million people.

Figure 5. The Netherlands Compared to Sea Level

Conclusions? Large portions of the World's coasts are very vulnerable to major storm surges, tsunamis, and rising sea levels. Over the next 50 years many authorities believe there will be a massive loss of life and many trillions of dollars in economic losses from these threats unless the vulnerable coasts are protected by strong high barriers. LISA coastal ice barriers can provide such protection as described in chapter 7, together with other very important applications, including low-cost off-shore housing, airports, seaports, and hazardous industrial plants.

Chapter 6 describes LISA ice island structures floating in the open ocean, at substantial distances from coastlines, with sizes up to 1 kilometer or greater. Chapter 7 describes LISA ice structures that are located either directly on the coastline or near off-shore to the coast.

A very important application of floating LISA islands will be the production of large amounts of fresh water by desalination of seawater, and large amounts of electric power.

A floating LISA ice island 1 kilometer in size can produce 1 Billion gallons of fresh water daily and 2,000 megawatts of electric power, using the OTEC (Ocean Thermal Energy Conversion) cycle that operates using the temperature difference between warm ocean surface water and deep cold ocean water.

Other important applications of floating LISA ice islands include, among many:

- Independent off-shore cities and nations for individuals and communities desiring to be "sea-steaders."
- Mining platforms for ocean resources
- Ocean research platforms.

Large ice structures, both natural and man-made have been around a long time and function extremely well as long as they are kept cold. Natural ice islands break off the ice caps in Antarctica and Greenland and survive for many years in cold environments. The biggest natural ice island is the Pobeda ice island (5) which breaks off from the Shackleton Ice Shelf in Antarctica every 40 years or so, and floats around, lasting for 15 to 20 years before melting away even though it is not thermally insulated.

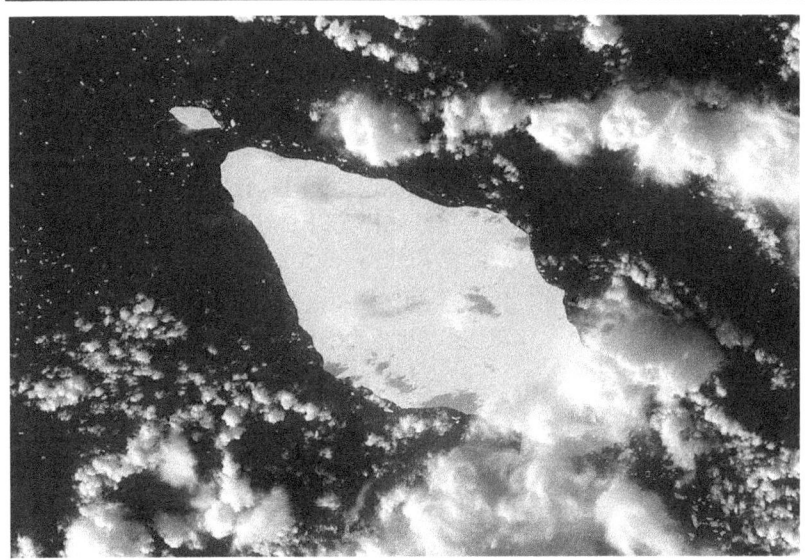

Figure 6. Iceberg A22A, South Atlantic Ocean (2002).

The last Pobeda ice island was 43 miles long and 22 miles wide, with a total area of 579 square miles. Thermally insulated, one could keep it around indefinitely. However, 579 square miles is a bit big for most applications. One could insulate smaller natural ice islands like the one shown in Figure 6. For the applications described in Chapters 6 and 7, however, it is preferable to manufacture the LISA structures by freezing fresh water into the desired shape and size, with particular features appropriate to the application. The cost of freezing is very small. Including both the cost of the refrigeration energy needed for freezing plus the amortized cost of the freezing equipment, for ocean applications, the cost of freezing is about 1 dollar per cubic meter of ice, more than 100 times cheaper than a cubic meter (1.3 cubic yards) of concrete. For freezing in coastal locations where deep cold water is not available as a heat sink for the refrigeration cycle, freezing cost is about 2 dollars per cubic meter (1.3 cubic yards).

Natural ice can be very strong and tough. Figure 7 shows a photo of iceberg with a smear of red paint on it, taken a few miles south of where the Titanic went down on April 15, 1912.

The photo was taken by the Chief Steward of the liner Prinz Adelbert. The steward didn't know that the Titanic had gone down – what attracted his attention was the red paint on the iceberg. (6)

Titanic vs Iceberg! Iceberg won, with virtually no damage. Ice is tough and strong. Not only natural ice, but also man-made ice.

Figure 7. The Iceberg Suspected of Having Sunk the RMS Titanic.

First, the Ice Hotels, that delight tourists who sleep, dine, and party in them. The Ice Hotels are beautifully designed with clear transparent ice pillars and ceilings. The ice structure is very strong and stable, with no collapse or problems (7,8,9). Figures 8 and 9 shows photos of the interior of the Hotel DeGlace in Quebec, Canada.

Ice hotels are erected at a number of sites in Finland, Norway, Sweden, and Canada, using natural transparent ice during the winter season. They have dozens of spacious rooms and large areas with vaulted ceilings for meetings, dining, weddings, and other functions. The tables and chairs are made of ice. The Ice Hotels are not insulated, and melt when spring comes and air temperatures rise. They then are built anew the next winter. If insulated, they could operate year around for many years. Because ice is very low cost, it is practical to build

Ice Hotels every year. It would be impractical to rebuild a concrete hotel every year.

The Hotel De Glace is built with 30,000 tons of snow and 500 tons of ice. Its ice walls are 4 feet thick (8). At last building, it had 51 double beds, with 16 foot high ice arches above the furniture and beds, which are also ice – but they do have deer furs, mattresses and Arctic sleeping bags.

Figures 8 & 9. Scenes from the Interior of Typical Ice Hotel, the Hotel De Glace, Quebec, Canada

The Hotel has lots of nice spots for tourists to hang out – an art gallery, the "Ice Café", a nightclub (the N'Ice Club), movie theater, chapel, indoor heated washrooms and outdoor hot tubs. First opening was on January 1, 2001. It cost $350,000 to build and operate through April. After its 13th season, it had a total of 1,000,000 visitors and 43,000 overnight guests (8).

As delightful as the Ice Hotels in Canada, Sweden, Norway, and other countries are, the pinnacle of large man-made ice structures is the Harbin International Ice and Snow Sculpture Festival in China (10, 11, 12). The Festival covers an area of approximately 600,000 square meters (150 acres), with a magnificent array of towers, buildings, and sculptures. At the 2013 Festival, the Crystal Palace Tower topped out at 160 feet above ground level. 200,000 tons of ice are used to construct the Festival.

Figures 10 (13) and 11 (14) provide a tiny sample of the wonderful variety of ice structures at the Harbin Festival. They shine with a dazzling variety of colored lights. The structures are made using 3 foot thick blocks of clear crystal ice cut from the frozen surface of the Songhua River. Interested readers can access a multitude of photos and descriptions of the Festival through a Google search of the China Harbin Ice Festival.

Figure 10. Tower at Harbin Ice and Snow Festival 2012

Figure 11. Harbin Ice Festival

It is unbelievable that until now, the concept of Ice World has not been proposed. Geoffrey Pyke is to be thanked for 2 very important first steps towards Ice World:

Pyke was the first person to propose a real mission for a large ice structure. During the Second World War, cargo transport ships delivering supplies to Britain and Russia were being decimated by the German U-Boats operating in the North Atlantic. Allied aircraft did not have enough range to protect the ships, and it was not practical to build conventional aircraft carriers so that the planes could land and refuel in the Mid-Atlantic.

In her fascinating book ICE, Mariana Gosnell (15) describes how Geoffrey Pyke, the director of programs for Lord Mountbatten, proposed making aircraft carriers out of ice. It's attraction? It couldn't be blown up and sunk. The proposed bergships, code named "Habbakuk", greatly excited Winston Churchill. In his note to his chief of staff, Churchill wrote, "I attach the greatest importance to the prompt examination of these ideas. The advantages of a floating island or islands, even if only used for refueling depots for aircraft, are so dazzling that they do not at the moment need to be discussed. There would be no difficulty in finding a place to put such a 'stepping-stone' in any of the plans of war now under consideration."(15)

That was Pyke's first contribution to the world of ice structures – a practical, desperately needed application. His second contribution was to adopt for the bergship a very important discovery by two chemists at the Polytechnic Institute of Brooklyn. They found that by adding a small amount of wood pulp and freezing it in the ice, the ice became much stronger and non-brittle. The new ice-wood pulp material was named pykrete after Geoffrey Pyke.

Pykrete is amazing stuff. Bullets bounce off it and it rivals concrete in strength (but much cheaper). Pyke's pykrete bergship was designed to be 2000 feet long and 300 feet wide, with a displacement 26 times that of the Queen Elizabeth passenger ship. There was room for 100 Spitfires and 1500 crewmen (15). The hull was a square hollow box beam with 35 foot thick walls covered with cork for insulation. Cold air was to be circulated through cardboard tubes frozen in the hull to keep the pykrete frozen, even in the tropics (15).

Alas, the bergship was never built. The range of aircraft increased and there was no longer a need for it. The U-boat menace faded, and the cargo ships were able to deliver their supplies. Even so, the concept of large ice islands, and the very important benefits of pykrete were established, and not lost in the dustbin of history. Thank you, Geoffrey Pyke! When the Ice World comes into being, there should be a memorial to you.

Interests in pykrete has continued, and people still do small scale demonstrations and experiments on it. Figure 12 shows a photo of water and sawdust that can be used to make pykrete (16). Paper and other fibrous materials can also be used. Charles Nichols (17) made a slab of pykrete and fired rifle bullets into it. The first 7.62 x 39 mm rifle bullet bounced off the surface of the pykrete slab. It took 7 additional rounds fired from a distance of 5 meters (15 feet) to penetrate the slab.

Figure 12. Water and Saw Dust to Make Pykrete.

For many of the ice structure applications described, pykrete will be an important and integral part of the structure. The mechanical properties of pure ice and pykrete are discussed later in the book

We now turn to the building of large ice structures in the ocean for the production of fresh water and electric power and other applications (Chapter 6) and along the coastline for protection against storm surges, tsunamis, and rising sea levels, together with low-cost housing and other applications (Chapter 7).

As mentioned earlier, detailed description of the various LISA structures and applications are given in the book, <u>The Ice World Cometh, Life in the Wild Wet</u>, by James Powell, Jesse Powell and John Powell, soon to be available on Amazon.com. Also, for those interested in reading the original paper on LISA ice structures, "Large Scale Ocean Based Ice Structures for Habitats, Storm and Flood Protection, and Industrial, Energy, and Transport Applications", by James Powell and John Powell, presented at Oceans 07 Conference, Vancouver, Canada, September 29-Oct 4, 2007, it is in the conference proceedings.

Finally, many of the LISA applications can be rapidly developed and implemented. Its basic technologies, freezing ice, thermal insulation, and refrigeration, are well developed and commercially available, and can be quickly adapted for use in LISA structures.

Moreover, manufacturing and testing the performance of LISA does not have to be done at full scale. LISA structures can be constructed and proven modularly, and expanded in size and numbers for application. For example, the investment to build and test a short segment of a coastal ice barrier would be modest, and could be accomplished in a short time. Once demonstrated, coastal ice barriers would be rapidly built at many sites around the World. LISA floating ice island modules could similarly be developed quickly for a low investment, and rapidly implemented for many applications. Though for certain applications, like OTEC-ICE electric power generation and desalination, the technology that would be used on the LISA island would require substantial separate development.

Chapter Six

LISA Islands for Power Generation & Desalinization Using Ocean Thermal Energy Conversion (OTEC)

James Powell, Jesse Powell, John Powell and James Jordan

When I get my new house done
Sail away lady, sail away
Give the old one to my son
Sail away lady, sail away

LISA ice islands[1] will open up a whole new world of ocean applications as outlined in the previous **Introducing LISA** section. Figure 6.1 illustrates the LISA ice island that is 1 square kilometer in area. Its top surface is 20 meters (66 feet) above the ocean surface, high enough to prevent any wave from washing over it.

The top, bottom, and side surfaces are the walls insulated and refrigerated to keep LISA's ice interior at -30°C (-22 degrees Fahrenheit). The thermal insulation on the side surfaces of the LISA island is armored to protect it from wave impact. The refrigeration electric power to keep the ice interior is modest – for the 1 square kilometer LISA island shown in Figure 6.1, it is 4.5 megawatts, a little less than 1/6th of the 30 Megawatts of electric power per square kilometer consumed by the residents of Manhattan island in New York City.

---

[1] More detailed descriptions of LISA islands are given in the book, "The Ice World Cometh-Life in the Wild Wet", by James Powell, Jesse Powell, and John Powell, soon to be available on Amazon.com

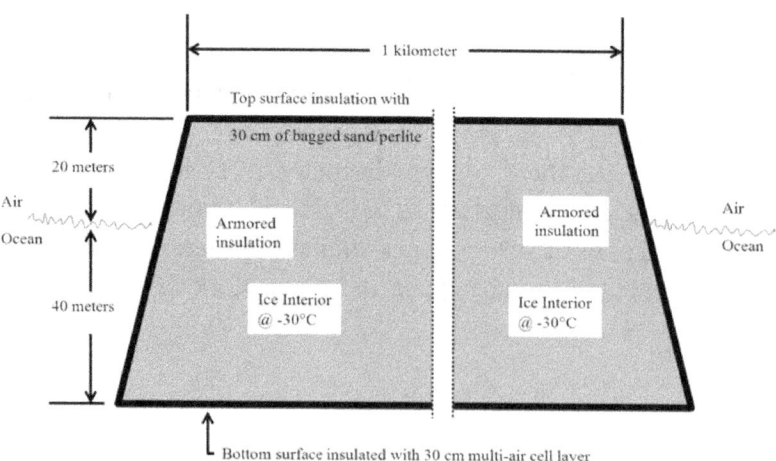

1 kilometer

Top surface insulation with
30 cm of bagged sand/perlite

20 meters

Air
Ocean

Armored
insulation

Armored
insulation

Air
Ocean

40 meters

Ice Interior
@ -30°C

Ice Interior
@ -30°C

Bottom surface insulated with 30 cm multi-air cell layer

Figure 6.1. Typical Large Ice Island Structure

If you are one of 2,000 persons – 1/10th of the population density of Manhattan, living on the 1 square kilometer ice island, your monthly power bill to keep the ice interior at -30°C at 5 cents per KWH is only $80 dollars. For your share of a 1 square kilometer ice island, e.g., 500 square meters or approximately 5400 square feet, you pay only $80 per month. Compare that to today's rent of $3,000 per month for a 400 square foot apartment in Manhattan.

The capital cost of the 1 square kilometer LISA ice island is also modest. Total capital cost, including thermal insulation, energy for freezing and freezing operations, is projected to be 112 million dollars. The property value per square kilometer of Manhattan is much greater. The total property value of Manhattan's 60 square kilometers is 914 Billion dollars (1) per square kilometer of Manhattan, that's 15 Billion dollars, 130 times greater than the cost of a 1 square kilometer LISA ice island.

Amortizing the capital cost of the LISA one square kilometer island with 2,000 inhabitants over a 30 year period, the annual cost per person is 112 million dollars divided by 30 x 2,000, or 1,900 dollars – much less than persons now pay for property taxes, mortgages, and rentals.

We think that individuals, groups, communities, corporations and government are going to be very interested in owning LISA ice islands located in the deep ocean, especially if new ice island communities can create their own regulations, laws, and taxes structures.

In the quest for their own independent new islands, many individuals and groups are already, today, seeking ways to construct floating cities on the ocean. Not only can the cities be independent nations that are not controlled by existing governments, but they would be able to move to any ocean location in the World that the islands owners chose.

This idea, termed seasteading, has already captured the imaginations of people, entrepreneurs, and corporations. To date, various designs have been proposed based on second hand cruise ships, new ships, barges, floating pontoons, and platforms mounted on semi-submersible columns (2).

However, there are problems with the present concepts. They tend to be:

- Subject to rolling and bobbing waves
- Vulnerable to big storms
- Very expensive
- Very crowded

Most of floating city concepts are basically smaller versions of Manhattan Island – crowded buildings and apartments, lots of steel and concrete, with no big parks and no beautiful surroundings. In no way do these cramped designs capture the island paradises of our imagination. They simply do not provide enough buildable area to allow humans to live in the ways they would prefer.

Ice islands probably cannot fully match the wonder and beauty of the islands of those pictured in exotic travel magazines, but they can have forests, parks, comfortable and spacious dwellings, and be quiet and uncrowded. And ice islands can be built in a wide variety of shapes to suit the needs of the owners. Figure 6.2 shows a few basic shapes possible with ice islands, but the architectural and landscape possibilities are really only limited by the imagination.

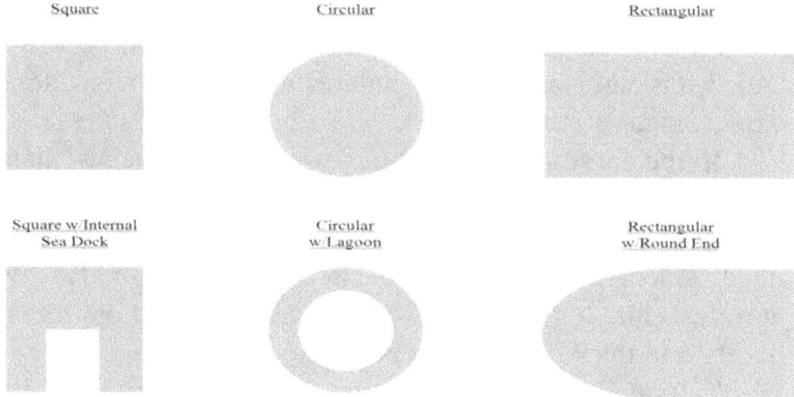

Figure 6.2. Top view of possible Ice Island Shapes.

The top surfaces of ice islands will be sufficiently above the sea surface, e.g., 20 meters (66 feet) or more, that waves can never wash over them. They will be unsinkable. Unlike the case with natural islands, the owners of ice islands will be able to relocate their island paradise to another location or even another nation if they wish to – no need to stay in one place. Ice islands will have lots of internal cavities for storage, equipment, manufacturing activities, power production, hydroponic gardens, if they choose to, and even interior living quarters – essentially ice hotels on the ocean.

What kinds of various applications are possible with ice islands? A lot that we can imagine, and a lot that we can't. Once ice construction becomes an established technology, however, we're sure that myriad ice islands devoted to a wide variety of applications will be found floating around the seven seas. Below, we list 10 potential applications.

1. OTEC-ICE plant ships for producing large amounts clean, low cost electric power and large amounts of desalinated fresh water
2. Habitats and communities – religious, refugees, and others
3. Industrial sites
4. Mining platforms – minerals, oil and gas
5. Resorts, hotels, and casinos
6. Banks and financial centers
7. Ocean research centers
8. Large scale fish-farming
9. Prisons
10. Military bases

We believe that OTEC_ICE plant ships will prove to be the most important application for LISA ice islands. Imagine the benefits of generating thousands of megawatts, per island, of clean, non-greenhouse gas emitting electric power! Imagine, as well, the benefits of producing billions of gallons of desalinated fresh water per island per day. Later in this chapter we describe in detail how ice islands can manage this feat.

However, other LISA applications will also be very important, in both beneficial as well as detrimental ways. A major challenge for humanity in the 21st Century will be to transform our global economy and governance structures into something that works for all people and our global ecosystem as a whole. Time and time again we have seen that our civilization seems poorly equipped to come to terms with the tragedy of the commons, be it climate change, air pollution, overfishing, or whatever. Ice World represents a transformational moment in global history. The prospect of cheap "land," unfettered by outdated tax structures and burdensome regulations, will appeal to many. Perhaps even more appealing will be the fact that new communities and industries won't have to submit to corrupt officials and institutions, nor submit to paying graft as a cost of doing business. In this sense, the prospect of Ice World represents an enormous leap forward in human freedom. It is a true Libertarian paradise.

On the other hand, unless we solve the tragedy of the commons, LISA and the resultant ICE world will simply get us to our current likely destination that much quicker – a degraded biosphere in which only the Rich are able to escape the daily misery brought on by human greed. We need to move beyond the tired Libertarian versus Statist debate, and approach our common destiny with open eyes.

The use of LISA ice islands for mining and oil and gas drilling is a case in point. Let's put aside the debate of whether we will need oil and gas in the foreseeable future. We believe we will. Let's ask the question, if we need to extract oil and gas, how should we do it? LISA ice islands will provide much stronger and more stable than present platforms, greatly reducing the possibility of oil & gas spills and explosions, and the need for undersea pipelines to transport oil & gas. And if we are to drill in the arctic, as it appears we are determined to do, which would be better to protect the Arctic Ocean from oil spills: a giant, immovable and indestructible LISA platform, or comparatively small tin can of a platform which is susceptible to the relentless forces of arctic ice flows and intense storms. LISA can help protect the arctic from the inevitable pollution that is attendant with conventional oil and gas technology, but it can also help to vastly expand the total amount of arctic drilling and thereby prolong our dependence on oil and gas.

Turning now to the very important first LISA application, **OTEC-ICE plant ships** can be of tremendous benefit in helping to generate the terawatts of electric power needed for the world's inhabitants without resorting to the consumption of coal, gas, and oil fossil fuels, and their consequent greenhouse gas emissions and global warming. OTEC-ICE plant ships can also supply enormous amounts of fresh water to drought stricken areas to grow food, provide water for people's personal needs, fight wildfires, etc.

For a decent standard of living, substantial electric power is required. Americans use on average about 1.5 kilowatts of electric power per capita for personal use, industry, government, etc. That's equivalent to 500 million kilowatts (500,000 megawatts) of average power generation for the US - less during off peak periods, and more during high demand peak power periods. To meet just the American power generation load would require 250 OTEC-ICE plant ships, each generating 2000 megawatts of power. And our demand for electric power will only grow in the future, as we switch from fossil fuel powered transport to electrically powered transport, from oil and gas heated homes to electrically heated homes and so on.

The world's future demand for electric power that is not based on fossil fuels will be much greater still. If China and India, with 1/3 of today's world population, and other developing countries, continue to increase their living standards, electric power generation will also increase. World population is expected to grow from today's 7 billion people to exceed 9 billion in 2050. Assuming that the world average per capita electric power consumption is ½ of today's American per capita consumption, would electric power generation would be 7.5 million megawatts in 2050 AD – four times greater than today. If it all came from OTEC-ICE plant ships (which it won't) one would need 3750 plant ships, each generating 2000 megawatts of power.

However, OTEC-ICE plant ships will not be the only source of renewable electric power for the World. Chapter 4 describes how beamed power from space solar power satellites will provide a major portion of the World's future power needs. Operating together, beamed power from space and OTEC-ICE plant ships can provide the bulk of the World's future alternative power needs. 1,000 plant ships, for example, could supply 30% of the world's power, plus a large fraction of the fresh water needs for drought affected regions. 1000 OTEC-ICE plant ships with dual electric power/fresh water capability could supply 2,500 cubic kilometers per year of desalinated water, 50 times greater than the total annual water consumption of California – enough for everybody in the planet.

Will OTEC-ICE plant ships be owned and operated privately or publicly, as independent for profit entities, or controlled by governments? Probably a mixture of all. One interesting question is whether an existing land based nation and a 200 mile Exclusive Economic Zone (EEZ) out from its coastline prevent independent OTEC-ICE plant ships from operating in the EEZ. Would nations allow them if they paid taxes? Would rival OTEC-ICE corporations and plant ships fight each other legally and illegally for location? Would they try to sabotage rivals? Time will tell. Based on human history, it's a real possibility. Overall, though, we believe OTEC-ICE will be of great benefit to humanity. OTEC-ICE plant ships will also supply the fresh water and refrigeration power to construct new ice islands, provide 8 million cubic meters of fresh water per day, and 2000 megawatts of electric power, sufficient to deliver a new one square kilometer ice island every week. (The construction time for a given ice island would be several months, but by having multiple units in construction, the output rate could be one per week.)

At 50 units per year, over a 10 year period, one OTEC-ICE plant ship could construct 500 units. With several construction sites, there could be many ice island floating on the ocean in only a decade. The revenues for the OTEC-ICE construction plant ships would be very large. At 150 million dollars for a 1 km² ice island, that would be an annual of revenue of 7.5 billion dollars. For 15 million dollars for a 1 kilometer long coastal ice barrier, with the capacity to turn out 600 one km units per year, that would be a revenue of 9 billion dollars per year. A big market with lots of profit.

Before describing the design and economics of OTEC-ICE plant ships, we first describe how to construct and insulate the large LISA ice islands to be used for OTEC power generation and desalinization, and the other applications described above. A very important factor in constructing and insulating LISA ice islands is the availability of deep cold water hundreds of meters below the ocean surface. This in addition to serving as the energy input between the warm surface water (Figure 6.3) and the deep cold water (Figure 6.4), enabling the deep cold water to reduce the temperature difference across the thermal insulation barrier which insulates the -30° C ice interior of LISA islands. Without the 5° C deep water, the temperature difference, ΔT, would be 60° C (30° C minus -30° C). With the 5° C deep water, the temperature difference, ΔT, is reduced to 35° C, which greatly reduces the thermal input requirements of the refrigeration cycle. Deep cold water (Figure 6.4) is available all over the World, often only a few miles off-shore, as is the case off the US West Coast.

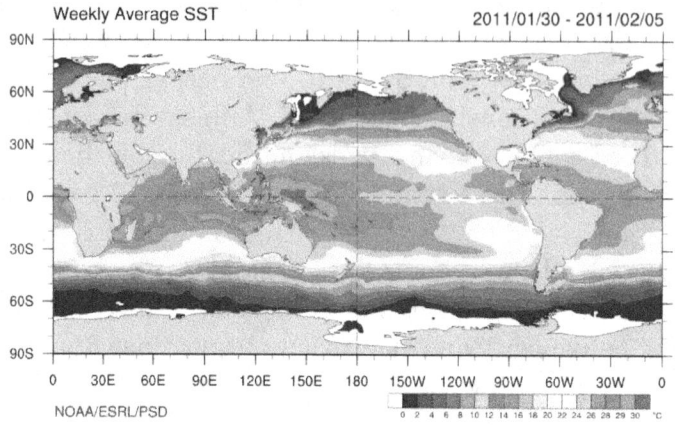

Figure 6.3. A Map of World Ocean Surface Temperatures.

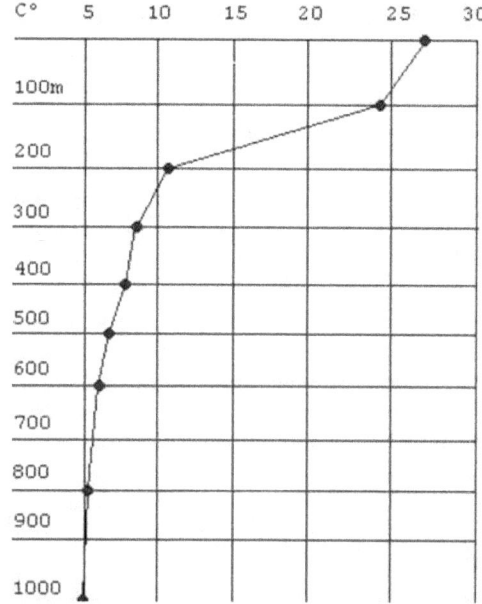

Figure 6.4. Ocean Thermocline Obtained During the International Geophysical Year.

Insulation and Refrigeration of LISA Structures

Insulating LISA structures is not high tech nor expensive. Cheap, efficient insulation already exists. To put LISA insulation in perspective, if the reader has camped, hiked, snowshoed, skied, or climbed mountains in cold weather, as the authors have, he or she will understand how important thermal insulation is, and how comfortable one can be with good sleeping bags, air mattresses, and down jackets.

Snowshoeing in the Adirondacks many years ago, with daytime temperatures of -30 degrees Fahrenheit, the temperature difference ΔT between my (Jim) body temperature of 98.6 degrees F and -30 degrees F was 128 degrees F, equal to 71 degrees centigrade. This is considerably greater than the ΔT across LISA's insulation, yet I was quite comfortable. At night in my sleeping bag on an air mattress, the ΔT was probably even greater than 71 degrees centigrade.

For LISAs that operate in the deep ocean where deep cold water at +5 degrees centigrade is available as the first stage of refrigerant, the ΔT between +5°C and LISA's ice interior at -30°C is only 35°C, ½ of that in the Adirondack winter. For LISA's operating on land and in coastal waters where deep cold +5°C water is not available and the ambient atmospheric temperature is on the order of +30°C (86 F), the ΔT will be greater, about 60°C – still less than the 71°C between Jim and the Adirondack surroundings.

The heat loss between Jim and the surrounding Adirondacks is also relevant. 2500 calories of food consumed daily is equivalent to an average body thermal output of 120 watts. Think of it – each human being is a bit brighter, thermally anyway, than a 100 watt light bulb.

At about 6 feet, 4 inches tall, Jim's body surface area is about 3 square meters, including arms, legs, torso, neck, head, ears, fingers, and toes. Inside a sleeping bag on an air mattress, his thermal heat loss rate through the insulation is approximately 120 watts(th) divided by 3 square meters, or about 40 watts (th)/square meter.

And that's with a relatively thin insulation layer. With thicker insulation like that insulating LISA structures, and a smaller ΔT, the thermal heat leak rate would be reduced by a factor of 10, down to the very practical level of 3 to 4 watts/m².

The important factors for insulation and refrigeration are:

- Low thermal conductivity of the insulation – one is warmer with a down comforter than an ordinary blanket
- Thickness of the insulation – one is warmer with 3 down comforters than just one
- The temperature difference across the insulation layer – one is warmer in sleeping bag if the outside temperature is +30 degrees F, rather than - 30°F, because the temperature difference between you and the outside is less.
- Strength and robustness of the insulation - one does not want insulation that will crumble and degrade under loads
- Cost of the insulation – cheaper is better. Buy your insulation on-line, rather than from a high end boutique
- Coefficient of Performance (COP) – the higher the better

That last factor, COP, needs more explanation. People generate heat, and need insulation – sleeping bags, down jackets, thick underwear, etc. – in cold weather to stay warm. In contrast, refrigerators that house the food we eat need insulation to keep the food cold. They also need electric or mechanical power to remove the thermal energy that leaks through their insulation and dump it into the warmer surroundings.

The energy efficiency of the refrigeration system is measured by its COP, defined as the watts(th) of thermal energy removed from the cold region divided by the watts(e) of electric or mechanical power used to remove it. The COP is expressed as

COP = watts(th)/watts(e)

There are two COP's for the refrigeration cycle – the perfect one with no irreversibilities and losses, and the actual one, where the refrigerant and refrigeration equipment have losses and irreversibilities. The perfect COP is based on the Carnot cycle while the actual COP is a fraction of the perfect COP, where the value of the fraction depends on the type of equipment used and the operating temperature range. In the Carnot cycle, thermal energy transfers without irreversibilities between two temperature levels, T_1 the higher temperature, and T_0, the lower temperature. The Carnot efficiency is defined as

$$\eta_c = \frac{T_1 - T_0}{T_1}$$

where the temperatures are given in degrees Kelvin, defined as the temperature above absolute zero. The freezing point of water is at 0 degrees centigrade (32 degrees Fahrenheit). Expressed in degrees Kelvin, water freezes at 273.16 K. If one runs a Carnot power cycle between the boiling point of water at 1 atmosphere pressure which is 100°C (212 degrees Fahrenheit) or 373.16 K, the Carnot efficiency is

$$\eta_c = \frac{373.16 - 273.16}{373.16} = 0.268$$

Putting 100 watts of thermal energy into this Carnot cycle, one would get out 26.8 watts of electrical or mechanical power, and dump 73.2 watts of thermal energy into a heat sink at 0 degrees centigrade (273.16 K). Conversely, if one ran a perfect Carnot refrigeration cycle between T_0 and T_1, one could remove 73.2 watts of thermal energy from the 0 degrees centigrade region using 26.8 watts of electric or mechanical power, and dump 100 watts of thermal energy into a heat sink at 100 degrees centigrade.

The perfect COP for the refrigeration cycle would then be

COP = 73.2/26.8 = 2.73 watts(th)/watts(e)

Expressed as formula, COP is

$$COP = \left. T_0 \middle/ (T_1 - T_0) \right. = \frac{273.16}{(373.16 - 273.16)} = 2.73$$

Based on experience with actual refrigeration cycles for the temperature range encountered in maintaining the ice interior of LISA structures at -30°C, actual COP's will be ½ of the perfect COP, i.e., COP(actual) = 0.5 COP(perfect). That is, it will actually take twice as much electric or mechanical power to remove a given thermal leakage into LISA's ice interior as a perfect Carnot cycle would.

There is a major energy efficiency advantage in maintaining the ice interior of LISA structures at a low temperature, e.g., -30°C, when LISAs operate in deep ocean locations, compared to LISAs operating in shallow water coastal locations.

The reasons for the deep ocean advantage are illustrated in Figure 6.5. Deep cold water at +5°C is present at depths of 500 meters or more. This results in two benefits.

First, the insulation for LISA can have two layers. Deep cold water at +5°C cools the bottom surface of the top layer, removing the thermal energy that leaks through the top layer. The deep cold water is essentially a free refrigerant. The flow rate is relatively low, and the pumping energy used to bring up the deep cold water through a pipe from the depth where it is available is very small compared with the extra refrigeration energy needed when the +5°C water is not present.

Refrigeration Circuits for LISA Structures

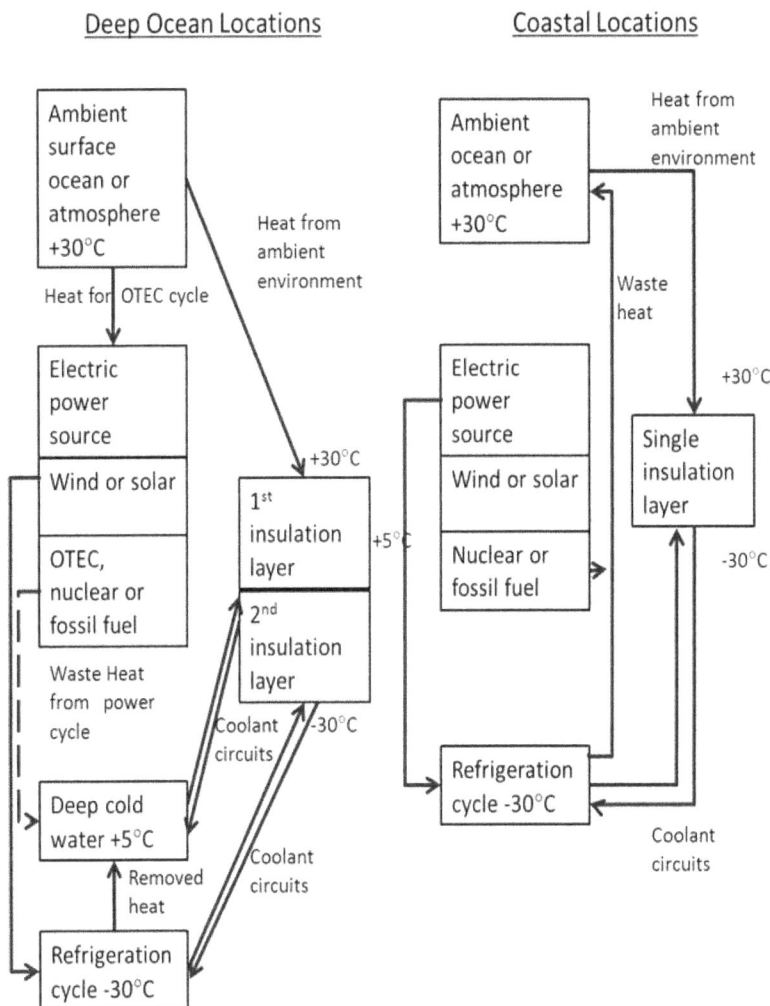

Figure 6.5. Refrigeration Circuits for LISA Structures.

As an example, consider a layer of dry sand 30 centimeters thick (0.3 m) as the top layer of insulation cooled at its base by a layer of +5°C deep cold water. Assuming a surface temperature of +30°C, the thermal leakage through the sand layer would be

$$g_1 = \frac{k_{sand}\Delta T}{\Delta X} = k\text{sand } \frac{(30°C - 5°C)}{0.3m} \quad watts\ (th)/m^2$$

For k_{sand} = 0.14 W m^{-1} K^{-1},
  g_1 = 0.14(25)/0.3 = 11.7 watts (th) per square meter

For a system that constrained the temperature rise in the +5°C water coolant to 2°C, the flow rate per square meter is only 1.4 cubic centimeters (1/700th of a liter) per second. For 1 square kilometer (1 million square meters, ⅓ of a square mile) of surface area, total flow rate is only 1.4 cubic meters per second, equal to 370 gallons per second – a very small amount.

The ΔT across the bottom layer, of insulation is now considerably smaller than if the -5°C cold water were not available. Without the cold water, the ΔT across it would be

$$\Delta T = T\ ambient - T_{ice} = +30°C - (-30°C) = 60°C$$

With +5°C cold water coolant, the ΔT across the bottom layer of insulation is

ΔT = +5°C – (-30°C) = 35°C. So the thermal leakage into LISAs -30°C, i.e., ice interior, using +5°C coolant is only 58% of the leakage without it.

$$\frac{g(\Delta T = 35°C)}{g(\Delta T = 60°C)} = \frac{35}{60} = 0.58$$

The second advantage is that operating the refrigeration cycle between T_0 = -30°C and T_1 = +5°C results in a considerably higher COP than operating between $T_0 = -30°C$ and $T_1 = +30°C$.

Operating at 50% of Carnot efficiency,
COP @ +5°C reject temperature = 3.49
COP @ +30°C reject temperature = 2.02

Combining the advantages of a smaller ΔT across the insulation and a higher COP made possible using +5°C deep cold water, the ratio R of electric or mechanical power required to remove the thermal leakage through a LISA insulation layer of a given thickness and thermal conductivity for operation in coastal water to the power in deep ocean water is

$$R = \frac{Power\ in\ coastal\ waters}{Power\ in\ deep\ waters}$$

$$= \left(\frac{\Delta T}{COP}\right)_{coastal} \Big/ \left(\frac{\Delta T}{COP}\right)_{deep\ water}$$

$$= \left(\frac{60}{2.02}\right) \Big/ \left(\frac{35}{3.49}\right) = 2.96$$

That is, it takes almost 3 times as much power to refrigerate LISA's ice interior if it operates in shallow coastal water rather than in deep water, based on the same insulation material and thickness.

LISA structures operating in deep ocean and coastal waters have 3 kinds of surfaces – top, sides, and bottom – that require thermal insulation as shown in Figure 6.6.

The top surface is 20 meters (66 feet) or more above the sea surface, sufficient height that it will not be subject to wave action. The surface is shown as flat, but can be shaped to whatever profile is desired.

A detailed analysis of various insulation options for the top surface – choice of material, thickness, method of manufacture, etc. – has been carried out. Thus far, the favored insulation option seems to be a 30 centimeter thick layer of sandbags filled with a 50/50 mixture of compacted dry sand and perlite, shown in Figure 6.7. Dry sand is cheap, on the order of $10 per cubic meter and has a reasonably low thermal conductivity, about 0.14 W m^{-1} K^{-1} (watt per meter per degree Kelvin). Perlite is expanded volcanic rock, with a very low thermal conductivity (0.031 W m^{-1} K^{-1}), very low density (50 kilograms per cubic meter), and very cheap ($50 per tonne, equivalent to $2.50 per cubic meter).

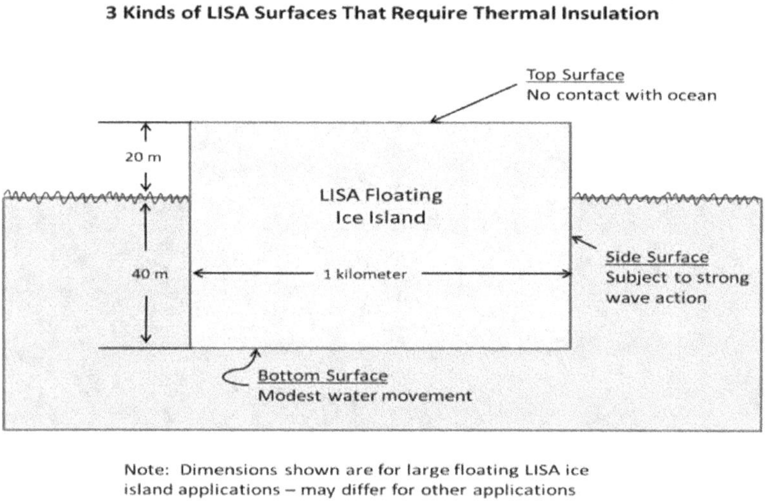

3 Kinds of LISA Surfaces That Require Thermal Insulation

Note: Dimensions shown are for large floating LISA ice island applications — may differ for other applications

Figure 6.6. Three Kinds of LISA Surfaces that Require Thermal Insulation.

Pure perlite would be the ideal insulation material for insulating LISA's top surface, except for its limited strength. At a compressive strength of about 20 psi, it's much better than Styrofoam and air in plastic bubbles, but limited in strength if buildings and other structures are to be erected on LISA's top surface.

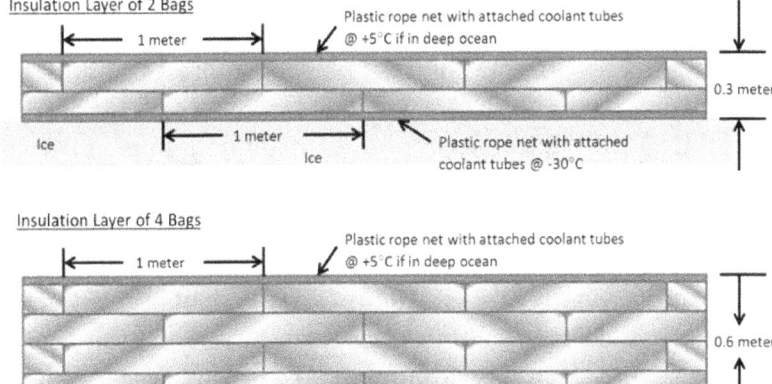

Figure 6.7. Deployment of Bagged Sand or Perlite Insulation
on Top of LISA Structures.

A 50/50 mixture of dry sand/perlite will be much stronger than pure perlite, with a good average thermal conductivity (0.085 W m^{-1} K^{-1}) and a low average cost of material ($6.25 per cubic meter). The insulation performance of a 30 and 60 centimeter thick layer of 50/50 mixture of dry sand and perlite are shown below, per square meter of insulation area.

	Deep Ocean Location ($\Delta T = 35°C$; COP = 3.49)		Coastal Location ($\Delta T = 60°C$; COP = 2.02)	
	30 cm	60cm	30 cm	60cm
Thermal leakage, w(th)/m^2	9.9	4.95	17.0	8.5
Refrigeration power, w(e)/m^2	2.84	1.42	8.41	4.20
Annual refrigeration power cost @ 5 cents per KWH, $/m^2	$1.23	$0.62	$3.64	$1.82

As discussed above, the annual cost of refrigeration for LISA structures in coastal waters is almost a factor of 3 greater than in deep ocean locations, because of the use of deep cold (+5°C) water for refrigerating the outer layer of insulation, and as a low temperature heat sink for the refrigeration power cycle. This suggests that a thicker layer of sand/perlite insulation, e.g., 60 centimeters, may be more optimum for LISA structures operating in coastal waters. To evaluate this possibility, the capital cost of the insulation must also be considered, since a thicker layer will have a higher capital cost.

The capital cost of bagged top surface insulation has been analyzed. The bags of sand/perlite would be manufactured at the location where LISA structures were constructed. The construction facility would produce many LISA's over a period of at least 30 years, enabling its own capital cost to be amortized and allocated to the cost of each LISA produced. On the FAST (Floating Assembly Station Table) facility floating platform, plastic bags are filled with dry sand/perlite material, sealed, and then attached to a floating plastic rope net to form the insulation layer for LISA's top surface. (Airbags are attached to the rope net holding the sand/perlite bags to maintain positive buoyancy.) LISA's ice interior is then frozen beneath the insulation layer.

Included in the capital cost of the sandbag layer (Table 6.1) are the costs of the dry sand/perlite material, the plastic liner for the sandbags, the rope nets attached to the sandbags, the amortized capital cost of the FAST facility and its operating cost. The capital cost of the top surface insulation layer is the same for both deep ocean and coastal locations of LISA structures. Also included is the capital cost of a 30 centimeter thick sand layer above the insulation layer for protection. For deep ocean locations, there would be coolant tubes carrying +5°C coolant water located between the outer sand layer and the insulation layer reducing the ΔT across it to +35°C.

Table 6.1. Capital Cost of Sandbag Insulation for Top Surface of LISA.

| | Top Surface Insulation Thicknesses | |
	30 Centimeters	60 Centimeters
Capital cost of insulation, $/m²	$12.45	$21.90
Capital cost of outer sand layer, $/m²	$3.00	$3.00
Total capital cost, $/m²	$15.45	$24.90
Amortized annual capital cost $/m² per year (30 year amortization)	$0.52	$0.83

Adding together the annual cost for refrigeration power and amortized capital cost, one obtains total annual cost for insulating LISA's top surface (Table 6.2).

Table 6.2. Total Annual Cost for Sandbag Insulation

Cost component	Deep Ocean Location [ΔT = -35°C; COP = 3.49]		Coastal Location [ΔT = 60°C; COP = 2.02]	
	30 cm	60 cm	30 cm	60 cm
Annual refrigeration cost, $/m²/year	$1.23	$0.62	$3.64	$1.82
Amortized annual capital cost, $/m²/year	$0.52	$0.83	$0.52	$0.83
Total annual cost, $/m² year	$1.75	$1.45	$4.17	$2.65
Refrigeration power, megawatts(e) for surface area of 1 km² (1 million square meters)	2.84 MW(e)	1.42 MW(e)	8.41 MW(e)	4.20 MW(e)
Total annual cost, million $/year for surface area of 1 km² (1 million square meters)	1.75 M$	1.45 M$	4.17 M$	2.65 M$

The above costs indicate that a 30 centimeter insulation layer is very acceptable for deep ocean locations. The 60 centimeter layer has a slightly lower total cost, but involves more manufacturing, and the savings are small. For coastal locations, however, there is a definite cost advantage in a thicker insulation layer, because of the greater ΔT across the insulation and the smaller COP for the refrigeration cycle. The 60 centimeter layer has a 35% lower total annual cost than the 30 centimeter layer, and the refrigeration power is reduced by a factor of 2.

These costs are very acceptable. For a 50 meter wide, 100 kilometer long ice barrier protecting a coastal region from storm surges, tsunamis, and rising ocean levels, the annual insulation cost for the top surface would be 13 million dollars, 1/6000th of the 85 billion dollar cost for Hurricane Sandy, and 1/40,000th of the 250 billion dollar cost for the Fukushima disaster, both of which costs would not incurred if LISA ice barrier had been in place.

The bottom surface of the LISA structure is on the order of 30 to 50 meters below the ocean surface. In the deep ocean, the bottom surface is in contact with ocean water. In shallow coastal water, LISA's bottom surface can be in contact with the ocean water below it for some applications. In other applications, like ice barriers that protect against storm surges, tsunamis, and rising ocean levels, the LISA structure will rest on the seabed so that it cannot be moved by storms and have water flow beneath it to flood shorelines.

The bottom surface of floating LISA structures is thermally insulated by an inflatable air mattress, as illustrated in Figure 6.8. Two configurations of LISA's bottom surface are shown – a flat surface and a square grid of air filled chambers formed by ice walls that enclose the chambers. The air pressure inside the chambers excludes water from them, so that the ocean water only contacts the bottom surface of the ice walls. For a grid 50x50 meters, ice walls 2 meters thick, only 8% of LISA's flat bottom surface insulation is in actual contact with ocean water, with the rest in contact with air.

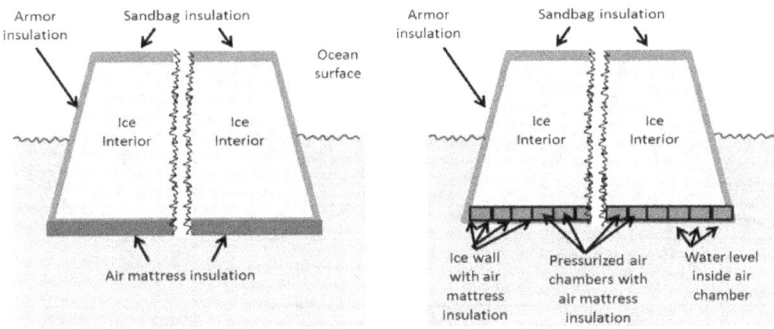

Figure 6.8. Air Mattress Insulation Geometries for Bottom Surface of LISA Structures.

The air mattress insulation would be cellular with multiple thin plastic films inside each cell to inhibit convective and radiative heat transfer so that the heat transfer is purely conductive. The air mattress insulation is inflated with carbon dioxide (CO_2) gas which has a lower thermal conductivity than air (0.0155 w/mK compared to 0.024 w/mk for air). Thermal leakage, refrigeration power, and annual cost for a 30 centimeter (12 inches) thick air mattress insulation are:

The total annual cost for air mattress insulation is a factor of almost 3 cheaper per unit area than the sandbag dry/sand insulation in deep ocean locations, and almost a factor of 4 cheaper per sandbag insulation in coastal locations.

By way of comparison, one can currently buy air mattresses online at very low prices - $11.97 for a 39 inch wide, 75 inch long and 8.75 inches thick twin size mattress. Retail price of the 1.9 square meter air mattress = $6.30 per square meter. Deducting profit and sales costs, the manufactured cost per m² would be considerably less. The very conservative projected cost of $15/m² for LISA air mattress insulation is roughly 3 times greater than the manufactured cost of commercial on-line mattresses.

Table 6.3. Total Annual Cost of Air Mattress Insulation for Bottom Surface of LISA Structures.

	Deep Ocean Location [$\Delta T = 35°C$; COP = 3.49]	Coastal Location [$\Delta T = 60°C$; COP = 2.02]
Thermal leakage, watt(th)/m²	1.8	3.1
Refrigeration power, watts(e)/m²	0.52	1.54
Annual power cost @ 5 cents/KWH, $/m² year	$0.22	$0.65
Capital cost, $/m²	$15	$15
Annualized capital cost, $/m² year	$0.50	$0.50
Total annual cost refrigeration plus amortized capital, $/m²year	$0.72	$1.15

The third and last kind of LISA insulation is the insulation that covers the side surfaces of LISA structures. The top surface sandbag insulation is high enough above the ocean surface that it never comes in contact with ocean water, even in major storms. The bottom surface air mattress insulation is far enough below the ocean surface that water movements are gentle and there are no floating objects that could impact the insulation.

The situation is very different for the insulation on the side surfaces of LISA structures. The insulation will be subject to high waves during severe storms and possible impacts by floating objects. Sandbag and air mattress insulation would be seriously damaged during storms. For LISA's side surfaces, armored insulation is required that can withstand impacts by high waves and floating objects. Figure 6.9 shows a promising approach for insulating LISA's side surfaces. It has an outer surface of steel plate, similar to that used for the steel hulls of conventional ships. The exterior steel plate covers a 10 centimeter (4 inch) thick internal layer of asphalt concrete, a mixture of asphalt and stone aggregate, the same material that is used as paving on hundreds of thousands of miles of US highways. Asphalt concrete is tough and strong. It withstands the heavy loads and impacts of many thousands of big 18 wheeler trucks weighing 40 tons or more. According to the US Department of Transportation, one big truck does as much damage to US highways as 9400 automobiles.

Asphalt concrete pavement can degrade – witness the annoying potholes we periodically encounter. However the potholes are caused by bad weather – freeze thaw cycles, floods, etc. Such conditions will not apply in LISA applications. There will not be freeze thaw cycles and the asphalt concrete is not in contact with water. In turn, the asphalt concrete layer rests on a 30 centimeter layer of dense fully compacted bags of dry sand – pure dry sand, not the sand/perlite mixture used for insulating LISA's top surface. Underneath the dry sand layer is a network of coolant tubes in LISA's -30°C ice interior. The ice on which the armored insulation rests is pykrete, ice containing on the order of 10-15% by volume of wood pulp. Pykrete is extremely strong and tough as described earlier. Rifle bullets will bounce off it.

Steel flanges on the underside of the steel plate that armors the insulation are connected to a square grid of fiber reinforced plastic sheets that enclose the asphalt concrete and bagged dry sand material. Plastic rope cables extend from the bottoms of the sheets into the -30°C pykrete ice zone, anchoring the insulation to LISA's very strong ice structure. LISA's side surfaces will be much stronger than the hulls of conventional ships – there is no empty space underneath its steel armor plate, no possibility of denting or buckling the hull, and no leakage of ocean water into the interior.

Thermal leakage, refrigeration power, and costs of both capital and refrigeration for the armored insulation are summarized below (Table 6.4). The projected capital cost of $153 per square meter is based on unit costs of $1 per kg for steel plate (current market price is $0.70 per kg), $105 per tonne for asphalt concrete (California Department of Transportation), $3 per kg for plastic sheet, and $14 per square meter for the 30 centimeter thick bagged dry sand layer.

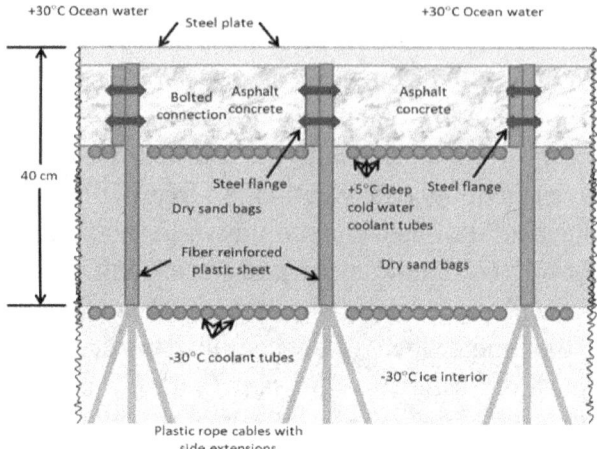

Armored thermal Insulation for Side Surfaces of LISA Structures Subject to Wave Action

Figure 6.9. Armored Thermal Insulation for Side Surfaces of LISA Structures Subject to Wave Action.

Table 6.4. Capital and Refrigeration Costs for LISA Armored Insulation.

	Deep Ocean Location (COP = 3.49)	Coastal Location (COP = 2.02)
ΔT across dry sand layer	35°C	60°C
Thermal leakage into -30°C interior, watts(th)/m²	16.3	27.9
Refrigeration power for -30°C interior watts(e)/m²	4.67	13.8
Annual refrigeration cost, $/m²	$2.02	$6.07
Capital cost, $/m²	$153	$153
Annualized capital cost, $/m², over 30 year period	$5.10	$5.10
Total annual cost, refrigeration plus amortized capital, $/m²	$7.12	$11.17

Per unit area, the armored insulation is considerably more expensive than the sandbag and air mattress insulation. However, the side area to be covered by the armored insulation is generally much less than the top and bottom surface areas of LISA structures. For example, a LISA structure that is 1 kilometer square in shape has a top surface area of 1 square kilometer (1 million square meters) and a bottom surface area also of 1 square kilometer.

The thickness of the LISA structure would be on the order of 60 meters = 20 meters above the ocean surface and 40 meters below - , yielding a side surface area of 4 x 1 x 0.06 = 0.24 square kilometers (240,000 square meters).

The annual cost for each surface in a deep ocean location is then, for 30 centimeter thick sandbag and air mattress insulation including the cost of refrigeration plus amortized capital cost:

- Top surface, (10^6) ($1.75) = 1.75 million dollars
- Bottom surface, (10^6) (0.72) = 0.72 million dollars
- Side surface, (0.24×10^6) (7.12) = 1.71 million dollars
- Total annual cost for insulation = 4.18 million dollars

A very reasonable cost, considering the very large size, 1 kilometer square, of the LISA structure. Refrigeration power requirements are also very reasonable.

- Top surface, (10^6) (2.84) = 2.84 megawatts(e)
- Bottom surface (10^6) (0.52) = 0.52 megawatts(e)
- Side surface, (0.24×10^6) (4.67) = 1.12 megawatts(e)
- Total refrigeration power = 4.48 megawatts(e)

The estimated cost of power is 5 cents per kilowatt hour. Wind or solar sources on the LISA structure or a small on-board OTEC plant can readily meet the refrigeration power needs. Energy storage is not needed, and the refrigeration power supply can be intermittent. LISA's ice interior has a very high thermal inertia. The increase in temperature of the ice beneath the insulation layer would be less than 0.1°C even if the refrigeration power shut down for a day.

Energy for Freezing LISA Structures

To turn water at +30°C (86 F) into ice at -30°C (-20 F) requires removing the same amount of thermal energy, 478 kilojoules per liter of ice, from the water, no matter where or how you do it. However, if you locate your construction facility in the deep ocean, you can use deep cold water at +5°C is to pre-cool the water that will eventually become structural ice. In other words, let nature do the work for you. The deep cold seawater can cool down the fresh water that is to be frozen to ice to about 6°C, removing approximately 20% of the 478 kilojoules per liter of ice at much lower cost than a refrigeration cycle could. Also, the +5°C cold water can act as the heat sink for the refrigeration cycle, rather than +30°C in coastal locations, reducing refrigeration power and cost by almost a factor of 2.

Thus, freezing in deep ocean really matters, because it greatly reduces refrigeration power and cost. Besides where you do it, it also really matters how you do it. Rather than using a single -30°C refrigeration system to freeze water to ice, it is much more energy efficient to use an intermediate refrigeration system at -5°C to freeze water to an ice/water slush at 0°C, and then freeze the slush using the -30°C system.

Freezing 0°C water to 70% ice/30% water slush removes 215 kilojoules of thermal energy from the total of 478 kilojoules per liter of ice, almost half of the total thermal energy extracted. The COP for the -5°C refrigeration system is much higher than that for the -30°C system. For deep ocean locations, the -5°C COP is 13.4, compared to 3.49 for the -30°C system. Per watt(th) removed the -5°C system uses ¼ as much electric power as the -30°C system.

Summarizing the options, the KWH of electric energy required to freeze a cubic meter of ice, and the corresponding cost at 5 cents per KWH are:

	Deep Ocean Locations		Coastal Locations	
	KWH/m³ Ice	$/m³ Ice	KWH/m³ Ice	$/m³ Ice
Freeze water	30.6	$1.53	58.3	$2.92
Freeze 70%/30% ice/water slush	16.6	$0.83	44.3	$2.22

Clearly, the refrigeration cost per cubic meter of ice to be frozen is much lower, $0.83/m³, if it is done in the deep ocean using ice/water slush than if it is done in coastal locations, by almost a factor of 3. The energy and cost advantages of freezing in the deep ocean are so advantageous, and the costs of towing LISA structures so low that even the LISA structures that operate in coastal waters would likely be manufactured at a facility in the deep ocean to keep costs down. Towing a large LISA structure (1 square kilometer in top surface area) at a speed 1 meter per second (2.3 mph) 1000 kilometers (600 miles) would take 12 days and cost about 1 cent per cubic meter of ice, based on 10 cents per KWH for propulsion energy. At 45 million cubic meters of ice, that is a towing cost of 0.5 million dollars. The propulsion power of 16 megawatts would be a small fraction of the propulsion power for large container ships, which is on the order of 100 megawatts.

Conventional ships travel at high speeds because their cargo is valuable. Sellers do not want to pay interest fees on the money invested in their product, and they do not want buyers to buy from someone else. And, the owners of the ship want to make as much revenue as possible. The same priorities will not apply for LISA structures – they can travel at low speeds and save money.

As a result, manufacturing LISA structures in the deep ocean and towing them to coastal locations appears very practical and advantageous. The cost of the freezing energy is greatly reduced, and the added cost of towing is much smaller than the savings in manufacturing cost. The cost of freezing to -30°C is low, only $0.83 per cubic meter of ice, with a cost savings of $1.39 per cubic meter if coastal LISA structures were first manufactured in deep water and then transported to coastal waters. The towing cost of $0.01 per m³ is negligible compared to the cost savings.

Methods for Freezing LISA Structures

Now for the next step. What is the best method for carrying out the freezing process? Two promising approaches have been analyzed. Summarizing, the two options are:

1. A distributed array of -30°C coolant tubes that freeze water or ice/water slush to form LISA's ice interior. The coolant tubes remain in place after the freezing process has been completed.

2. Option #1, but almost all of the coolant tubes are removed after freezing is complete, and re-used in the next LISA structure to be manufactured.

The freezing process would be carried out at a manufacturing facility located in the deep ocean, as illustrated in Figure 6.10. The facility is a floating LISA ice island with insulated ice walls that enclose and protect a central fresh water pond in which the new LISA structure is manufactured. The enclosing ice walls are armored and thick, and sufficiently high, e.g., 20 meters or more, above the ocean surface that they prevent any ocean waves from acting on the fresh water pond. The bottom surface of the facility is an insulated ice sheet that prevents any contact of the fresh water in the central pond with salty ocean water. The facility would produce new LISA structures to be towed to their destined operating locations when completed. Typical manufacturing times for each new LISA structure would be 3 to 4 months. With a 4 month manufacturing period, the facility would produce 3 LISA structures annually – 90 over a 30 year period. The amortized cost of the facility would be spread out over roughly 100 new LISA structures during the 30 year period, adding a few percent to the cost of each new LISA.

Figure 6.11 shows how options #1 and #2 would be deployed in the fresh water pond. The flexible coolant tubes would hang from a floating rope net, with lengths equal to the vertical thickness of the completed structure. For most LISA applications, thickness would be in the range of 40 to 60 meters.

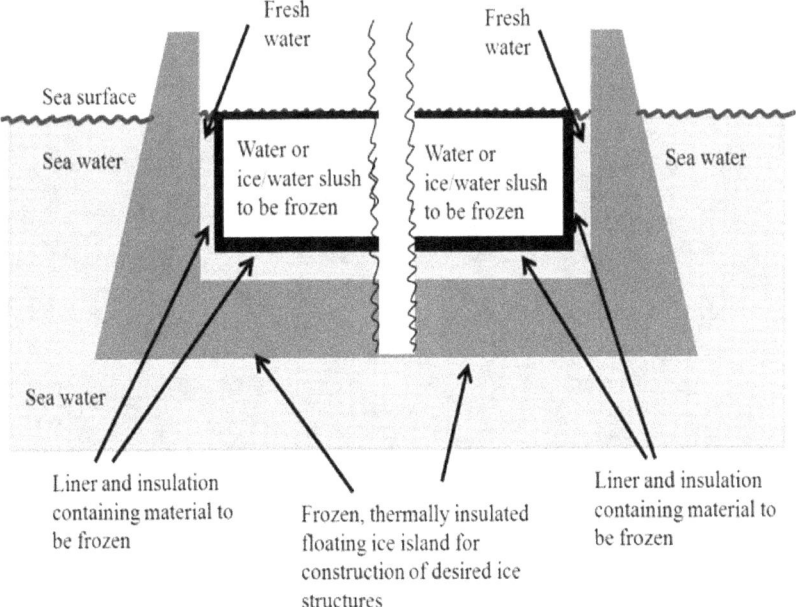

Fresh water

Fresh water

Sea surface

Sea water

Sea water

Water or ice/water slush to be frozen

Water or ice/water slush to be frozen

Sea water

Liner and insulation containing material to be frozen

Frozen, thermally insulated floating ice island for construction of desired ice structures

Liner and insulation containing material to be frozen

Figure 6.10. Ocean Based Facility for Construction of LISA Ice Structures.

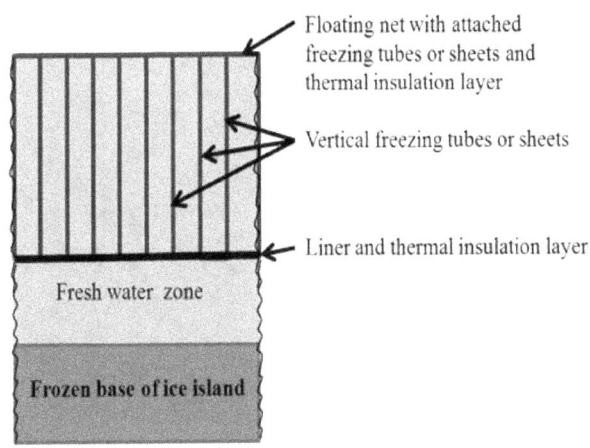

Floating net with attached freezing tubes or sheets and thermal insulation layer

Vertical freezing tubes or sheets

Liner and thermal insulation layer

Fresh water zone

Frozen base of ice island

Figure 6.11. Arrangement of Freezing Tube or Sheets Inside Ice Structure for Options 1 & 2

Figure 6.12 illustrates how ice layers would grow in thickness around the tubes once -30°C coolant started to flow in them. After sufficient time the individual ice layers would meet and refreeze together to form a solid block of ice, with a central water filled cavity remaining in the center of the tube grid array. Coolant flow would continue until the remaining small water cavity was frozen.

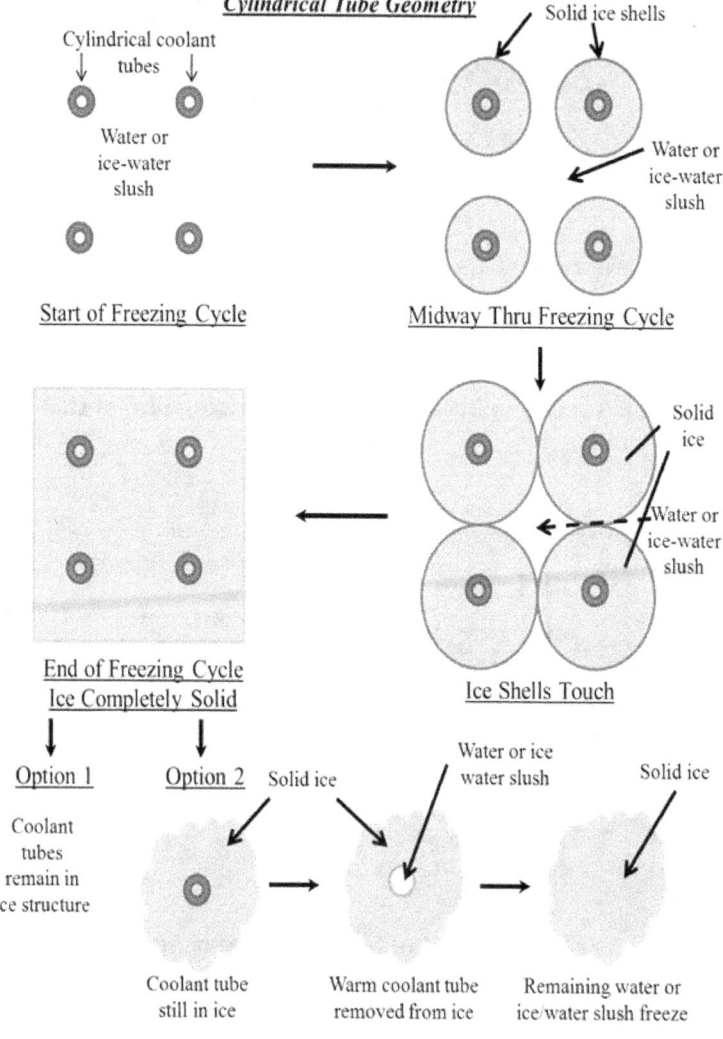

Figure 6.12. Examples of Heat Transfer Freezing Geometries for Construction of LISA Ice Structures.

The coolant tubes are fabricated as flexible cylinders from multi-layers of plastic sheets that form the walls of the tubes. When filled with pressurized coolant, the tubes expand to their full diameter. Typical expanded diameter is on the order of 10 centimeters (4 inches). Tube wall thickness is typically about 0.08 centimeter (30 mils).

Inside each coolant tube is a small diameter central tube that carries the inlet coolant flow downwards to the bottom of the coolant tube. The inlet flow then exits from the central tube and flows back upwards though the annular channel between the outer wall of the coolant tube and its inner small central tube. As the coolant flows upwards, it absorbs the latent and sensible heat released by the water or ice/water slush that surrounds the coolant tube, turning it to solid ice. The temperature of the coolant rises slightly during its upwards flow by 1 or 2 degrees centigrade. The warmed coolant exits at the top of the coolant tubes and flows back to the refrigeration system to be re-cooled to its design inlet temperature. There is a layer of thermal insulation between the inner tube and the annular channel to inhibit heat transfer between them.

In option 1, all of the coolant tubes remain in place after the freezing of LISA's ice interior has been completed. A small fraction remains active, providing continuing refrigeration of LISA's ice interior.

In option 2, warm coolant flows through the coolant tubes for a short time, e.g., a few minutes, causing in a thin layer of water to form on the surface of the tubes. The tubes are then depressurized so that the pressure inside them is less than the hydrostatic pressure outside of the tube. The flexible plastic tubes then collapse to a small diameter and are withdrawn from the ice interior. A small fraction, e.g., a few percent, remain in the ice interior, providing continuing refrigeration.

The time required to fully freeze the ice interior depends on the spacing between the coolant tubes. Shown below is an example of the parameters for a spacing of 1.5 meters (4.9 feet) between coolant tubes (center to center of the tubes) in a square array, using 70%/30% ice water slush.

- Diameter of coolant tubes = 10 centimeters (4 inches)
- Freezing time until 70 centimeter thick cylindrical ice layers on tubes touch each other = 18.1 days
- Freezing time until central cavity in square array is completely frozen = 36 days
- Fractional tube area in square array = 0.0035
- Cubic meters of ice produced per meter of tube = 2.25 m^3
- Cost of tube per meter of tube length = $0.92/meter (30 mil wall $4/kg of HD polyethylene)
- Cost per cubic meter of ice if left in = $0.41/m^3 (option #1)
- Cost per cubic meter of ice if withdrawn = $0.13/m^3 (option #2, 10 withdrawals, $10/per tube operating cost per withdrawal)

Increasing the tube to tube spacing to 2 meters would increase the time for complete freezing to approximately 70 days, but cut the cost per cubic meter of ice by a factor of 2: $0.20/m^3 for option 1 and $0.07/m^3 for option 2.

The cost of the process to freeze LISA's ice interior using option #1 and #2 are low, about 1/10th to 1/5th of the $0.83 per cubic meter refrigeration energy cost.

Both freezing options are able to create as desired a wide variety of localized structures inside the ice interior of LISA, including buoyancy cavities, reinforcement sections, chambers for work, storage, and residence, etc. Buoyancy cavities are desirable for all structures. Completely solid iceberg and ice islands are mostly underwater. A natural ice island with its top surface 20 meters above the ocean surface will extend down to 184 meters below the ocean surface. If the same ratio of height above surface/depth 20/184 = 0.11 below surface were to apply for LISA structures, their cost would be much higher.

To achieve a more reasonable height/depth ratio, it is necessary to incorporate air filled buoyancy cavities inside LISA's ice interior. By having 25% of the ice interior volume be air filled buoyancy cavities, the height/depth ratio would increase to 0.5. For a height above water of 20 meters, the depth below water would be 40 meters, a much more reasonable and lower cost situation.

The cavities would be filled with low cost cold air. No thermal insulation of the cavities would be required. Their diameter would be relatively small, on the order of 1 to 2 meters. Their effect on ice strength would be small, as evidenced by the experience with the man-made structures built from natural ice. For example, Ice Hotels and the structures built at the Harbin Ice Festival have much larger cavities and a much larger fraction of their total structure is air filled than would be the case for LISA structures. Despite their large cavities, however, the ice walls and ceilings do not come crashing down on guests and visitors. People sleep, dine, walk around, and party in complete safety inside Ice Hotels and sculptures.

At temperatures of -30°C, the creep rate of ice under loads is very small, enabling LISA ice structures to maintain their configurations over many years without significant deformation. For locations that may have high stress levels, 10 percent wood pulp can be incorporated in the ice to form pykrete which further increases the strength and toughness of ice. Alternatively, sediment e.g., sand, soil, garbage, etc., can be incorporated to form permacrete, which also increases strength and toughness.

The economics of LISA ice structures is discussed in the following section taking into account their capital costs for thermal insulation, refrigeration energy, and freezing equipment, as well as their annual cost for refrigeration to remove thermal leakage through their insulation.

Design and Economics of OTEC-ICE Plant ships for Electric Power and Fresh Water.

The world's oceans absorb a tremendous amount of energy from the sun, many thousands of times the primary energy that humans obtain from the combustion of fossil fuels. It is clean, renewable energy with no emissions of pollutants and greenhouse gases that cause global warming. And it won't run out in a hundred years or so, like fossil fuels will, but go on forever.

Converting the solar thermal energy input to the world's oceans into useful electric energy for humanity has been a dream for the last 140 years. In Jules Vernes' "Twenty Thousand Leagues Under the Sea", published in 1870, Captain Nemo proposed generating electricity from the difference in temperature between the ocean's warm surface and the cold deep water a kilometer beneath it, using a thermoelectric device. Stimulated by Verne, eleven years later the French inventor D'Arsonval, devised a more practical approach. Instead of thermoelectric wires, he proposed vaporizing ammonia with thermal energy from the warm surface water, expanding the ammonia through a turbo-generator to produce electricity, and condensing the ammonia in heat exchangers cooled by deep cold water drawn from the ocean depths. D'Arsonval's concept is termed Closed Cycle Ocean Thermal Energy Conversion (CC-OTEC) and is the favored OTEC approach today. Subsequent to D'Arsonval, his student, Georges Claude, proposed a different approach. Instead of vaporizing ammonia with warm ocean water and expanding the ammonia through a turbo-generator, the Claude cycle produces water vapor from the warm surface water and expands it through the turbo-generator. The exhaust water vapor from the turbine is then condensed using cold deep ocean water.

Claude's concept is termed the Open Cycle Ocean Thermal Energy Conversion (OC-OTEC) cycle. It requires a much larger turbine diameter than the closed cycle OTEC, because the very low pressure of the water vapor that enters the turbine, about

0.025 atmosphere compared to approximately a pressure of several atmospheres if ammonia is the working fluid. The turbines for open cycle OTEC plants will be much bigger and more expensive than those for closed cycle OTEC plants. Over the decades since D'Arsonval and Claude there has been continuing interest in OTEC. Small pilot plants have been built and demonstrated net power production (3,4,5,6). The largest plant tested to date generated a gross power output of 255 kilowatts(e) with a net power output of 103 kilowatts after deducting power for operating equipment (pumps, etc.). The open cycle OTEC plant operated for six years (1993 to 1998) in Hawaii (3).

The only currently operating OTEC plant is a 50 kilowatt(e) plant in Okinawa, being carried out by Saga University and Japanese industry (6). Proposals continue to be made by various organizations, however, typically to build 10 megawatt(e) size OTEC plants (6). In spite of its great attractiveness, OTEC has not moved ahead to commercial implementation, but remains in the R&D phase. This is due to a number of important issues that remain to be resolved, and its projected high capital cost, on the order of $10,000 per kilowatt(e) of capacity. OTEC issues and costs are discussed in the rest of this chapter, together with ways of how large ice structures can help resolve OTEC issues and greatly lower costs.

To convert ocean thermal energy to electric power, power cycles must operate between two temperatures that are very close together, ocean surface temperatures of 20 to 25° C, (Figure 6.3) and the +5° C temperature of deep cold water 1000 meters or more below the surface (Figure 6.4). The warm surface water heats a working fluid, which then expands through a turbine or other device to generate electric power. After expansion, the working fluid dumps the portion of the input thermal energy that was not converted to electric or mechanical energy into a cool heat sink, and circulates back to the warm energy sources for its next pass through the expansion device.

As a result of the small temperature difference, 15 to 20° C between warm ocean surface water and deep cold water, only about 2 to 3% of the oceans thermal input energy can be converted to useful electrical or mechanical energy – rather than 30% to 50% conversion for power cycles that operate on fossil fuels.

OTEC has a number of major issues that require resolution before it can become a major source of renewable energy for the world. These issues are discussed below

1. Ocean Thermal Energy Conversion (OTEC) cycles require very large thermal inputs to produce useful energy.

2. OTEC cycles require much more heat exchanger area and turbine flow area than conventional power cycles that operate on fossil fuels.

3. OTEC power plants require moving tremendous volumes of warm and deep cold water to generate power. To match the energy output from a 40% efficient conventional power cycle using 1 barrel of oil fuel, the 1.6% efficient OTEC cycle would have to move 3000x40/1.6 = 75,000 barrels of warm water and 58,000 barrels of cold water. The power and cost needed to move such massive volumes of water is very challenging.

4. The very large volumes of OTEC equipment constrain how much power can be produced from practical size conventional ships. Parameters for the 50 megawatt(e) closed cycle OTEC plant ship designed by Luis Vega and Dominic Michaelis (4) are summarized in Table 6.5.

Table 6.5

Parameters for 50 MW(e) Closed Cycle OTEC Plant ship

Source: First generation 50 MW(e) OTEC Plant ship for the production of electricity and desalinated water, L. Vega and D. Michaelis (4)

- Ammonia working fluid
- 198 meter plant ship length
- 39 meter plant ship beam
- 16 meter draught
- 24 meter depth
- 120,600 tonnes displacement
- 5 modules of 16 MW(e) gross capacity
- 5 seawater/ammonia evaporator modules, each 7100 m^3 volume (34 m long, 13 m wide, 16 m high)
- 5 ammonia turbo-generator units, each 580 m^3 volume (12 m long, 8 m wide, 5 m high)
- 5 seawater/ammonia condenser modules, each 7100 m^3 volume (34 m long, 13 m wide, 16 m high)
- 73,900 m^3 total volume of 5 modules on 50 MW(e) plant ship
- Dimensions of RMS Titanic
  - 269 meter length
  - 28 meter beam
  - 10.5 meter draught
  - 19.7 meter depth
  - 52,000 tonnes displacement
- 246 m^3/sec warm water flow rate
- 138.6 m^3/sec cold water flow rate

It carries 5 power modules, each generating a gross electric output of 16 MW(e), with a net electric output of 10 MW(e), for a total of 50 MW(e) output from the ship. Total volume of all the modules is 73,900 cubic meters.

The Vega and Michaelis 50 MW(e) plant ship is bigger than the RMS Titanic was. The plant ship's deadweight tonnage is 120,600 tonnes, over 2 times greater than the deadweight tonnage of the Titanic.

Total world electric generation capacity (7) in 2010 was 2.3 million megawatts(e), producing 20.2 trillion KWH per year. To generate it using Vega and Michaelis 50 MW(e) OTEC plant ships would require 46,000 ships, an impossibly high number. To just generate 1000 MW(e), standard output from one coal fired or nuclear power plant, would take 20 Vega and Michaelis OTEC plant ships.

In contrast, ice islands of the type could hold much more OTEC equipment to generate much greater amounts of electric power. One OTEC-ICE plant ship could generate 2000 MW(e), equivalent to 40 of the 50 MW(e) OTEC plant ships. 575 OTEC-ICE plant ships could generate 1.15 million MW(e), 50% of the present total world generation.

Large ice structures as platforms for OTEC power and desalinization systems have major advantages, including:

1. Significantly lower cost of the OTEC platform. A conventional ship structure for the 50 MW(e) OTEC plant ship would cost on the order of 100 million dollars. 40 such ships for an OTEC output of 2000 MW(e) would cost at least 4 billion dollars. A large ice structure platform for 2000 MW(e) OTEC plant would cost only 1/10th as much.

2. Maximum OTEC plant ship output using conventional ship platforms will be limited to approximately 50 MW(e) because of construction limitations for conventional ships. With ice platforms the OTEC output can be much greater, resulting in economies of scale for equipment and personnel. OTEC-ICE plant ships will need fewer operating and maintenance personnel per MW(e) of output. It also will be simpler and cheaper to deliver power and desalinated water from a large OTEC-ICE platform than from multiple ships having much smaller outputs.

3. Large OTEC-ICE platforms will be much more stable and secure against high waves, hurricanes, and storms than will OTEC systems based on conventional ships. OTEC-ICE platforms cannot sink – conventional ships can.
4. Construction of large, low (~0.02 atm) pressure chambers for evaporation and condensation of water vapor for desalination and/or open cycle OTEC will be much simpler and cheaper using ice structures than on conventional ships. Ice as a material of construction is much cheaper than steel or concrete.
5. Warm and cold water intake and discharge pipes will be easier to construct and much cheaper than conventional pipes, enabling longer pipes that will not affect local ocean temperatures.

The design and economics of 3 OTEC-ICE systems are described below:

System 1:   OTEC-ICE Closed Cycle 2000 MW(e) Electric Power Plant
System 2:   OTEC-ICE Open Cycle 1 Billion Gallons Per Day Desalination Plant
System 3:   OTEC-ICE Closed Cycle Electric Power Plant Combined With Open Cycle Desalination Plant.

The OTEC-ICE Closed Cycle 2000 MW(e) design is based on Vega and Michaelis 50 MW(e) closed cycle design (4). Table 6.5 gives the parameters for the 16 MW(e) module. Five of the modules generate 50 megawatts of net electric power after deducting power for the pumps and other equipment in the cycle.

Figure 6.13 shows the arrangement and dimensions of the 5 modules in the 50 MW(e) unit for the OTEC-ICE plant ship. Forty such units would generate 2000 megawatts. Each 50 MW(e) unit has 5 ammonia (NH_3) evaporator modules, 5 turbogenerator modules, and 5 NH_3 condenser modules.

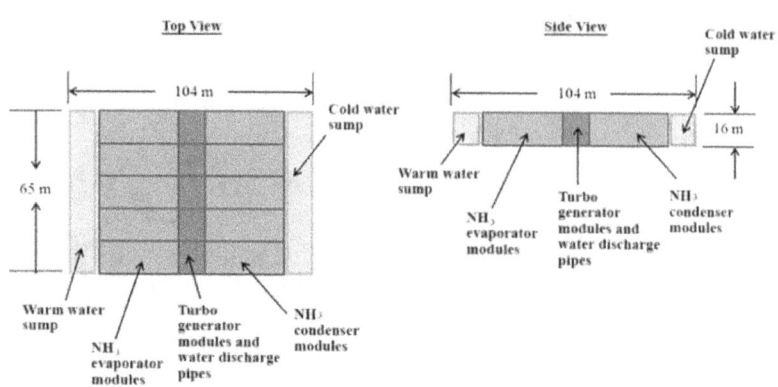

Figure 6.13. Layout of Closed Cycle OTEC Modules for 50 MW(e) Power Unit.
Basis: Vega and Michaelis (4).

The overall dimensions of the 50 MW(e) unit are
- 104 meters long
- 65 meters wide
- 16 meters high

The 50 MW(e) OTEC units are arranged on the ice plant ships as shown in Figure 6.14. Each unit is installed in its own individual section of the plant ship with a 50 meter thick ice wall between the ends of each section and a 65 meter thick ice wall between the sides of each section. Each section is covered by a conventional roof for protection from storms and weather.

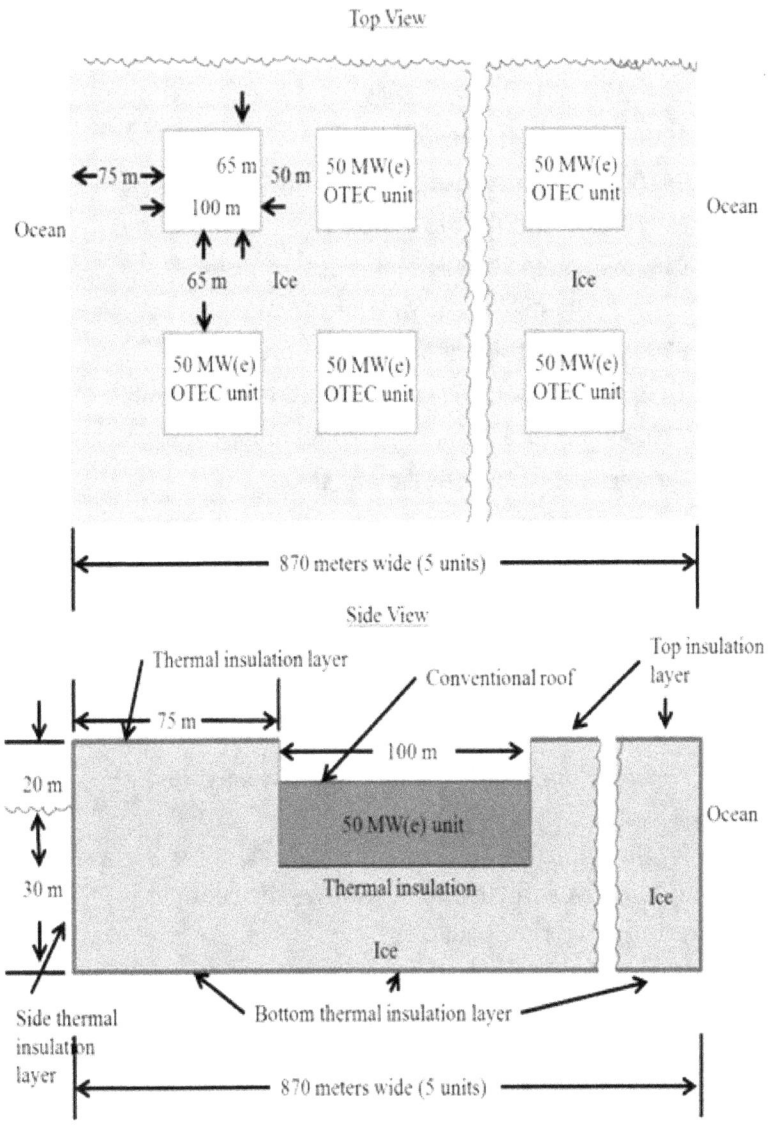

Figure 6.14 Arrangement of 50 MW(e) Units on 2000 MW(e) OTEC Ice Plant ship.

Table 6.6

Design Parameters for 2000 MW(e) Closed Cycle OTEC-ICE Plant ship.

- 2000 MW(e) net output
- 3420 MW(e) gross output
- Net/gross output ratio = 0.62
- Dimensions of 50 MW(e) net output unit
    - 104 m long, 65 m wide, 16 m high
- 6800 m^2 floor area of 50 MW(e) unit
- 40 units (50 MW(e) each) on OTEC-ICE plant ship
- 270,000 m^2 total floor area for 40 units
- Dimensions of OTEC-ICE plant ship
    - 1150 m long, 870 m wide, 50 m deep
- 1×10^6 m^2 bottom surface area of OTEC-ICE plant ship
- 10,500 m^3/second warm water input at 26° C
- 5500 m^3/second cold water input at +5° C
- 50/50% sand/perlite insulation for top surface
- CO_2 gas cell insulation for bottom surface
- Armored dry sand insulation for side surface

Table 6.6. Summarizes the design parameters for the 2,000 MW(e) ICE Plant ship. Table 6.7 shows the plant ship refrigeration power and cost.

Table 6.7.

Refrigeration Power and Cost for -30° C Ice Interior of 2000 MW(e) OTEC-ICE Plant ship

Basis: +26° C warm surface water temperature

+5° C deep cold water temperature

Deep cold water used as intermediate coolant between 26° C and -30° C

-30° C ice interior insulation has outer temperature

of   +5° C and inside temperature of -30° C; ΔT

= 35° C

3.49 COP

Parameter	Plant ship Top Surface Under 50 MW(e) Units	Plant ship Top Surface and Unit Walls	Plant ship Bottom Surface	Plant ship Side Surface
Insulation type	Dry sand bags	Sand/perlite bags	CO_2 gas cells	Armored dry sand
Area, m^2	270,000	944,000	1,000,000	202,000
Thickness, cm	30	30	30	30
k, w/mK	0.14	0.08	0.0155	0.14
qth, w(th)/m^2	16.9	9.9	1.8	16.9
P(e), w(e)/m^2	4.67	2.84	0.52	4.67
Total power, MW(e)	1.26 MW(e)	2.66 MW(e)	0.52 MW(e)	0.93 MW(e)
Annual cost $M/year (5 cents/KWH)	0.54 M$/year	1.66 M$/year	0.22 M$/year	0.40 M$/year
Total power for all surfaces = 5.37 MW(e)				
Total refrigeration power cost = 2.32 million $/year				

Finally, there is the capital cost of the ice structure plant ship – the cost of freezing the ice, the capital cost of the insulation and refrigeration equipment and the ice pipes that bring up deep cold water from a depth of 1000 meters or more. Table 6.8 gives the capital costs of the various components listed above. The capital costs of freezing the ice structure and the unit costs of the various thermal insulations used – bags of dry sand, bags of 50/50 dry sand and perlite, CO_2 gas cell insulation, and armored dry sand insulation for the side surfaces of the ice plant ship – are described, previously.

Table 6.8. Capital Cost of 2000 MW(e) OTEC-ICE Plant ship

Basis: Included are capital costs of energy and equipment for freezing ice structure, insulation layers on surface of ice structure and refrigeration equipment, and ice pipes to draw up cold water. The Capital cost of the NH_3 closed cycle power equipment, heat exchangers, water pumps and anchor cables are assumed to be $5,000/KW(e).

Parameter	Top Surface	Bottom Surface	Side Surface
Insulation cost/m^2	15.46 $/$m^2$	15 $/$m^2$	153 $/$m^2$
Area m^2	1.21 x 10^6	1 x 10^6	202,000
Capital cost (million dollars)	19	15	30
Total capital cost for insulation = 65 million $			
Total ice volume = 37 million m^3			
Capital cost for freezing (@ $2/$m^3$) = 74 million dollars			
Total length of cold water ice pipes = 12 kilometers (8 pipes, 22 m ID, 1.4 km long pipe)			
Total capital cost of cold water pipe @ 12.5 M$/km = 150 million $			
Total capital cost of 2000 MW(e) OTEC-ICE plant ship = 290 million $			
Equivalent cost per KW(e) = 145 $/KW(e)			
30 year amortized cost/KWH = 0.055 cents/KWH			
30 year amortized cost/KWH for $5000/KW(e) OTEC cycle equipment = 1.9 cents/KWH			

Adding the 65 million dollars capital cost for insulation and 74 million dollars for freezing the ice structure, plus $150 million dollars for the cold water ice pipes, the total capital cost for the OTEC-ICE plant ship is 290 million dollars (Table 6.8). Per KW(e), this amounts to $145/KWH for the 2000 MW(e) plant ship. Amortized over 30 years, the ice plant ship only costs 0.055 cents per KWH for the electric power it produces. By far, the major cost for the OTEC-ICE plant ship will not be the plant ship, but rather the capital cost of the OTEC power cycle equipment. Estimates of OTEC capital cost per KW(e) of capacity vary, with a value in the range of $4000 to $5000 per KW(e) appearing frequently. At $5000 per KW(e) for the OTEC cycle portion of the 2000 MW(e) OTEC-ICE plant ship, 30 year amortized capital cost would be 1.9 cents per KWH (Table 6.8), 35 times greater than the cost of the ice plant ship itself.

In addition to generating electric power in the OTEC cycle, surface water and deep cold ocean water can also be used to desalinate ocean water to produce large quantities of fresh water. The OTEC-ICE system desalination process is illustrated in Figure 6.15 for a production rate of 2 billion of fresh water daily. The intake rate of warm ocean water is 7500 cubic meters per second. As it passes through a set of evaporation chambers, a small portion (1.2%) of the ocean water evaporates and flows through ducts to a set of condensation chambers, where it condenses by transferring its latent heat of condensation to deep cold water flowing inside a set of heat exchangers.

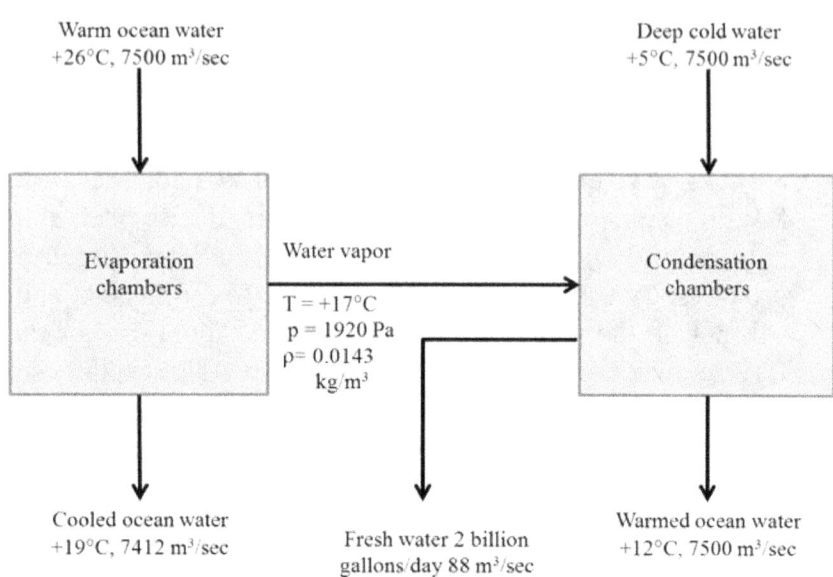

Figure 6.15. Flow Sheet for OTEC-ICE Plant ship to Produce 2 Billion of Fresh Water Per Day.

In the process, the temperature of the warm ocean surface water drops from its inlet temperature of +26°C down to +19°C as the fresh water evaporates from the ocean water. After reaching +19°C, the sea water is discharged back into the ocean. Similarly, the deep cold water temperature increases from its inlet temperature of +5°C to +12°C as it condenses the water vapor to fresh liquid water. After reading +12°C, the deep cold water is discharged back into the ocean. Two billion gallons per day of desalinated water sounds like a lot of water, and it is a lot of water. But people need a lot of water for personal use – growing food, industrial production, etc. California's 38 million population use 35 billion gallons of water daily, over 17 times more than a 2 billion gallon per day OTEC-ICE plant ship would produce. Per capita, Californians use almost 1000 gallons per person per day.

Today there are 7 billion people in the world. Tomorrow, in 2050 AD, there will be 9 billion, needing sufficient clean, safe water. And the situation is steadily getting worse as droughts caused by global warming increase, aquifers are over-pumped, and water sources become more polluted with toxic chemicals. There is a tremendous need for massive amounts of low cost water. Present desalination technology cannot meet this need. The cost of desalinated water using fossil fuel or electrically powered systems is several dollars per 1000 gallons, far too high for farmers and low income countries.

To desalinate ocean water by evaporating warm surface seawater and condensing it on a heat exchanger cooled by cold deep water is a major engineering challenge. First, one has to handle an enormous flow rate of warm and cold ocean water. Second, and even more challenging, is the problem of the vast volumetric flow rate, low pressure, low density water vapor. At 17°C, the nominal vaporization temperature, water vapor pressure is 1920 pascals, 1.9% of the standard atmosphere pressure of the air we breathe in and out 10 times a minute. The density of water vapor at 17°C is only 0.014 kilogram per cubic meter, slightly greater than 1/100,000th of the density of liquid water. For a fresh water production rate of 2 billion gallons per day, the required volumetric flow rate of water vapor based on an OTEC cycle with vapor at 17°C is 6.2 million cubic meters per second. To put this flow rate in perspective, the normal human breathing rate is about 9 liters of air per minute, or 0.15 liters per second. A volumetric flow rate of 6.2 million cubic meters per second is equivalent to the breathing rate of 40 billion humans, 6 times greater than the 7 billion people on Earth today. Because of the very low pressure of the water vapor, 1.9% of normal atmospheric pressure, the flow velocity of the water vapor as it moves through ducts on the OTEC desalination plant ship must be kept low to avoid excessive pressure loss. This, and the low density of the water vapor, 1/100,000th that of liquid water, necessitates very large duct area. For a maximum velocity of 50 meters per second and 6.2 million cubic meters per second, total duct area on the plant ship would be 124,000 m^2.

To achieve practical OTEC based desalination, the plant ship must

- Be very large in size to accommodate the very large volumes of equipment and water and vapor flows
- Have very efficient, very low pressure drop evaporators and condensers
- Have very low pressure drop for the water vapor as it moves from the evaporation chambers to the condenser chambers
- Very low cost, very reliable evaporators and condensers

The design concept described below, based on a large ice plant ship, aims at achieving the above objectives. Table 6.9 summarizes the design approach. The design is based on existing correlations for fluid flow and heat transfer of thin films, and is intended as a preliminary design to assess feasibility. It should not be viewed as an engineering design. Experimental testing of the fluid flow and heat transfer performance will be necessary to develop the engineering design for actual implementation. Since the design is highly modular, it will not be necessary to test and entire OTEC-ICE desalinization plant ship – only a few of the many thousands of its modules.

Table 6.9 Summary of Evaporator/Condenser Design Approach for Sea Water Desalination

- This films of warm sea water flow down each side of multiple vertical plastic sheets, 1 meter in height, inside a low pressure evaporation/condensation chamber
- Approximately 1.17% of warm sea water input vaporizes. Remaining sea water is discharged back into the ocean.
- Water vapor flows from the set of evaporator sheets to a set of heat exchanger plates that are cooled by deep cold sea water brought up from a depth of ~1000 meters

- Water vapor condenses as fresh water on cold heat exchanger plates and is collected for users. Deep cold water warms by +7°C and is discharged back into the ocean.
- Warm ocean water enters at +26°C and is discharged at +19°C. Temperature drop of +7°C is due to latent heat of vaporization
- Water vapor temperature in low pressure (1720 Pa) chamber is +17°C
- Cold deep water enters chamber at +5°C and is discharged at +12°C. Temperature increase of +7°C is due to latent heat of water vapor condensation.
- Fresh water output is 2 billion gallons per day (88 m³/sec)
- Warm sea water input is 7500 m³/sec, cold sea water input is 7500 m³/sec
- Equivalent thermal power evaporator rate is 2.2x10¹¹ watts

Figure 6.16 shows a view of a low pressure evaporator/condenser Chamber that contains the evaporator and condenser modules. The chambers are located on the OTEC-ICE plant ship above sea level.

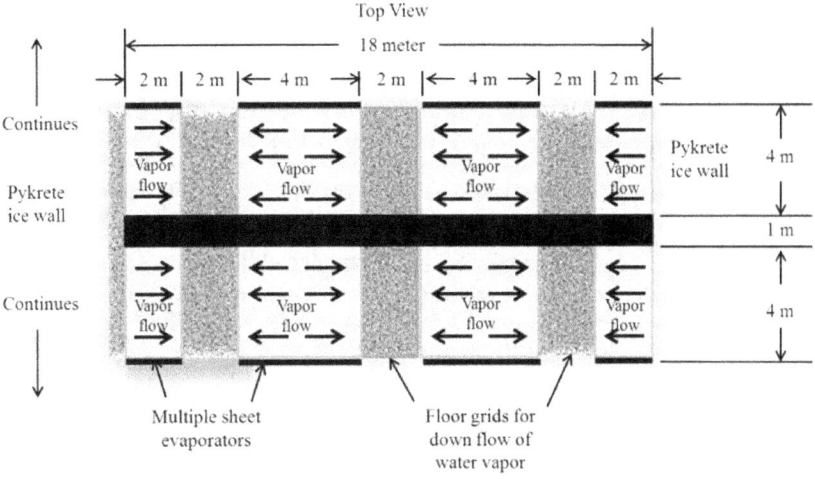

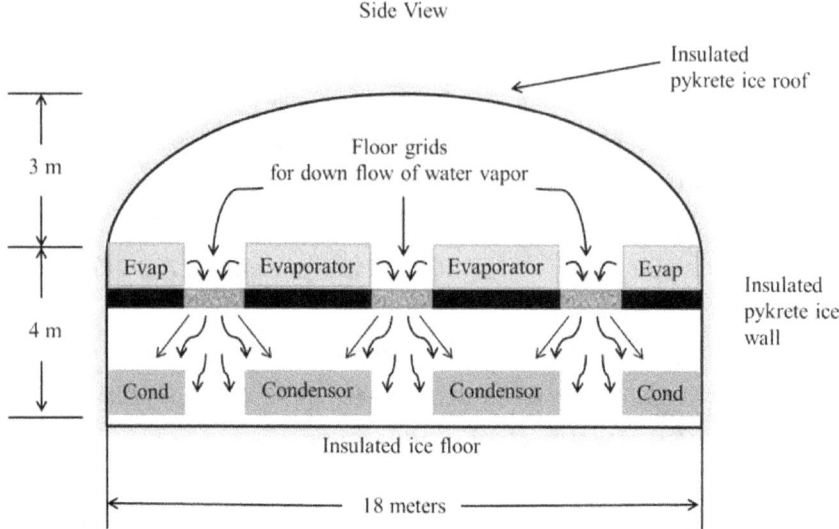

Side View

Insulated pykrete ice roof

Floor grids
for down flow of water vapor

3 m

Evap   Evaporator   Evaporator   Evap

Insulated
pykrete ice
wall

4 m

Cond   Condensor   Condensor   Cond

Insulated ice floor

18 meters

Figure 6.16. Layout of Water Vapor Evaporators and Condensers.

Water vapor from the falling water films flows outwards in the spaces between the multiple stacked plastic sheets inside each evaporator unit, as shown in Figure 6.16. The vapor flows outwards in both directions from the center of the unit to the openings between the units. The evaporation units are positioned on a floor above the condensation units. The portion of the floor between them is a grid structure through which the water vapor flows down to the condensation units. Per 4 m x 4 m unit, the flow area for water vapor out from the 4 meter wide, 1 meter height of the unit is approximately 3.6 m² taking into account the 1 millimeter widths of the plastic sheets, spaced 1 cm apart. Each 4 m x 4 m unit produces 7.3 kg/sec of water vapor. At the vapor density 0.014 kg per m³, this corresponds to a vapor volumetric flow rate of 525 cubic meters per second. One half of the 525 m³/sec vapor flow rate exits through the 3.6 m² flow area on the left side of the 4 m x 4 m unit with the other half exiting from the right side. The water vapor velocity at the exit is then (525/2)/3.6, equal to 73 meters per second. The corresponding Bernoulli ΔP is then ΔP

$= \frac{1}{2} \rho_{vapor} V_{exit}^2 = \frac{1}{2} (0.014 (73)^2 = 37$ Pascals, or only 1.9% of the static pressure of 1920 Pa. Since the water vapor velocity is less inside the evaporator module, the Bernoulli ΔP is even smaller there.

After exiting the module, the water vapor flow turns downward, descending through the floor grid to the condenser section of the low pressure chamber. The condenser modules are the same size, 4 m x 4 m, as the evaporator modules and located in the underneath them in the same pattern. The condenser module is made up of a set of parallel aluminum plate heat exchangers. The heat exchanger consists of 2 parallel aluminum plates each 1 millimeter thick and 1.4 meters in height separated by a 2 millimeter channel through which cold deep sea water flows. Water vapor condenses on the outer surfaces of the aluminum plates, with the latent heat of condensation conducted through the aluminum sheets and absorbed by the flowing water inside the heat exchanger. The aluminum sheets are bonded together by aluminum struts between the 2 sheets. The condensed water vapor flows as a thin film of water vertically downwards on the surfaces of the 2 aluminum sheets to the bottom of the 1.4 meter high heat exchanger, where it is collected and pumped to the fresh water reservoir on the plant ship. The fresh water films are much thinner than the sea water films on the plastic sheets in the evaporator modules, because the fresh water film flow rate is much smaller, i.e., 1.2% of the sea water film flow rate.

Figure 6.17 shows the layout of the low pressure evaporation/condensation chambers of the OTEC-ICE plant ship and Table 6.10 summarizes the design parameters. The desalinization plant ship is larger than the closed cycle OTEC-ICE plant ship for electric power production described earlier (Table 6.6). The length of the electric power plant ship is 1150 meters compared to 1400 meters for the desalinization plant ship, and its beam is 870 meters, compared to 1080 meters for the desalinization plant ship.

The desalinization plant ship could be made smaller by having thinner ice walls between the low pressure chambers, a possibility that can be explored in future design studies. The ice wall thicknesses of 30 meters might be reduced to 20 meters, for example. The low pressure chambers occupy only a small fraction, 4%, of the plant ships volume. This is a consequence of having to have the warm sea water enter the low pressure chamber at a height of 10 meters above sea level.

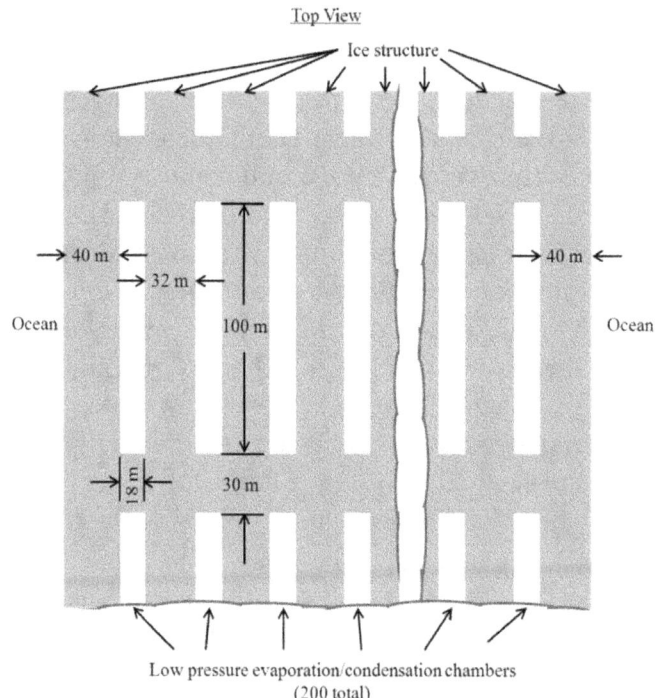

Top View

Ice structure

40 m

32 m

Ocean

100 m

Ocean

18 m

30 m

Low pressure evaporation/condensation chambers
(200 total)

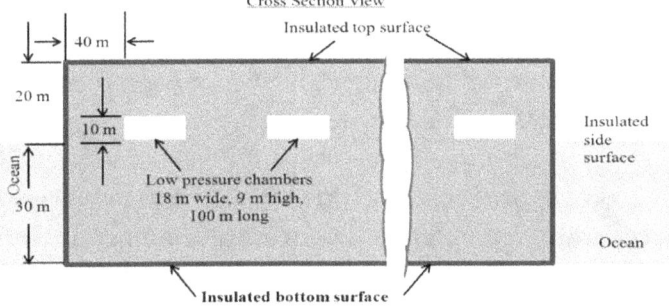

Overall dimensions of OTEC-ICE Fresh Water Plantship
- 1400 meter length
- 1080 meter width
- 20 meter height above ocean surface
- 30 meter depth below ocean surface

Figure 6.17. Layout of OTEC-ICE Plant ship for Fresh Water Production of 2 Billion Gallons Per Day.

Table 6.10. Summary of OTEC-ICE Plant ship Design Parameters for Fresh Water Desalinization

- Evaporation/condenser chambers
  - 200 low pressure chambers on OTEC-ICE plant ship, 18 meters wide, 100 meters long, 9 meters high
  - Water vapor in chamber at 17°C, 1920 Pa pressure
  - 60 module evaporator units in each chamber plus 60 module condenser units
  - 3 module units across width of chamber, 20 lines of 3 units along 100 meter length of chamber
  - 32 meter thick ice walls between 18 meter wide adjacent chamber, 30 meter thick ice walls between ends of 100 meter long chambers
  - Top of evaporator plastic sheet feed points are 10 meters above ocean surface
- OTEC-ICE plant ships
  - 1400 meter length
  - 1080 meter beam
  - Top surface of plant ship is 20 meters above ocean surface
  - Bottom surface is 30 meters below ocean surface

Raising the input warm sea water to a height of ~10 meters lowers its hydraulic pressure to ~2% of sea level pressure, essentially matching the water vapor pressure of 1920 Pa. If it entered at a lower height above sea level, there would be a much larger pressure difference, and the ΔP loss would be large, necessitating much greater pump power. If the warm sea water entered the low pressure chamber at sea level, there would be an irreversible ΔP at 10^5 Pa. Pumping 7500 m³ per second flow rate with a ΔP input of 10^5 Pa so that it could be discharged to the ocean would then require a pump power of 7500 x 10^5, or 750 megawatts – an impractically large power requirement. By entering the low pressure chamber at 10 meters, the only ΔP loss experienced by the warm sea water input is its vertical flow of 1 meter downwards on the plastic evaporator sheets, equal to 10^4 Pa. The remaining hydraulic pressure of 0.9 x 10^5 Pa, plus pump power input of 10^4 Pa, brings the pressure of the discharge sea water to 10^5 Pa, so that it can flow back into the ocean.

The refrigeration power required for the OTEC-ICE desalinization plant ship is a bit greater than that for the closed cycle OTEC-ICE plant ship that would generate electric power for two reasons. First the desalinization plant ship outer insulation surface area is about 50% greater. Second, the inner insulation surface area on the ice walls of the low pressure chambers is greater than the insulation area beneath the closed cycle OTEC power generator units. As with the closed cycle OTEC-ICE plant ship for electric power production (Table 6.7) the refrigeration power and cost for the OTEC-ICE desalinization plant ship are very low, 6.78 MW(e) and 2.96 M$/year. In both cases, the refrigeration power and cost will be much smaller than the electric power and cost for pumping the warm and deep cold sea water streams. For the desalinization plant ship with a 7500 m³/sec flow of warm sea water at a ΔP of 10^4 Pa, the pump power would be 75 MW(e), more than 10 times the total refrigeration power. The corresponding power cost for the warm sea water input would be more than 10 times the refrigeration power cost. Added to the power cost for the warm water input would be an equal amount for the cold water input, also at 7500 m³/sec with a ΔP of 10⁴ Pa.

At 2 billion gallons per day of fresh water production, the refrigeration power cost of 2.96 million dollars per year corresponds to only 0.4 cents per 1000 gallons. Even if refrigeration power and cost were 3 times greater than projected, it would still be negligible – a little bit over 1 cent per 1000 gallons. Turning to the capital cost of the OTEC-ICE plant ship, the costs for the following components have been analyzed.

- Thermal insulation
- Freezing the ice structure
- Cold water pipe
- Evaporator sheet modules
- Condenser heat exchanger modules

Not included are the cost of the water pumps, maintenance equipment, crew quarters and plant ship refrigeration and propulsion equipment. These will be relatively small. Total capital cost for the above components for the OTEC-ICE the desalinization plant ship is 886 million dollars. The most expensive component 450 million dollars for the aluminum heat exchangers in the condensation modules, accounts for over ½ of the total capital cost. The amortized capital cost corresponds to only 4.0 cents per 1000 gallons of fresh water. Conclusions: Adding the 0.4 cents per 1000 gallons for the refrigeration power cost to the 4.0 cents per 1000 gallons amortized capital cost results in a total cost of 4.4 cents per 1000 gallons. This is a factor of 100 less than present desalinization costs. Even if the OTEC-ICE desalinization costs are double or triple the values projected here, it will be an extremely attractive approach for producing many billions of gallons of fresh water per day.

Turning to system 3, the OTEC-ICE plant ship that produces both electric power and fresh water. The closed cycle

50 MW(e) OTEC power generation units would be located underneath the low pressure evaporation/condensation chambers, as illustrated in Figure 6.18. The OTEC-ICE power/desalinization plant ship would be the same size as the desalinization plant ship described previously, 1400 meters long, 1080 meter beam, 20 meters above the ocean surface, and 30 meters below the surface.

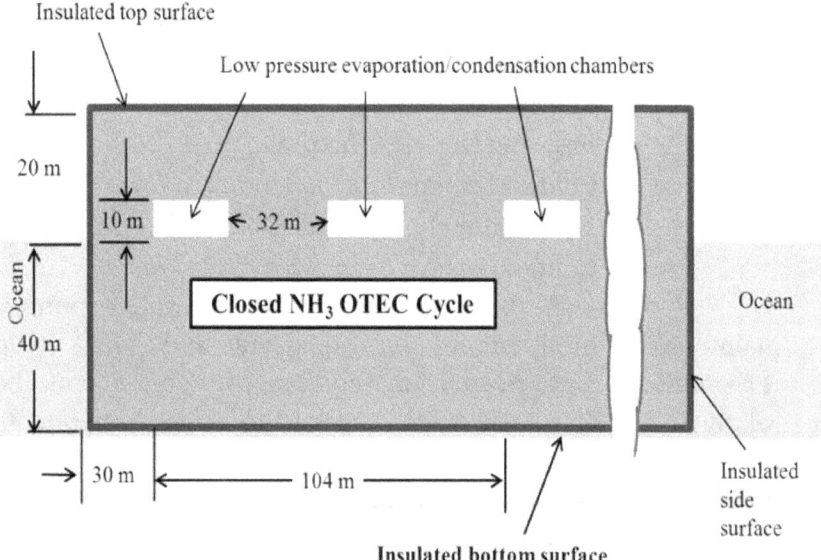

Cross Section View

Figure 6.18. Layout of OTEC-ICE Plantship for Production of 1.9 Billion Gallons Per Day of Freshwater Plus 2000 Megawatts of Electrical Power Using Closed Ammonia OTEC Cycle, Author, Powell

It is substantially larger than the dimensions of OTEC-ICE 2000 MW(e) closed cycle power plant ship,which are 1150 meters long, 870 meters wide, 20 meters above the ocean surface, and 30 meters below the surface. With the larger size of the combined power/desalinization plant ship, it will be simple to locate the 40 OTEC module power units, each generating a net power of 50 MW(e), beneath the low pressure chambers.

The power generation and desalinization modular units use the same warm sea water and cold deep sea water flows, rather than separate flows. This reduces the sea water flow rate required, as compared to having separate flows. The input warm sea water flows first to the closed cycle OTEC power generation units at +26°C, at a flow rate of 10,500 m³ per second, and then at a lower discharge temperature to the desalinization units. Similarly, the input deep cold water flows first to the heat exchangers that condense the ammonia working fluid in the OTEC power cycle, and then slightly warmer, to the heat exchangers that condense water vapor to liquid desalinated water.

The outlet sea water temperature from the 50 MW(e) closed cycle power modular units is not specified in the design by Vega and Michaelis. In their design for the open cycle OTEC system, the warm sea water input temperature is also +26°C, with the outlet temperature of 23.3°C. Taking the warm outlet temperature from the 50 MW(e) closed ammonia cycle OTEC units to be the same as the open cycle OTEC, +23.3°C, the ΔT available for the desalinization portion of the OTEC-ICE plant ship decreases from ΔT = 26°C - 19°C = 7°C, down to ΔT = 23.3°C - 19°C = 4.7°C. However, the warm seawater input increases from 7500 m³/sec to 10,500 m³/sec. Scaling the fresh water production rate to account for the decreased ΔT corresponding thermal input per cubic meter of warm water input to the evaporator modules and the increased warm water flow rate, the desalinated water output from the combined power/fresh water OTEC-ICE plant ship.

$$Fresh\,water\,output = 2\,billion\frac{gallons}{day}\left(\frac{4.7}{7}\right)\left(\frac{10,500}{7500}\right) = 1.9\,billion\,gallons\,per\,day$$

So the fresh water output from the combined power/desalinated plant ship is almost the same as for the desalinization only plant ship. Conclusions. The combined power/desalinization OTEC-ICE plant ship is the favored approach. It will produce 2000 MW(e) of electric power plus 1.9 gallons per day of fresh water. As an example of the economic benefits from OTEC-ICE power/desalinization plant ship, based on a delivery price of 2 cents per kilowatt hour – far less than other renewable sources – and a delivery price of 10 cents per 1000 gallons – again, far less than the cost of other desalinization systems - , total revenues from the plant ship over a period of 30 years would be 12.6 billion dollars. The comparison below shows that the projected revenues, even at the very low prices of 2 cents per KWH and 10 cents per 1000 gallons, are much greater than the projected costs for the OTEC-ICE plant ship, a factor of 13 greater.

30 Year Revenues
12.6 billion dollars

30 Year Costs
0.866 billion dollar capital cost
0.090 billion dollar refrigeration cost
0.96 billion dollar total cost

The 30 year costs include the capital costs of the insulation, freezing the ice structure, cold water ice pipe, evaporator, and condenser modules. They do not include the cost of the closed ammonia OTEC cycle equipment, and the cost of personnel, crew quarters, pumps, and maintenance, refrigeration, and propulsion equipment. Even if these additional costs doubled capital cost to 1.7 billion dollars, which is unlikely, 30 year costs would rise to 1.8 billion dollars, still a factor of 7 less than the revenues.

Summarizing, OTEC-ICE plant ships have the potential to provide enormous amounts of electric power and fresh water at very low cost to countries all around the World. The cost of the OTEC plant ship for electric power generation is only 2 cents per KWH, while the cost of the OTEC plant ship for fresh water production is only 10 cents per 1,000 gallons.

Chapter Seven

LISA COASTAL SECURITY

Coastal LISA Ice Structures – Protection from Storm Surges, Tsunamis, and Rising Sea Levels, and Off-Shore Habitats, Airports, and Seaports

James Powell, Jesse Powell, John Powell and James Jordan

We are tied to the ocean. And when we go back to the sea, whether it is to sail or to watch - we are going back from whence we came.

John F. Kennedy

A few hundred millions of years ago our ancestors crawled out of the ocean onto land. Curious, but cautious, they began to explore the New World. Fortunately for us, they decided it was a good place, and settled down to live there; otherwise, we would not be here.

Over the millennia, countless land dwelling species have transformed the lifeless desert they crept onto into today's gorgeous world we and our fellow species share. Sadly, however, humans are trashing Eden – pouring 30 billion tons of greenhouse gases into the atmosphere every year, cutting down rain forests, poisoning soils with toxic wastes, and mass extermination of our fellow species. By the end of the 21st Century, the Sixth Extinction, created by humans, will have wiped out half of the species on Earth.

Global warming, the product of the enormous amounts of greenhouse gas emissions from the fossil fuels humans have burned, is melting ice caps and glaciers, raising sea levels and increasing the severity of storms and hurricanes.

We thought we had escaped from the ocean, but now the ocean is coming back for us – particularly for the large portion of humanity that lives close to it in coastal regions. And it's coming back in a big way. Storm surges are getting bigger, rising sea levels are flooding low lying islands and shorelines, and giant tsunamis are devastating coastal populations.

How serious are the threats from storm surges, tsunamis, and rising sea levels to the more than 1 Billion people living in and near the Word's coastlines? Very serious. Let's review the history of storm surges, tsunamis and ocean flooding. The disasters that have happened and the much worse disasters that lie ahead in the coming decades, unless LISA ice barriers are built to protect to protect the human populations that are at risk.

Storm Surges

Hurricanes (typhoons, if you live in the Pacific) have been with us forever, from ancient times until the present, causing tremendous damage and deaths. They will go on forever into the future, most likely getting stronger and deadlier as global warming goes on. Let's look at four of the deadliest storm surges that have hit America, starting with the 1900 Galveston Hurricane and its storm surge of over 15 feet (1). Unfortunately, the highest point in the city of Galveston, located on the coast of Galveston Bay, was only 8.7 feet above sea level. Galveston was completely destroyed, and between 8,000 to 12,000 people died. The cost of the damage in today's (2010) dollars is estimated to be $104 billion (1), more than the $85 billion damage from Hurricane Sandy.

It was impossible to bury all the dead bodies. After retrieving from the rubble (Figure 7.1), they were at first weighted down and dumped at sea (1), but the ocean currents washed many of them back onto the beach. The smell of the decaying bodies was horrible and could be detected from miles away. As a result, the dead were then burned on funeral pyres. Following the 1900 catastrophe, the citizens of Galveston decided, "Never again". They raised the elevation of the city by as much as 17 feet using dredged sand, including tearing down 2100 buildings plus the 3,000 ton St. Patrick church (1). And they built a 17 foot high seawall, starting in 1902 (Figure 7.2). By the time the next storm hit in 1915 with a 12 foot storm surge, Galveston was ready. The storm claimed only 53 lives, far fewer than the more than 8,000 that died in 1900, and far

Figure 7.1. Carrying Bodies, Galveston Hurricane, 1900.

less than would have died if they had not built the seawall. Today, Galveston is behind a 10 mile long, 17 feet high seawall (1).

However, storms and storm surges are getting bigger. Katrina storm surges were as high as 26 feet. When LISA barriers begin to be implemented, Galveston should install a much higher seawall.

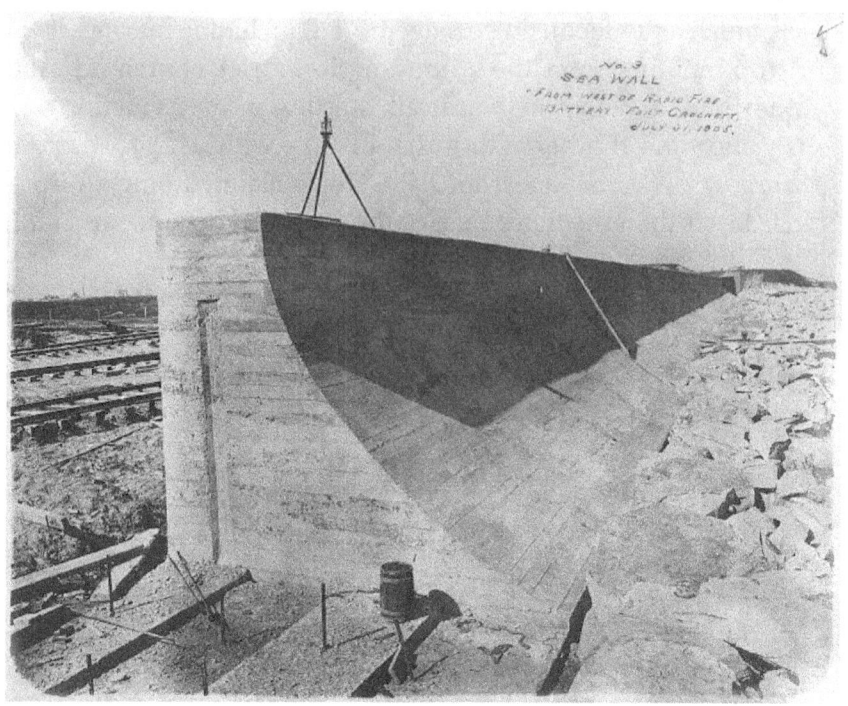

Figure 7.2. Sea Wall from West of Rapid Fire Battery, Fort Crockett – NARA.

Let's go now to the next great American storm surge. The Great Hurricane of 1938 that struck Long Island and New England on September 21, 1938. The storm surge was 10 to 12 feet in height (2). The estimated death toll was between 682 and 800 people with over 57,000 homes damaged or destroyed and property damage estimated at 4.7 billion dollars, adjusted to current (2014) dollars (3). If the hurricane had hit in 2005, the damage would have been much greater estimated at about 39 billion dollars, due to the increased population and property values (3).

Eastern Long Island was particularly hard hit with wave heights of 25-35 feet. Montauk became a temporary island, and 10 new inlets were created in the sand dune barrier island that runs along Long Island's ocean-facing south shore. But the damage didn't stop at Long Island. Even though Long Island offered some protection to Connecticut, the storm surges on Long Island Sound destroyed many small towns along the Connecticut coast. It was the worst natural disaster in the 350 year history of Connecticut (3). Nor did Rhode Island or Massachusetts escape the 1938 Hurricane unscathed. Downtown Providence was flooded with more than 13 feet of water in some areas (3). Beach communities on Rhode Island's coast were swept away, houses and people, along with the Whale Rock light house and its keeper. In Massachusetts, Falmouth and New Bedford were under 8 feet of water. A 50 foot wave, the biggest during the storm, was recorded at Glouster (3). If Long Island, Rhode Island and Massachusetts had coastal ice barriers with top surfaces 20 meters (66 feet) above the ocean surface, they would not have been flooded, the surge would not have demolished thousands of homes and buildings, and many lives would have been saved. Damage from the hurricane's high winds would still have occurred, but the damage and loss of life would have been much less.

Now on to America's next great storm surge, the Katrina disaster. Figure 7.3 is a view from space of Hurricane Katrina approaching New Orleans. It landed on Louisiana on August 29, 2005, causing tremendous flood damage. When its levees failed, over 80% of New Orleans was flooded (4). Figure 7.4 shows a section of New Orleans with flooded streets. Storm surges of up to 28 feet (2) reached 6 to 12 miles inland (4) leaving boats and ships high and dry, far from the ocean (Figure 7.5). There were 1,833 confirmed deaths with a total damage of 108 billion dollars (2005 dollars).

Figure 7.3. NOAA Photo of Hurricane Katrina Approaching New Orleans, Louisiana, August 28, 2005.

There were 53 breaches of the levees that supposedly protected New Orleans, resulting in the mass flooding of the city. If they had held, the flooding would have been much less, though there still would have some flooding due to flood gates that were not closed (4). If coastal ice barriers had been in place to protect New Orleans, the deaths and damage from Hurricane Katrina would have been much less.

Figure 7.4. Katrina Storm Surge Flooding New Orleans.

Figure 7.5. Boats Ashore from Hurricane Katrina Damage in Bayou
La Batre, Alabama.

Now for the final example of major surges that have devastated the coasts of America, Hurricane Sandy. In fact, Hurricane Sandy not only hit New York, New Jersey and states all along the East Coast, but also did great damage in the Caribbean (5). It was a very big storm (Figure 7.6) affecting Jamaica, Haiti, the Dominican Republic, Cuba, the Bahamas, and Bermuda. Cuba had waves up to 29 feet high. Storm surges in Long Island Sound, Raritan Bay, and New York Harbor were in the range of 6 to 11 feet (6). Large areas in lower Manhattan were flooded. Battery Park had a water surge of 13.88 feet. Seven subway tunnels under the East River were flooded. Governor Cuomo estimated the damage to New York at 42 billion, while New Jersey estimated its damage at 36.8 billion. Plus lots of damage to other US States. Total damage in the US? Estimates range from 65 billion to 85 billion dollars, depending on the estimator. At least 286 people were killed by Hurricane Sandy (5).

As we have said with regard to the previous 3 examples of major US storm surges, had the affected regions been protected by coastal ice barriers with surfaces 20 meters (66 feet) above sea level, the deaths and damage from Hurricane Sandy would have been greatly reduced. Moreover, none of the countless historic storm surges from hurricanes and typhoons, not even the 2013 Typhoon Haiyan (7) in the Philippines that resulted in 6,241 confirmed deaths and 1785 missing, presumed dead, could have washed over protective coastal ice barriers.

Figure 7.6. Satellite Image of Hurricane Sandy, October 28, 2012.

We now turn to an even bigger threat than storm surges, tsunamis.

Tsunamis

While most (80%) of tsunamis happen in the Pacific (8), they don't happen only there. They happen all over the world throughout human history. The list (9) is extremely long (Table 7.1), and accounts date back thousands of years. In 463 BC, the Greek historian Thucydides in his <u>History of the Peloponnesian War</u> was the first to recognize that earthquakes were the cause of tsunamis (8). The Roman historian, Ammianus Marcellinus, described the 365 AD tsunami that devastated Alexandria in Egypt – first the earthquake, then the sea retreats from the shore, and then the BIG wave!

Table 7.1 Partial List of Major Historic Tsunamis

Location/Date	Maximum Wave Height, Meters (Feet)	Cause	Deaths	Notes
1. Norwegian Sea ~6200 BC	NA	290 km landslide off coastal shelf	NA	Floods went 80 km inland
2.Helike, Greece 373 BC	NA	Earthquake	NA	Submerged city of Helike, 2 km inland
3 Alexandria, Egypt 365 AD	30 (100)	Earthquake	Thousands	Ships flung 3 miles inland
4 Keicho, Japan 1605 AD	30 (100)	Earthquake (8.1 Mag)	5000	---
5 Bristol Channel, Britain 1607 AD	NA	Earthquake	2000	Flooded 200 square miles
6 Hoei, Japan 1707 AD	20 (66)	Earthquake (8.4 Mag)	30,000	---
7 Mt. Unzen, Japan 1792 AD	100 (330)	Earthquake and landslide	15,000	---
8 Hawaii, US 1868	18 (60)	Earthquake (7.5 – 8.0 Mag)	77	---
9 Messina, Italy 1908	NA	Earthquake	123,000	---
10 Lituya Bay, Alaska 1958	524 (1720)	Earthquake and landslide	NA	Highest wave ever recorded
11 Krakatoa, Indonesia 1883	40 (135)	Volcanic eruption	36,000	High tsunamis all over Pacific Ocean
12 Meiji Santiku, Japan 1896	30 (100)	Earthquake	27,000	---
13 Valdiva, Chile	25 (75) 3m (10 ft) in Japan	Earthquake (9.5 Mag)	6000	Strongest earthquake ever recorded
14 Alaska,US 1964	30 (100)	Earthquake (9.2 Mag)	121	Killed 11 people in Crescent City, California
15 Indian Ocean, 2004	33 (108)	Earthquake (9.1 – 9.3 Mag)	230,000	Deadliest tsunami ever
16 Fukushima, Japan 2011	40.5 (133)	Earthquake (9.0 Mag)	18,000	Nuclear power plants destroyed

The tsunamis kept on coming. Memorable ones include the 1755 Lisbon Earthquake (10). Estimates of the number of the dead in Portugal and other countries are in the range of 40,000 to 50,000 people. Tsunamis from that event struck locations all over the Atlantic. 20 meter (66 feet) high tsunamis hit the coast of North Africa. A 3 meter (10 foot) tsunami hit Cornwall in Britain. The earthquake and tsunami struck Lisbon on the morning of November 1, 1755, All Saints Day. (Figure 7.7) Theologians saw the disaster as divine judgment (10). Eighty-five percent of Lisbon's buildings were destroyed including its major churches. The disaster profoundly impacted European thinking with the question, "How could God let this catastrophe happen?" that became part of the Enlightenment.

Tsunamis can come from other causes than earthquakes. The volcanic eruption of Krakatoa in 1883 created a tsunami that resulted in tens of thousands of deaths (11). Then there are the landslides that cause tsunamis. The landslides can be triggered by earthquakes or volcanic eruptions. Two historic tsunamis resulting from landslides:

- Mt. Unzen in Japan, 1792. The volcanic eruption caused a massive landslide into the ocean. The resulting tsunami was 100 meters (330 feet) high, killing 15,000 people in the local villages (11).

- Lituya Bay, Alaska, 1958. An earthquake triggered a massive landslide at the head of Lituya Bay, creating a 91 meter (300 foot) high tsunami with a peak wave that reached a land height of 5.24 meters (1,720 feet) above sea level. Fortunately, it was confined to Lituya Bay with only minor impact beyond and killed only 2 people (11).

Figure 7.7. Copper Engraving Shows the Lisbon Earthquake, Nov 1, 1755

Finally, there are the asteroid impacts that create truly gigantic tsunamis. Two terrifying examples:

- The Chicxulab asteroid that wiped out the dinosaurs 66 million years ago when it hit the Yucatan (11). There were no measurements of the resulting tsunami, but it is thought it would have been as high as 5 kilometers (3.1 miles).

- An asteroid impact in the southeastern Pacific 2.5 million years ago generated a 200 meter (660 feet) high tsunami, striking Chile and the Antarctic Peninsula and much of the Pacific (11).

Asteroid impacts and their effects are especially interesting today. Three quarters of Earth's surface is ocean and the most likely place for an asteroid to hit. On Friday, April 13, 2029, the asteroid Apophis is scheduled to fly by Earth at a distance of 18,000 miles (12). The prediction is that it will miss us, thank goodness. However, if Apophis-like asteroid hits Earth, it will impact with an explosive energy equivalent to 500 million tons of TNT 2 times the explosive energy of Krakatoa. If as is probable, it hits the ocean, it will generate a giant tsunami. Where would it hit, and what coastal areas experience the tsunami? Nobody knows.

Coming back to the present, let's examine in more detail the recent 2004 Indonesian (13) and the 2011 Fukushima-Daiichi tsunamis (14). The 2004 Indonesian tsunami was generated by a magnitude 9.1 earthquake off the west coast of Sumatra, Indonesia on December 26, 2004. The resulting waves were up to 30 meters (100 feet) high. It killed an estimated 220,000 people in 14 countries located in the Indian Ocean, including Indonesia (167,000), Sri Lanka (35,000), India (18,000), Thailand (8000) and others. 1.69 million people were displaced and 45,000 people missing.

Figure 7.8 shows a photo of the 2004 tsunami striking AoNang in Thailand (13). The tsunami was detected in South Africa at a distance of 8,500 kilometers (5,300 miles) from its origin. Its wave height there was 1.5 meters (5 feet). In Banda Acen, Indonesia, close to the tsunami's origin, a 2,600 ton ship the Apung 1, was washed 3 kilometers inland. It has become a tourist attraction.

Figure 7.8. 2004 Tsunami, Ao Nang, Thailand.

In addition to the deaths, displacements, and injuries to the affected populations, the 2004 tsunami has caused extensive environmental harm including infiltrating soil and aquifers with salt, and major damage to coastal wetlands, forests, and thousands of plantations that grow rice, mangoes and bananas (13).

The March 11, 2011, Fukishima-Daiichi tsunami was extremely devastating. The death toll was lower than from the 2004 Indonesian tsunami, but the environmental and economic damage was even greater. Approximately 18,500 people died from the earthquake and tsunami. Another 300,000 people were evacuated after the earthquake, and about 1600 people died due to the evacuation conditions (14). In November 2011, 8 months after the Fukushima-Daiichi disaster, the Japanese Science Ministry reported that radioactive cesium had contaminated 11,580 square miles in Japan's mainland. 4500 square miles had radiation levels that exceeded Japan's allowable limit of 1 millisievert per year (15). However, 1 month after the disaster, on April 19, 2011, Japan increased its "safe" exposure level 20 fold to 20 millisieverts per year, reducing the area to be evacuated (15). The US safe exposure limit remains at 1 millisievert per year. Nonetheless, even with the higher "safe" limit, a very large area still had to be evacuated. All the land within 12 miles of the nuclear power plant (230 square miles) plus 80 square miles of land northwest of the plant were declared too radioactive for habitation. In September 2012, Fukushima officials stated that 159,128 people had been evacuated, losing their homes and possessions. Total economic loss is estimated to be in the range of 250 to 500 billion dollars (15).

The nuclear plant had a 10 meter (33 feet) high seawall, but the 14 meter high tsunami flowed over it (14). Figure 7.9 illustrates the scale of a human relative to a 10 meter tsunami (13). Visualize 4 meters (13 feet) more on top of the 10 meter scale, and one will have an idea of how high the Fukushima tsunami was. Various studies previous to the disaster had warned that the nuclear power plant was vulnerable to earthquakes and flooding by tsunamis, but were not taken seriously by the utility and regulators.

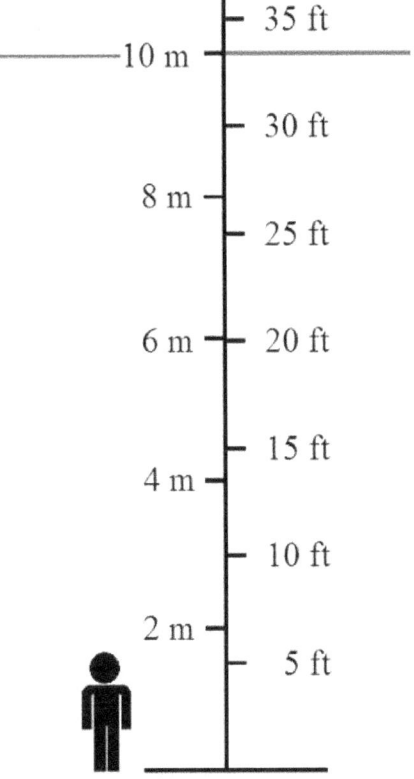

Figure 7.9. Tsunami Size Scale.

So, that's the past and present for storm surges and tsunamis. What can we expect in the future? More of the same. As global warming continues, storm surges will get bigger. Earthquakes under the sea floor and near the ocean will generate tsunamis. Submarine landslides will also generate tsunamis and the asteroids will continue to threaten us with disaster from deep space.

One future tsunami scenario in particular is of great importance to the United States – the tsunami that will be generated by inevitable earthquake on the Juan de Fuca Plate (16), which runs along the Northwest Coast of the US (Figure 7.10). The only question is when.

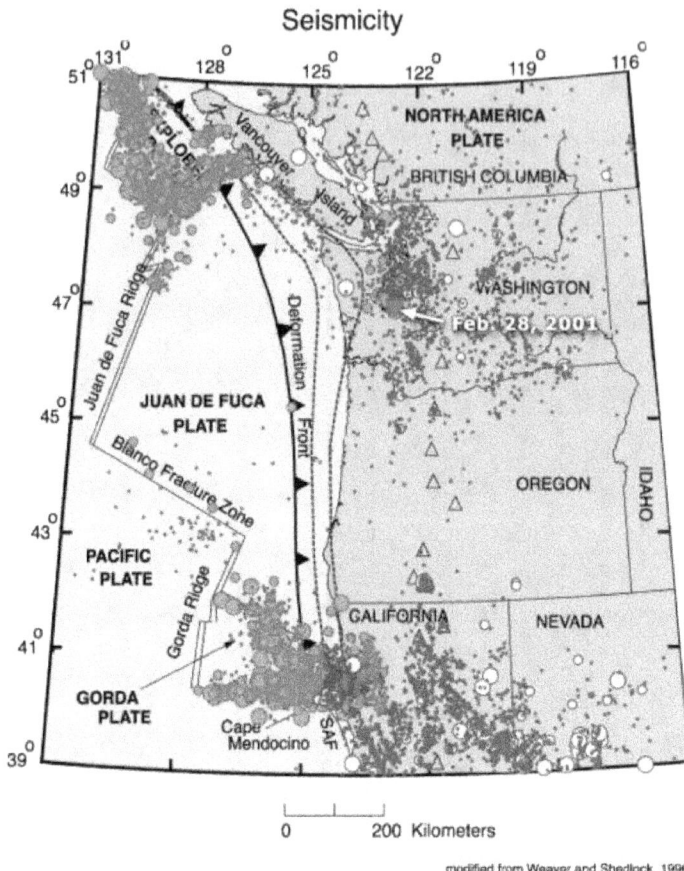

Seismicity

modified from Weaver and Shedlock, 1996

Figure 7.10. Earthquake locations along the Juan de Fuca Plate.

The earthquakes occur on the 600 mile long Cascadian fault, which is on the boundary between the Juan de Fuca Plate as it slides under the North American continental plate. So far, studies have identified 19 to 21 major earthquakes in the last 10,000 years. In at least 17 events, the entire fault zone appears to have ruptured causing magnitude 9 earthquakes and major tsunamis (17). The two most recent earthquakes and tsunamis occurred around 1500 AD and in 1700 AD. When the Cascadia fault ruptures producing an earthquake, the sea floor bounces, triggering a tsunami. The sea floor bounce can be as much as 20 feet setting off waves of up to 30 meters high (18). The July 26, 1700 earthquake and tsunami event on the Cascadian fault was a monster. Based on Japanese historical records, the tsunami was 5 meters (16 feet) in height (19) when it hit Japan's Honshu Island after traveling 5,000 miles at 500 mph. On the northwest coast close to its origin, the tsunami's height would have been much greater, 20 meters or more. It's been 314 years since the 1700 earthquake and tsunami. Scientists believe there is a 45 percent probability of an earthquake of magnitude 8.0 or more in the next 50 years and a 15% chance that it will be a magnitude of 9.0 or greater (19). When it happens, the devastation to the northwest coast will be tremendous.

In a paper published in the <u>Proceedings of the National Academy of Sciences</u>, a team of researchers concluded that a billion people are at risk from rising sea levels and that flood prevention actions should be undertaken (20). Their assessment of global flood damage examined projections from four different climate models, and projected the consequences without adaption to prevent flooding. They (20) found that, quoting, "without adaption, 0.2 – 4.6% of global population (20 million to 400 million) is expected to be flooded annually by the year 2100 under 25 – 123 centimeter (10 inches to 4 feet) of global mean sea-level rise, with expected annual losses of 0.3 – 9.3% of global domestic product (200 billion dollars to 6 trillion dollars)." (The annual losses in dollars are our calculation, assuming the 2100 global GDP remains at its present level of 65 trillion dollars, through it is likely to increase substantially, resulting in every greater economic losses.)

Adaptation to rising ocean levels is not new. The Dutch have done it very successfully for hundreds of years, starting around 900 AD. Their first dikes were low, about a meter high, to protect their growing crops (21). As the Dutch population grew, so did their dikes, both higher and more of them, to protect more and more land area. By 1250, most dikes were connected into a continuous defense (21). As the dikes grew, they also moved seawards, protecting more land. In the Middle Ages, the dikes relied on wooden timbers to reinforce them, but that changed when shipworms brought to the Netherlands by trading ships, ate the wood in the 1700's (21). Today's dikes have a core of sand covered by clay. Dikes subject to wave action have an outer layer of crushed rock below the waterline, and a layer of basalt stones or tarmac up to the high water line. Above that is grass and sheep. The Netherlands has 3,500 kilometers (2,100 miles) of dikes. Without them, roughly half of the country would be underwater, as illustrated in Figure 7.11 (21). Good as the dikes are, however, they will not protect the Netherlands from a major tsunami.

The Netherlands is not the only country vulnerable to rising sea levels, however. Countries all over the world are, including the United States. Studies (22) project that a 1 meter (3.3 feet) rise in sea level by 2100, a real possibility based on climate models, would put 3.7 million of the US population under water, more than 1% of our total population. Of particular concern is Florida. Studies by Frank Ackerman of the Stockholm Environment Institute (23) project that a sea level rise of just 2 feet would put 10% of Florida's land area underwater losing the homes of 1.5 million people. Quoting, "It's like our map of the area vulnerable to 27 inches of sea level rise looks like someone took a razor to the state right above Miami and sliced off everything below that".

Figure 7.11. The Netherlands Compared to Sea Level.

So, what can we conclude from this review of the flooding threats from storm surges, tsunamis, and rising sea levels?

Summarizing:

1. Large portions of the world's coasts are very vulnerable to major storm surges, tsunamis, and rising sea levels.

2. Over the next 50 years, the time span of many of those living today, many people will die and possibly trillions of dollars in damage will be lost due to storm surges and tsunamis if protective action is not taken.

3. The most damaging threats and highest waves come from tsunamis. Barriers built as protection against tsunamis will also protect against any possible storm surge and rising sea levels.

4. If based on conventional construction methods, the cost of building effective barriers against tsunamis, storm surges, and rising sea levels will be extremely high if large areas of the world's coasts are to be protected.

5. Ice barriers offer a practical, affordable way to protect large areas of the world against major tsunamis, storm surges and rising sea levels. Compared with the economic and human costs in lives lost and property damaged, the capital and operating costs of protective ice barriers will be far lower.

We now turn to the description of the proposed LISA barriers that would protect many of the world's coastal areas from storm surges, tsunamis, and rising sea levels.

We address the following issues:

- What is the design of the LISA ice barriers and where are they located on the coastline?
- How are they manufactured and how much do they cost?
- What is the required refrigeration power and cost to maintain them at temperature?
- Which coastal areas around the World would the ice barriers protect and what is the necessary investment?

First issue, design and location. Two types of coastal ice barriers are envisioned and illustrated in Figure 7.12, on-shore and off-shore. The on-shore barriers would be constructed on the existing soil or sand coastline, either at the water's edge as shown in Figure 7.12, or some distance back from the water's edge.

The off-shore ice barrier would be located at a distance from the actual water/land coastline, creating a water barrier between the existing coastline and the ice barrier, similar to the natural sand and soil barriers already present at many locations around the World. For example, the natural barrier island known as Fire Island in Long Island, New York, creates the Great South Bay between Fire Island and the principal land area of Long Island. While Fire Island does protect the Long Island from most storms, it has been breached and even washed over in very severe storms like the Great Hurricane of 1938 and Hurricane Sandy of 2012.

The distance between the coastline and off-shore ice barriers will depend on location. Factors that will affect the decision on distance include:

- How rapidly the ocean depth increases with off-shore distance
- Opposition from existing residents on or close to the coastline, to having an offshore barrier interfere with their ocean view.
- Environmental effects at locations with creation of the ocean Bay between the barrier and the coast.

Likewise the decision on whether to construct the barrier on-shore or off-shore will depend on the attitudes of local residents and the environmental effects.

Residents will have to assess the benefits of protecting their property, as well as their lives, from severe storms, tsunamis, and rising sea levels, versus the obstruction at some locations of their views of the ocean. Which is more important to residents living on the West Coast of the United States – surviving the tsunami after the Magnitude 9 earthquake when the Juan de Fuca Plate inevitably ruptures, or enjoying an unobstructed ocean view until it does rupture?

For both the on-shore and off-shore ice barriers, a height of the top surface above sea level of 20 meters (66 feet) seems reasonable. The storm surge during Hurricane Katrina (one of the worst on record) reached 28 feet, less than half of the proposed 66 feet height of the top surface. The Fukushima tsunami was 14 meters (46 feet) high. A 20 meter height would have protected the nuclear power plants and the surrounding towns and villages. The Indonesian tsunami had wave heights of up to 30 meters (100 feet). However, most of the affected regions would have been protected by a 20 meter high ice barrier.

We choose a nominal height of 20 meters above the sea surface, though it could be increased to 30 meters if further study warrants. Also, for certain locations near earthquake fault zones, it may be desirable to have heights greater than 20 meters. Heights of 20 to 30 meters should provide complete protection against tsunamis generated by earthquakes and most landslides. There remains the possibility of really big tsunamis, say 50 meters or more, from enormous landslides such as the Canary Islands, or a large asteroid impacting the ocean. Protection from such events does not appear practical, however. Fortunately, such events are extremely rare. It's 66 million years since the Chixulub asteroid wiped out the dinosaurs.

Figure 7.12 shows sketches of the on-shore and off-shore coastal ice barriers. The top surface height of the on-shore barrier is 20 meters, assuming that its bottom surface is at sea level. If built back from the shoreline on higher ground, its height could be less than 20 meters depending on how much its bottom surface is above sea level. For a total height of 20 meters and a width of 50 meters (166 feet), the total volume of ice would be 1000 cubic meters per meter of length, or 1 million cubic meters per kilometer (0.6 miles) of length. A 10 kilometer long on-shore ice barrier would have a capital cost on the order of 60 million dollars. Assuming that it protected 30,000 people, 3,000 per kilometer, from storm surges, tsunamis, and rising sea level, the investment per person would be only $2,000. Over an 80 year lifetime, that amounts to $25 annually, about 5 pounds of hamburger or 6 gallons of gasoline. This appears affordable.

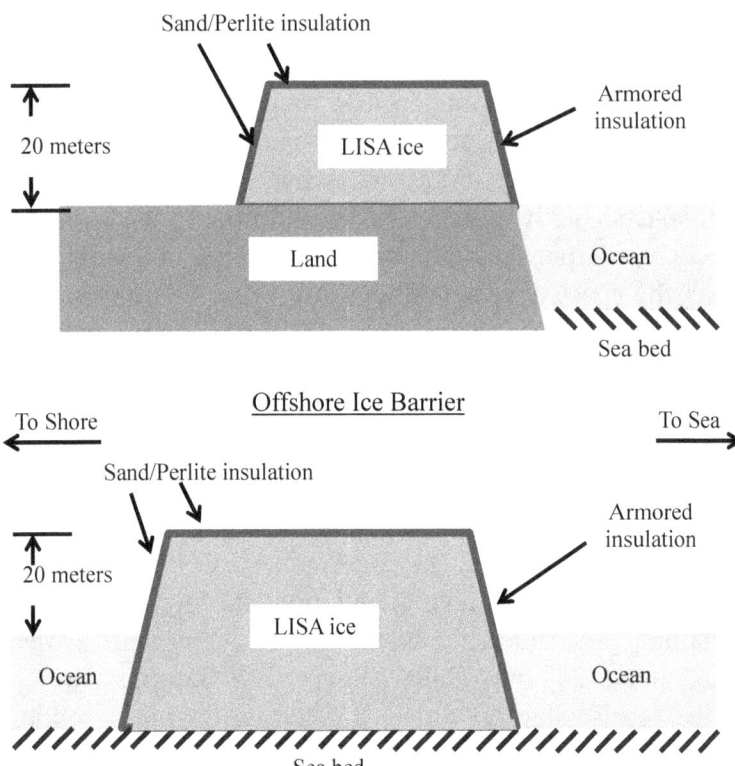

Figure 7.12
Features of On-Shore and Off-Shore Coastal Ice Barriers

On the order of 30 million people live in vulnerable areas along the 3000 kilometer long coast of the United States. On average, that is 10,000 people per kilometer. Protecting them would cost $8 per person annually, about 2 gallons of gasoline. Even more affordable. For the US as a whole, 123 million people, 39% of total US population, live in counties directly on the shoreline (24). The counties constitute less than 10% of US land area. An increase of 10 million people in the coastal counties is projected by 2020 AD. Not all of the population live directly on the coast, but even if they live some miles inland, they will be strongly affected if a disastrous storm surge or tsunami hits. The disruption, injuries and death of those persons directly impacted will affect them also.

World-wide, there are 620,000 kilometers of coastline with 2.4 billion people, one-third of the human population, living within 100 kilometers (60 miles) of the coastline (25). On average, that's 4,000 persons per kilometer of coastline. Many of the world's largest cities live directly live directly on the coasts, with much greater numbers of persons per kilometer. At particular risk are those living in low lying areas along the coast. A recent study (26) found that approximately 630 million people, about 1/10th of the world's population live in areas that are less than 10 meters (33 feet) above sea level. These people are especially vulnerable to storm surges, tsunamis and rising sea levels. The study found that the 10 countries with the most people in low coastal areas are China, India, Bangladesh, Vietnam, Indonesia, Japan, Egypt, United States, Thailand and the Philippines. Countries with the largest fraction of their population in low areas are the Bahamas, Suriname, Netherlands, Vietnam, Guyana, Bangladesh, Djibouti, Belize, Egypt, and Gambia. Two-thirds of the world's largest cities, defined as having 5 million or more people, are in low lying areas.

On-shore ice barriers should be acceptable at many locations around the world. However, they may not be acceptable at certain locations where the local coastal population wants to view the ocean, and doesn't want to have a 66 foot high ice barrier blocking their view, even though it protects them. Also, coastal city locations with harbors will probably want to continue having access for boats and ships, though some may be willing to establish off-shore ice island seaports like those described later. The off-shore seaports could be part of the coastal ice barrier constructed in sufficiently deep water, e.g., 30 meters or more, that ships could dock there. There are a lot of advantages to off-shore seaports – greatly reduced emissions of toxic gases and particulates from ship engines, greater safety to urban populations from accidents and terrorist attacks, and a quieter and cleaner environment.

Figure 7.12 shows a sketch of the off-shore coastal barrier. Its top surface would be a nominal 20 meters (66 feet) above sea level, same as the on-shore ice barrier, to ensure full protection from storm surges, tsunamis, and rising sea levels. Its bottom surface would rest on the seabed, at a nominal depth of 30 meters (100 feet), though it could be deeper or shallower, depending on local conditions. The practical limit to water depth is probably on the order of 60 meters (200 feet).

The off-shore ice barriers would probably have openings through which tidal water could flow in and out to minimize disturbance of the local ocean ecology in the waters between the barriers and the shoreline. These openings, which would occupy several percent of the barrier surface, could be closed during periods of high tide, to keep the sea water behind the barriers at a lower level than the outside sea water. For the projected sea level rises of a maximum on the order of 1 meter by 2100, this appears practical.

Ships going into or out of seaports and harbors protected by ice barriers could go through a lock system like those in various canals around the world, to prevent unwanted flows from storm surges, tsunamis, and rising sea levels. The distance between the off-shore ice barriers and the local shoreline will depend on location and how rapidly the sea water depth increases as one moves away from the shore. For some locations, it could be several miles, in others, a mile or less.

Off-shore ice barriers will be bigger in volume, with more surface area and greater refrigeration power, than on-shore ice barriers. As a result, they will cost more.

For an off-shore coastal ice barrier stationed at an ocean depth of 30 meters, with a width of 50 meters and a top surface height of 20 meters above sea level, total volume is 50x50 = 2,500 cubic meters per meter of length, or 2.5 million cubic meters per kilometer of length, 2.5 times greater than the on-shore barrier. A 10 kilometer long off-shore ice barrier would cost $100 million dollars if manufactured off shore where deep cold water at +5°C were available for the refrigeration cycle, and it was then towed to its operating location along the coastline. For 3,000 people per kilometer, and a projected 80 year lifetime, that would amount to $3200 per person total, or $41 per person per year. Still very affordable.

Table 7.2 summarizes the design features of on-shore and off-shore ice barriers. Both use a 50/50 mixture of sand/perlite insulation (Figure 7.12), as described in Chapter 6, for the top surface and inland side of the barrier and armored insulation for the side surface (S) subject to wave action. Both barriers rest on soil or sand on their bottom surface – dry land for the on-shore barrier and wet seabed for the off-shore barrier.

The side surfaces of the ice barriers are at an 80 degree angle relative to horizontal. This results in a wide base for the barrier, compared to the width of the top surface. For the 20 meter thick on-shore barrier, the top surface has a width of 46 meters, compared to a base width of 57 meters, with an average width of 50 meters. For the 50 meter thick off-shore barrier, the top surface is 41 meters wide, compared to a base width of 68 meters, with an average width of 50 meters.

The 80 degree angle for the side surfaces is nominal. Depending on location, construction method, etc., the side surface angle would probably be in the range of 60 to 90 degrees, e.g., vertical sides. The armored side insulation is much more expensive than the sand/perlite insulation, $153 per square meter of surface area, compared to $15 per square meter, as described in Chapter 6 and will dominate the capital cost of the insulation.

Table 7.2

Design Features of On-Shore and Off-Shore LISA Coastal Ice Barriers

Parameter	On-Shore Barrier	Off-Shore Barrier
Height of Top Surface Above Sea Level	20 meters	20 meters
Base Rests on Depth of Seabed	Dry Land	Sea Bed 30 meters
Total Height, Including Ocean Depth	20 meters	50 meters
Average Width	50 meters	50 meters
Total Volume/meter length	1000 m³	2500 m³
Angle of Side Surfaces	80 degrees	80 degrees
Top Surface Width	46 meters	41 meters
Top Surface Area/meter length	46m²/m	41m²/m
Bottom Surface Width	57 meters	68 meters
Bottom Surface Area (Sea Side)	57 m²/m	51 m²/m
Side Surface Area (Land Side)	20 m²/m	51 m²/m
Top Surface Insulation	50/50 Sand/Perlite	50/50 Sand/Perlite
Thickness of Top Surface Insulation	30 cm	30 cm
Side Surface Insulation (Land Side)	50/50 Sand/Perlite	50/50 Sand/Perlite
Thickness of Side Surface Insulation (Land Side)	30 cm	30 cm
Side Surface Insulation	Armored Insulation	Armored Insulation
Bottom Insulation	land	Sea Bed

Table 7.3 summarizes the capital costs of the insulation for the on-shore and off-shore coastal ice barriers, together with their refrigeration power and annual cost for a unit barrier length of 1 kilometer. The unit capital costs for the sand/perlite

and armored insulation, in dollars per square meter ($/m²), and the unit power requirements, watts per square meter (W(e)/m²), are described in Chapter 6.

The refrigeration cycle for the coastal ice barriers is assumed to operate between an ambient atmospheric/water temperature of +30°C (86 degrees Fahrenheit) and an ice interior temperature of −30°C (-22 degrees Fahrenheit). The COP (Coefficient of Performance for the refrigeration cycle is 2 watts (th)/watt (e). That is, to remove 2 watts of thermal power that leaks through the thermal insulation enclosing the +30°C ice interior, 1 watt of electric power to the refrigeration cycle is required. The 2 watts (th)/watt (e) COP is based on a refrigeration cycle that achieves 50% of the efficiency of a perfect Carnot cycle operating between +30°C and −30°C, a performance achievable with present commercial refrigeration equipment. If deep cold water were available at +5°C as a heat sink for the refrigeration cycle, as it would be for LISA ice islands operating in the deep ocean (Chapter 6), the COP would be greater, e.g., 3.5 watts (th)/watt(e), so that fewer electric watts would be needed to remove a watt of thermal leakage into LISA's ice interior at −30°C.

The thermal leakage through the insulation on the top and side surfaces of the on-shore and off-shore LISA ice barriers is constant and steady state assuming the outside ambient temperature stays constant at +30°C. At most locations of course, the ambient outside temperature will vary with the seasons and local weather conditions, being less in winter and more in summer. However, the effect on LISA's ice interior temperature will be very small even when the refrigeration cycle does not change its power level because of its very large thermal storage capacity. If, for example, the ambient temperature were to increase from +30°C to +40°C for 3 months, and the refrigeration cycle stayed constant in refrigeration power, the average temperature of the outer 2 meter thick ice layer next to the thermal insulation would only increase by 0.1 degrees centigrade. The refrigeration power for the top and side surfaces of the ice barrier can be sized to handle the average thermal input.

Table 7.3

Thermal Insulation Capital Cost and Refrigeration Power Requirements for LISA Coastal Ice Barriers

Basis: Refrigeration Cycle Operating Between +30^{0}C and - 30^{0}C, COP = 2 watts (th)/watt(e)

- $15/m^2 for 30 cm thick bagged Sand/Perlite Thermal Insulation
- $153/m2 for Armored Thermal Insulation
- 5 cents/KWH Power Cost
- 1 kilometer Length Ice Barrier
- 1 M$=1 Million dollars; 1 MW(e) = 1 Megawatt(e)

Parameter	On-Shore Barrier	Off-Shore Barrier
Top Surface		
Insulation Capital Cost	0.69 M$	0.61 M$
Refrigeration Power	0.39 MW(e)	0.34 MW(e)
Refrigeration Cost/year	0.17 M$/year	0.15 M$/year
Side Surface Facing Ocean		
Insulation Capital Cost	3.1 M$	7.8 M$
Refrigeration Power	0.28 MW(e)	0.70 MW(e)
Refrigeration Cost/year	0.12 M$/year	0.31 M$/year
Side Surface Facing Land		
Insulation Capital Cost	.30 M$	0.76 M$
Refrigeration Power	0.17 MW(e)	0.43 MW(e)
Refrigeration Cost/year	0.07 M$/year	0.19 M$/year
Bottom Surface		
Insulation Capital Cost	zero	zero
Refrigeration Power	0.14 MW(e)	0.16 MW(e)
Refrigeration Cost/year	0.06 M$/year	0.07 M$/year
Total Insulation Capital Cost	**4.1M$**	**9.3 M$**
Total Refrigeration Power	**1.0 MW(e)**	**1.6 MW(e)**
Total Refrigeration Cost/year	**0.44 M$/year**	**0.72 M$/year**

For both the on-shore and off-shore ice barriers, the bottom surface is not insulated and rests directly on land or the seabed. The analysis described below shows the temperature profile in the ground below the bottom surface of the ice barrier as a function of depth below the ground surface and time after the ice barrier is erected. The refrigeration power and costs shown in Table 7.3 are based on the time averaged thermal leakage over a period of 10,000 days (27 years). As discussed below, adding insulation on the bottom surface could reduce its refrigeration power and cost. However, they are already small compared to the refrigeration power and cost for the top and side surfaces. Further reduction probably is not worth it, since it would be offset by the increase in capital cost.

To be effective against the strong horizontal forces that act on them during storms and tsunamis, the ice barriers should rest land or on the sea bed. They should not allow any ocean flow through them. During non-storm conditions flow could be allowed using gates in off-shore barriers that would be closed during severe storms. This would also ensure non stagnant water conditions between the barrier and the coast that it protects. If protecting against rising ocean levels a limited amount of flow through the off shore barriers could be provided during low tide conditions.

Figure 7.13 shows two options for the bottom surface of the ice barrier structure. In Option 1, all of the bottom surface would directly rest on the sea bed, while in option 2, only a portion would rest on the sea bed. The other portion of the bottom surface would have a number of cavities, with the inside surfaces of the cavities lined with thermal insulation. The cavities would be pressurized with compressed gas, either air or CO_2, to prevent their filling with water or sediment from the sea floor.

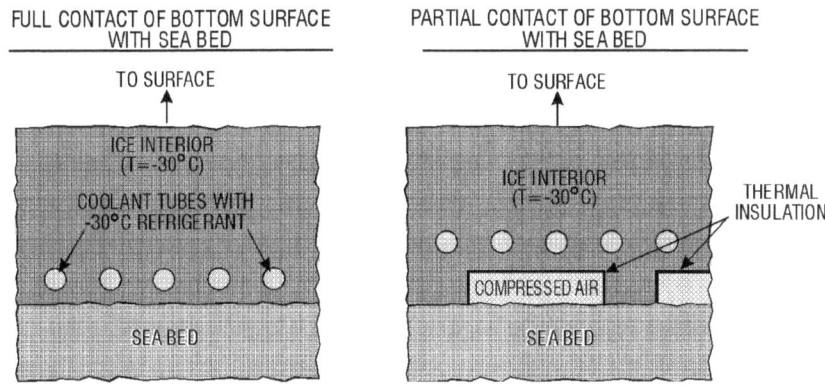

FULL CONTACT OF BOTTOM SURFACE WITH SEA BED

PARTIAL CONTACT OF BOTTOM SURFACE WITH SEA BED

TO SURFACE

TO SURFACE

ICE INTERIOR (T = -30° C)

COOLANT TUBES WITH -30° C REFRIGERANT

SEA BED

ICE INTERIOR (T = -30° C)

THERMAL INSULATION

COMPRESSED AIR

SEA BED

Figure 7.13. Potential Options for Insulation of Bottom Surfaces of LISA-2A Structures Resting on the Sea Bed.

Option 2 is more complex and higher in cost, but it would reduce the annual refrigeration cost compared to option 1. Using 50% coverage by cavities, for example, the refrigeration cost would be reduced by almost a factor of 2, compared to having the entire surface resting on the sea bed. With 50% of the surface on the sea bed, the ability to resist horizontal forces would still be very high. The thermal leakage into the sea bed from the -30°C ice structure above it will initially be high, but will decrease with time as the cold temperature wave moves downward into the sea bed. The time dependent temperature wave and the associated inside-the-sea bed thermal leakage can be determined using relationships developed in the Handbook of Heat Transfer by Rohsennow and Hartnett (p. 3-62, McGraw Hill, 1973 edition). The sea bed is treated as a semi-infinite solid slab with the temperature at the ice/sea bed interface suddenly reduced to -30°C, at t = 0. Figure 7.14 shows the time dependent temperature distribution as a function of depth in the sea bed below the ice/sea bed interface at times of 100 days, 1000 days and 10,000 days (27-4 years) after the LISA-2 structure is put in place.

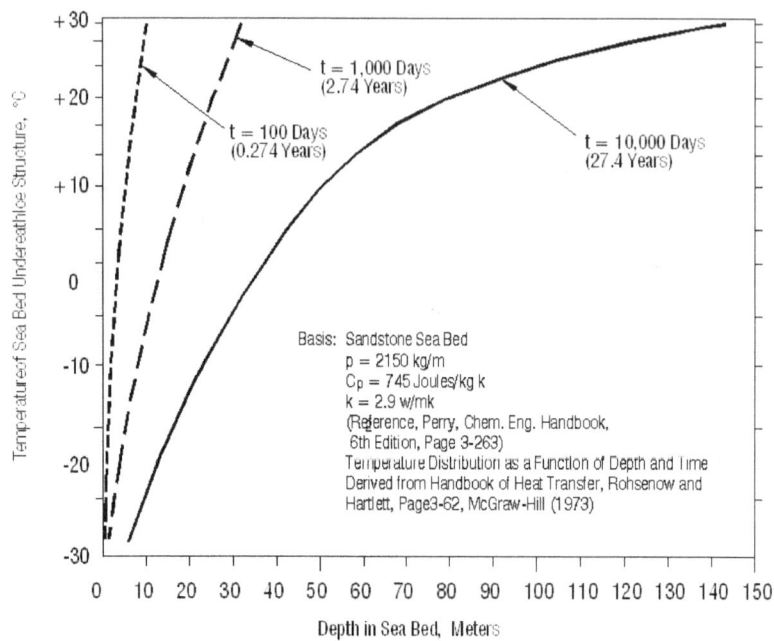

Figure 7.14. Temperature Inside A Sea Bed Underneath a Large Structure Resting on the Sea Floor as a Function of Depth and Time After Structure is Formed.

The properties of the sea bed are assumed to be the same as those for sandstone, with density = 2150 kg/m³, specific heat = 745 Joules/kg K, and thermal conductivity of 2.9 W/mK (Reference Perry, Chemical Engineers Handbook, 6th edition, p 3-263). As time goes on, the cold temperature wave penetrates more deeply into the sea bed. Table 7.4 shows the temperature at selected depths at times of 100, 1000, and 10,000 days. Also shown are the <u>average</u> thermal leakage rates into the sea bed from the -30°C ice interior over the three time intervals of 0 to 100 days, 0 to 1000 days, and 0 to 10,000 days.

For 100% of the ice structure resting directly on the sea bed (option 1), the average thermal flux into the sea bed is about 5 W/m² over a 10,000 day (27.4 years) operating period for the LISA-2 structure. With a COP (coefficient of performance) of 2 watts(th)/watt(e) for a refrigeration cycle operating between +30°C reject temperature and -30°C refrigeration temperature, the average annual refrigeration cost per square meter of bottom surface would be approximately $1.00/m², based on a cost of 5 cents per KWH for off-peak power. This assumes a cycle efficiency that is 50% of Carnot. If deep cold water at +5°C were available for heat rejection from an off shore service, the average annual refrigeration cost would be about $0.50/m².

For a 10 kilometer long ice barrier structure with a base width of 50 meters, the average annual refrigeration cost for the bottom surface of the ice barrier over 10,000 days would be only $500,000 at a unit cost of $1/m². This is very modest considering the protection that it would offer high value metropolitan areas like New York, New Orleans, Miami, etc.

Next question: are the ice barriers manufactured on-shore or off-shore, and if manufactured off-shore what are the towing costs? The on-shore ice barriers would be locally manufactured in place on the shoreline, using either distributed coolant tubes to freeze the ice as described in Chapter 6.

Off-shore ice barriers would probably be manufactured off-shore in the deep ocean and towed to their operating location, unless the location was in shallow water, e.g., less than 20 meters (66 feet) deep. In that case, the barriers would be manufactured locally similar to the on-shore ice barriers. The towing cost will depend on a number of factors, including towing speed, towing distance, length of the manufactured ice barrier, and number of ice barriers per trip of the towing tug. Based on a towing speed of 1 meter per second (2.25 mph) and a very conservative towing distance of 1000 kilometers (600 miles) from the deep ocean facility that manufactures the off-shore coastal barriers, the tow-time to the operational location would be 12 days. The return trip of the tug to the manufacturing facility would be at higher speed, e.g., 15 mph, and take only 2 days, for a round trip time of 14 days. Over 1 year the tug could easily make 20 round trips.

Table 7.4. Refrigeration Requirements for Bottom Surfaces of Ice Barrier Structures as a Function of Time After the Ice Structure is Formed.

Basis: Sandstone sea bed
ρ = 2150 kg/m³
C_p = 745 Joules/kg K
K = 2.9 W/mK

Temperature as a Function of Depth	Time After Ice Barrier Structure is Formed on Sea Bed: Full Contact		
	t^* = 100 Days	t^* = 1000 Days	t^* = 10,000 Days
T @ 5 meters	+8°C	-18°C	-28°C
T @ 10 meters	+28°C	-5°C	-22°C
T @ 20 meters	+30°C	+15°C	-12°C
T @ 40 meters	+30°C	+30°C	+2°C
T @ 80 meters	+30°C	+30°C	+20°C
$\bar{q}$ for t = 0 to t = t* interval, W/m²	46	15	5.0
	50% Contact Case (q through CO2 insulation = 0.4 W/m²)		

Nominal dimensions of a manufactured ice barrier unit to be towed for off-shore coastal application are:

- 50 meter height, with 30 meters below the ocean surface
- 50 meters wide, 1 kilometer long
- 50x30 = 1500 m² sub-surface frontal area when towed
- Drag coefficient = C_D = 0.4

The drag force on the towed ice barrier unit is

$$F_D = \frac{1}{2} C_D (A_F) \rho_w V^2$$

$$= \frac{1}{2}(0.4)(1500)(1000)(1)^2 = 3x10^5 \text{ Newtons}$$

The required propulsion power for the towed unit is

$$P_P = (F_D/\eta_P)(V) = (3 \times 10^5/0.6)(1) = 5 \times 10^5 \; watts$$

$$= 0.5 \; MW_{(e)} \; [670 \; shp] \text{ for propulsion efficiency} = 0.6$$

A tug towing four ice barrier units would require 4x0.5 $MW_{(e)}$ or 2 megawatts of propulsion power for the units, plus its own power. This is a very small propulsion power. The Emma Maersk container ship has a propulsion power of 100 megawatts, with a speed of 29 mph. Towing the four ice barrier units at a higher speed, say 2 meters per second (4.5 mph) would require more propulsion power, which scales with the cube of velocity. At 2 m/sec, propulsion power would be $(2) \times (2)^3 = 16$ megawatts. Perfectly feasible, but cutting the towing time from 12 days to 6 days does not seem that important. One tug making 20 round trips per year at 1 meter/sec, towing 4 units per trip, could deliver 80 units per year, enough for 80 kilometers of coast line.

The power cost per unit ice barrier delivered is

$$C_P = 0.5 \; MW \; (12 \; days)(24 \; hours/day)(\$/MW \; Hr)$$

at 10 cents per KWH ($100/MW Hr)

The propulsion power cost per unit is then

$$C_P = 0.5 \; x \; 12 \; x \; 24 \; x \; \$100 = \$1,440 \; per \; unit \; towed$$

A very small cost. The tug's capital cost per unit towed, plus the salaries of the crew are also very small. For an annual salary of $150,000 per crew member and 12 crew members (4 per shift), with 80 units towed per year, the crew cost per unit is

$$C_C = \$50 \; x \; 10^6/800 = \$62,500 \text{ per unit towed}$$

Total towing cost is $14,400 + $22,500 + $62,500 = $99,400, or about $100,000 per unit, very small compared with the 15 million dollar capital cost of the unit. If the towing parameters vary, e.g., a longer or shorter towing distance, a larger or smaller crew, or a higher or lower cost tug, it will make virtually no difference to the total cost of an ice barrier unit. There will be some minor additional costs for joining the towed units together when they reach their operational location to form a continuous coastal barrier. A separate tug and crew would move the delivered unit into place and freeze them together using coolant pipes. The cost of joining appears to be even less than the $100,000 towing cost and will not significantly affect the total cost of a unit.

What are the capital and operating costs for the ice barriers? The capital cost of the thermal insulation and the operating refrigeration cost are given previously in Table 7.3. Remaining are the capital cost for freezing the ice barrier and the operating, maintenance, and control costs.

The on-shore ice barriers could use distributed coolant tubes, as described in Chapter 6. They probably would use the ice/water slush freezing option, to reduce the power cost for freezing, particularly important at coastal locations where deep cold water at +5°C is not available. For the above conditions, the energy cost to freeze an on-shore ice barrier is $2.22 per cubic meter of ice (Chapter 6) at 5 cents per KWH. For a 20 meter high, 50 meter wide (1 kilometer long) ice barrier, the energy cost would be $(20)(50)(1000)(\$2.22) = 2.2$ million dollars. To this is added the cost of the distributed coolant tubes, which is on the order of 0.41 $/m^3$ of ice when left in place. For the 1 kilometer long on-shore unit, this amounts to 0.41 million dollars. The total capital cost for the 1 kilometer unit is then the sum of insulation capital cost (4.1 M$), freezing cost (2.2 M$), and coolant tubes (0.41 M$), equal to 6.7 million dollars.

The refrigeration operating power cost is 0.44 M$ per year. Added to this are the maintenance and control costs. Taking maintenance costs at 5% per year of capital cost, this amounts to 0.05x6.7 M$ = 0.33 M$ per year. This appears conservative since the ice barrier does not move and the internal coolant pipes are protected by the ice structure. They cannot corrode or be damaged. After the ice barrier is frozen, most of them are redundant. Since only a small fraction of the set is needed to maintain the ice at -30°C. If a tube fails, it can be simply shut off without increasing the ice temperature. The only damage that could require repair is local damage to the insulation from storm waves or some accident, e.g., a falling tree.

Control operating costs will be very low. For a 100 kilometer coastal on-shore ice barrier, with a conservative control budget of 10 million dollars annually, control costs would be only 0.1 million dollars per kilometer per year, making the total operating cost the sum of refrigeration cost (0.44 M$/year km), maintenance cost (0.34 M$/year km) and control cost (0.1 M$/year km) equal to 0.87 M$ per year per kilometer. A 100 km (960 mile) long on-shore coastal barrier would then have an annual cost of 88 million dollars, 1/1000th the damage cost from Hurricane Sandy. A 2500 kilometer long barrier along the US east coast would cost 2.2 billion dollars annually, 1/7000th of the US GDP of 15 trillion dollars per year.

Summarizing the capital cost and operating costs for on-shore coastal ice barriers that are manufactured locally, based on a 1 kilometer unit length

So, per kilometer of length, the on-shore ice barriers have a total capital cost of 6.7 million dollars and an annual operating cost of 0.87 million dollars. These costs see very acceptable in terms of the protection the ice barriers would provide against storm surges, tsunamis, and rising sea levels.

Capital Cost Component		Operating Cost Component	
Thermal insulation	4.1 M$/km	Refrigeration power (5 cents/KWH)	0.44 M$/km year
Freezing energy	2.2 M$/km	Maintenance (5% of capital cost/year)	0.33 M$/km year
Freezing equipment	0.41 M$/km	Control	0.10 M$/km year
Total capital cost	6.7 M$/km	Total operating cost	0.87 M$/km year

However, these already low costs can be further substantially reduced by a simple, environmentally attractive decision. The strength and toughness of ice can be greatly increased by incorporating additives, as described in Chapter 6. Adding on the order of 10% by volume of wood fibers or pulp creates "pykrete", doubling ice compressive strength and quadrupling its tensile strength. Rifle bullets can be fired at pykrete and it does not break or shatter. Similarly, dredged sediment, i.e., sand, etc., can be added to create "permacrete", strengthening and toughening ice.

We propose a new additive, garbage, to create "garcrete", with properties similar to pykrete and permacrete. Per capita, Americans dispose on the order of 3 kilograms per day of "garbage" consisting of residential and commercial waste, together with construction debris. That is about 1 million tons per day of garbage for the 310 million Americans, 365 million tons per year. Incorporating 10 percent by volume in the ice of coastal ice barriers, that corresponds to a length of 3600 kilometers of "garcretal" barrier that could be manufactured each year.

Why garcrete? First, as long as the ice barrier is maintained, it completely isolates garbage from the environment. Landfills, our current method, do not fully isolate garbage from the environment. Toxic substances leak into the ground water and aquifers, landfills collapse and wash away, and so on. Complete protection from the garbage is highly desirable.

Second, the cost of disposal in landfills can be very high. New York City, which generates 14 million tons of garbage per year (27), pays on the order of 100 dollars per ton to ship it to distant disposal sites. One kilometer of garcrete on-shore ice barrier that incorporates 10% by volume garbage would dispose of 100,000 tons of garbage. Per year that is 140 kilometers of on-shore ice barrier using New York City's garbage. At $100 per ton, that is 140 million dollars of garbage disposal cost avoided. Paying 10 million dollars towards the capital cost of 1 kilometer long ice barrier would make it virtually zero cost. Protecting the city at no cost!

There is a big caveat here, of course. A city or nation considering construction of coastal ice barriers with the use of garcrete need to ensure that power will be supplied to the structures for the indefinite future and proper maintenance is performed. Ice barriers are very resilient, compared to most infrastructure. The power requirements are modest and relatively inexpensive, and power can be disrupted for months before any serious damage is done to the structure. Better yet, the entire structure could easily generate its own power with the addition of solar panels or wind turbines to its top side. Furthermore, maintenance of the structure would be absolutely minimal.

We now turn to the capital and operating cost for off-shore ice barriers. The off-shore ice barrier is bigger in size and volume than the on-shore ice barrier, which increases the insulation capital and refrigeration costs. The capital cost for freezing the off-shore barrier will be about the same as for the on-shore barrier, even though its ice volume is 2.5 times greater. The lower cost per unit volume of ice for freezing off-shore offsets the larger volume.

Summarizing the capital and operating costs for the off-shore coastal barriers that are manufactured at a facility in the deep ocean with deep cold water at +5°C availability, based on 1 kilometer unit length for the barrier:

Capital Cost Component		Operating Cost Component	
Thermal insulation	9.3 M$/km	Refrigeration power (5 cents/KWH)	0.72 M$/km year
Freezing energy	2.1 M$/km	Maintenance (5%/year of capital cost)	0.60 M$/km year
Freezing equipment	0.15 M$/km	Control	0.10 M$/km year
Towing cost (1000 km)	0.1 M$/km		
Total capital cost	11.7 M$/km	Total operating cost	1.42 M$/km year

As with the on-shore coastal barriers, incorporating garbage in the ice will essentially make its capital cost zero, at a waste disposal cost of $100 per ton.

What coastal areas would ice barriers protect and what is the necessary investment? For the US, a large portion of the east and west coasts particularly these areas with substantial population, and virtually all of the Gulf Coast. Of special interest is the Pacific Northwest, which is vulnerable to a big tsunami when Cascadian Fault ruptures. For the world, major portions of its 672,000 kilometers of coastline are candidates for ice barriers – much of the coasts of Asian countries – Japan, China, Vietnam, Indonesia, Thailand, India, Malaysia, etc. Substantial areas in Europe, Africa, South America, North America, outside of the US and Australia. Also the many very vulnerable oceanic islands. Investment required? Assuming a 50/50 mix of on-shore and off-shore coastal ice barriers, the average capital cost is about 8 million dollars per kilometer. World annual GDP is 70 trillion dollars. Devoting 1/500th of that, 130 billion dollars per year, to coastal ice barriers would build 1600 kilometers per year, 160,000 kilometers in 10 years, 1/4th of the world's coastline. Seems affordable in view of the many trillions of dollars in damages and the deaths of millions of people that they would avert.

However, there are significant ecological challenges associated with creating unending coastal ice barriers. It is simply not wise to wall off an entire coast. Rivers need to flow to the ocean, estuaries need to be tidally flushed, marine life needs to migrate between coastal and deep ocean ecosystems. While it may be economically possible to wall off the entire Gulf Coast or Pacific North West, this obviously should not be done. At an aesthetic and ecological level, walling off every last inch of coastline would be horrific. Furthermore, it would not be in our own economic interests, nor even possible to build, build, build without considering what we are losing in the process. Natural ecosystems also provide tremendous economic benefit. One estimate of the economic value of so-called ecosystem services provided each year to humanity is $145 Trillion (Costanza et al., *Global Environmental Change,* v26. 2014). That is to say, that without plants recycling CO_2 and providing O_2, and wetlands filtering our water, and the myriad things we take for granted, we would have to pay $145 Trillion dollars per year to artificially reproduce all the benefits provided to us by the natural world.

Practically speaking, humans will need to find a balance between protecting the natural world and protecting their economic future and safety. Undoubtedly, some highly populated areas will need ice barriers in the years to come. New Orleans, is one obvious example. But it is our hope that planners will recognize the vast value provided by wild ecosystems not just on an aesthetic level, but also on an ecosystem services level.

What are the other benefits from coastal ice barriers? In general, the benefits will be mostly associated with off-shore coastal barriers, not on-shore barriers. The many potential benefits include:
- Installation of a high speed, low cost Maglev route along the east coast. Maglev vehicles can operate on the surface

of or a tunnel inside a continuous ice barrier along the coast.

- Location for off-shore wind turbines that generate large amounts of clean, non-greenhouse emitting, electric power.
- Creation of sub-surface artificial reefs that enhance plant and fish populations, and wild life parks on the top surfaces of the barriers.

Turning now to off-shore housing, and industrial applications, LISA off-shore islands hold great promise for low cost, healthier, better living conditions for urban and suburban populations, who today are crowded together into high population densities and very costly, barely affordable housing.

As cities grow in population, housing costs increase. Taking New York City as an example, the average apartment rental is more than $3000 per month (28). That's 3 times the average national rent of $1062 per month. San Francisco apartment rents lag New York City but are still high at $1998 per month. The sales price for apartments in New York City are pretty high. Elizabeth Harris in her July 8, 2013 *NY Times* story (29) on housing in New York City reported on the sale of a 408 square foot, 5th floor walkup apartment in Manhattan. Sale price? $595,000.

The average price of a Manhattan apartment in the Spring of 2013? 1.45 million dollars. The median price? $865,000. The highest price for a Manhattan apartment? Over 100 million dollars for a top-of-the-line penthouse apartment. The major reason for the high prices is a lack of supply. In the second quarter of 2013, there were only 4,795 listings. The population of Manhattan is 1.6 million people (30). There are 847,090 total housing units, 763,846 of which are occupied and 4,795 for sale listings, about 1/200th of the total units (31). No wonder many sellers are asking for all cash offers from the buyer and often get 100 or more people looking at the apartment. And New York City is wealthy. Per capita income was over $100,000 in 2005 for its 1.6 million inhabitants, with an estimated annual GDP of over 1.2 trillion dollars, 8 percent of total US GDP (30). The rich can afford the high cost of housing, but the not so rich cannot. They are moving out of Manhattan to try to find lower cost housing – Brooklyn, Queens, wherever. Trouble is, lots of people are also looking for apartments and homes there.

And high housing prices are not just hurting people in the boroughs of Manhattan, Brooklyn, Queens, Bronx, and Staten Island. People in the counties that surround New York City are also hurting. For rentals, 53 percent of the people in Nassau County on Long Island spend more than 30% of their gross income on housing (32). In Suffolk County, also on Long Island, it's 54%. The story is the same north of NY City. On the order of 55% of people pay more than 30% of gross income on rentals in Westchester, Orange, Greene, Ulster, Rockland, Putnam, and Monroe Counties (32). A majority of the population in the New York region spend more than 30% of their gross income before taxes, pensions, contributions, etc., on housing. That really hurts! And housing prices are rising – that will hurt even more.

So, $3,000 per month for a 5th floor, 408 square foot apartment in Manhattan. That's attractive? With 1.6 million people living on 59.5 square kilometers of land – that's a population density of 27,000 persons per square kilometer (30). It's hard to understand what that means in terms of personal space. To visualize how crowded that is, imagine all 1.6 million persons standing on 59.5 square kilometers of land. Each person has 59.5 million square meters of land divided by 1.6 million people, or 38 square meters (410 square feet) of personal space. In a square array of persons (Figure 7.15A) that's a distance of 20 feet to the 4 people nearest you, East, West, North and South. [The drawing of a person is by Leonardo DaVinci (33)].

A

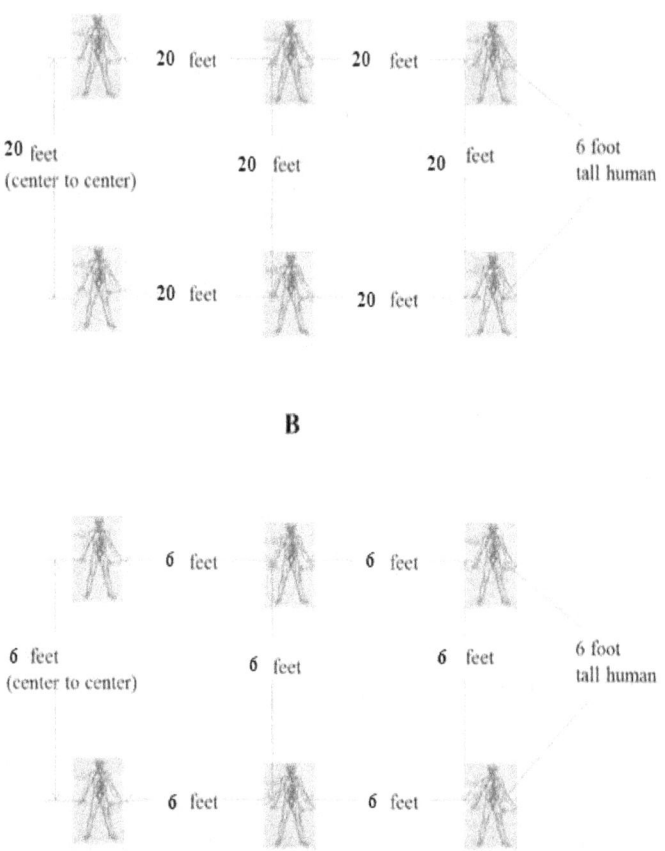

B

Figure 7.15 (A) Average Distance Between Inhabitants of Manhattan Island, United States (B) Average Distance Between Inhabitants of Central Mumbai, India.

Assuming that you are 6 feet tall, that's 3 body lengths to your nearest (and dearest?). During the day when workers come to town, and the night, when tourists and party goers come to the city, it gets even more crowded, and the distance to the 4 nearest would shrink to less than 20 feet.

However, New York City's population density is positively roomy compared to cities with even greater densities, where the person-to-person distance is much smaller. In India, Mumbai has a population of approximately 20 million people, with an average population density of 27,348 persons per square kilometer in Greater Mumbai (34), about the same as Manhattan. However, over 60% of Mumbai's population lives in high density slums. The Dharvai slum in central Mumbai is home to 800,000 to 1 million people in 2.39 square kilometers, with a minimum density of 335,000 people per square kilometer, 12 times the population density of Manhattan. The average distance between persons is reduced from the 20 feet in Manhattan, down to 6 feet – 1 body length. Your head is only inches away from your neighbor's feet (Figure 7.15B).

How would Manhattan residents feel about moving to an off-shore LISA island, with much less expensive apartment rentals, say $1000 per month instead of $3000 per month, with much lower population densities, say 1/4th of Manhattan's density – 7000 persons per square kilometer instead of 27,000, with 40 feet to the next person on average, not 20 feet? Not to mention much more attractive views of the coastline and ocean. To be fair, there are lots of wonderful places to see and visit in Manhattan, but the residents of an off-shore island could still go into Manhattan, taking only a few minutes for the trip, to work, dine, visit friends, and take in the sights they want to see. Imagine six off-shore islands, each 10 km^2 in area, the same total area as Manhattan, but with 1/4th of the population, just 400,000 people. 66,000 people live on each island, with population density of 6,600 persons per square kilometer, 1/4th of Manhattan's population density of 27,000 persons per km^2.

Before describing the shape and cost of the off-shore LISA islands, it is important to answer the question, why LISA ice islands? Why not just build artificial islands with landfill? The answer? Too expensive. Too vulnerable to erosion and settling, too vulnerable to storm surge, tsunamis, and rising sea levels, and too close to shore. Off-shore airports have been built using landfill. They are very expensive to construct, costing tens of billions of dollars, and are often troubled by settling of the landfill, as is the case with the Kansai off-shore airport in Japan. LISA off-shore islands with surface areas on the order of 10 km², the area proposed for off-shore habitats, are not subject to settling.

Dubai in the United Arab Emirates undertook in 2003 to construct a series of off-shore islands, the Palm Islands and the World Islands (Figure 7.16) using dredged sand and rock (35). The islands are very small, ranging from 0.014 to 0.04 km² in area. They were constructed from 321 million cubic meters of sand and 31 million tons of rock. Total surface area is well under 10 km², the area of 1 off-shore LISA island. Development costs were estimated at 14 billion dollars in 2005. The project was completed in 2008.

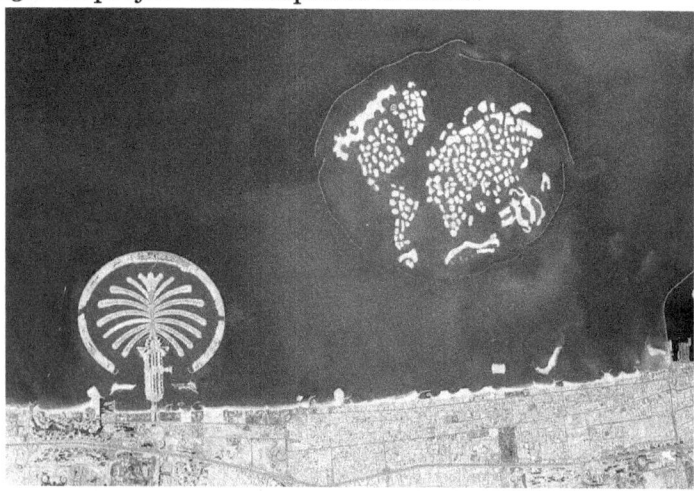

Figure 7.16. Artificial Archipelagos, Dubai, United Arab Emirates.

The project has had difficulties with erosion and the islands sinking back into the sea (35). As of 2011, only one of the islands had a building (a show home) on it and residential and commercial buildings were not being constructed on the other islands (35). Besides being very expensive and subject to erosion and settling, their elevation above sea level is constrained, rendering them vulnerable to strong storm surges, tsunamis, and rising sea levels. The top surfaces of LISA coastal ice barriers and the floating off-shore airports and seaports, are at least 20 meters (66 feet) above sea level, and not vulnerable to storm surges, tsunamis, and rising sea levels. Moreover, LISA off-shore islands can be located much further from shore than landfill artificial islands, which must be located in shallow water to minimize the amount and cost of landfill required. Since LISA off-shore islands can float, they can be anchored miles off-shore in deep water with no problems, making them acceptable to shoreline inhabitants since the off-shore islands won't interfere with their activities and ocean views.

What shapes are possible for off-shore LISA islands? A wide variety. Figure 7.17 shows four examples of possible shapes – square, circular, rectangular, and annular with a central seawater bay for aquatic activities. All have the same top surface area of 10 square kilometers.

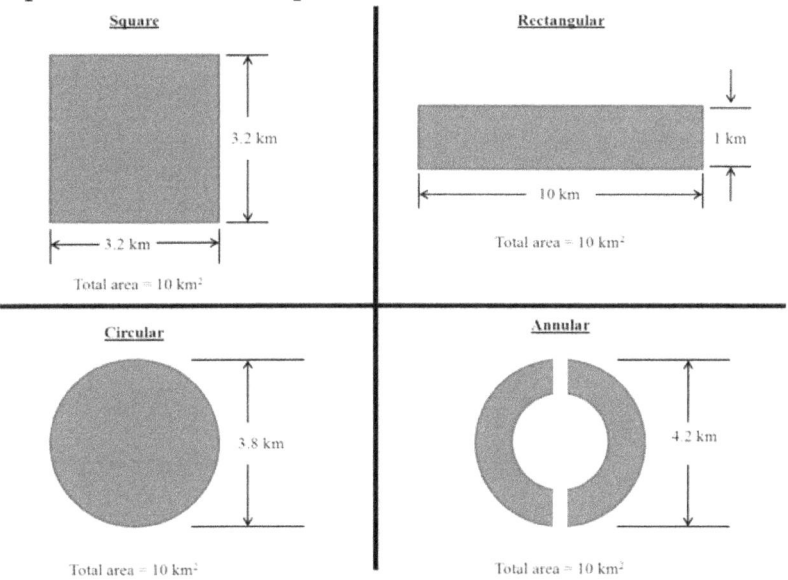

Figure 7.17. Types of LISA Off-Shore Islands.

The differences in capital and refrigeration costs between the four shapes are small as shown in Table 7.5. Each of the examples has 70,000 inhabitants, 7,000 per square kilometer, 1/4th of Manhattan population density of 27,000 persons per km².

Table 7.5. Capital and Operating Costs of Off-Shore LISA Floating Islands Habitats

Basis: 70,000 persons living on island (1/4 of Manhattan density)
10 square kilometer top and bottom surfaces
Off-shore, deep water @ +5°C available
30 cm bagged sand/perlite insulation @ top
30 cm air mattress insulation @ bottom
Armored insulation on side surface
-30°C ice interior, 3.49 COP, 5 cents/KWH

Type of Shape				
Parameter	Square	Circular	Rectangular	Annular
Dimensions	3.2 km	3.6 km	1x10 km	4.12 km (outer) 2.06 km (inner)
Side surface area (50 meter height)	0.63 km²	0.56 km²	1.1 km²	0.64 km² (outer) 0.32 km² (inner)
Capital Cost (M$) of Insulation				
Top @ $15.5/m²	155	155	155	155
Bottom @ $15/m²	150	150	150	150
Side @ $153/m²	97	86	168	102
Total Insulation	402	386	473	407
Freezing & Towing	450	450	450	450
Total Capital Cost (M$)	852	836	923	857
Annual Refrigeration Cost (M$/year)				
Top surface	12.3	12.3	12.3	12.3
Bottom surface	2.2	2.2	2.2	2.2
Side surface	1.3	1.1	2.2	1.7
Total refrigeration cost (M$/year)	15.8	15.6	16.7	16.2
Cost Per Inhabitant				
Capital	$12,200	$11,900	$13,200	$12,200
Refrigeration	$225/year	$223/year	$238/year	$231/year
30 year amortized	$407/year	$398/year	$440/year	$407/year

capital, $/year				

The capital cost of the LISA Island is in the range of $12,000 to $13,000 per inhabitant. The annual refrigeration cost per inhabitant is about $230. Amortizing the capital cost over a 30 year period corresponds to a cost per inhabitant of approximately $400 per year. Added together that's a total cost, amortized capital plus refrigeration, of only $600 per year – about 5 days of the average $3000 per month, $36,000 per year they now pay for a 400 square foot apartment in Manhattan.

Of course, this is only the cost of manufacturing and refrigerating the LISA off-shore island. It does not include the cost of constructing an apartment, roads, service facilities, etc. However, construction costs will be much cheaper in areas with lower population densities and fewer tall buildings. Average construction costs of new housing in the US are on the order of $100 per square foot, but considerably more in density populated areas like Manhattan. At $100 per square foot on a LISA island, a 600 square foot apartment would cost $60,000 – a real bargain for inhabitants with Manhattan's average income of over $100,000 per person (30).

Even if the capital and refrigeration costs of a LISA off-shore island are 2 or 3 times greater than the estimated values given here, living on a LISA island would be very affordable. At 2 times the projected amortized capital and refrigeration cost of $600 per year per inhabitant, it would be $1200 per year, 1 percent of annual income, or 10 days-worth of the present apartment cost. At 3 times the projected cost, it would be $1800 per year, less than 2% of annual income, and equivalent to 15 days of present apartment rental costs. The capital and refrigeration cost of a LISA off-shore island will not be an important factor in deciding whether to move there or not.

What will be the most attractive shape for an off-shore LISA island? Of the four examples shown, the rectangular and annular shapes appear especially attractive. On the rectangular LISA with the long, 10 kilometer, side running parallel to the coast line, many of its inhabitants will have waterfront views of the coast, if they live on the shore facing side, or on the ocean if they live on the ocean facing side. To get to either side from any point on the island will only be a short walk. The annular LISA island has a very attractive central lake of seawater, 2 kilometers in diameter, with beaches to swim in and docks for boats. There are two open passages to the outside ocean through which boats can travel into or out of the islands central lake. Again, it will be only a short walk from any point on the island to the view the outer ocean and coast, or to walk to the inner lake for a swim or boat ride.

Ten off-shore LISA islands with a capital cost of only about 9 billion dollars, could house almost half of Manhattan's present population, with a population density 1/4 that of today's Manhattan. New York City's 5 boroughs have an annual GDP of 1.2 trillion dollars (31). Over a 30 year period that is 36 trillion dollars, 4000 times the cost of 10 off-shore LISA islands. Seems affordable and a real bargain. Put in a different context, 9 billion dollars for 100 km² of off-shore LISA islands is less than construction cost of the 3 mile Long Island Railroad tunnel under the East River to the Grand Central Railroad Station.

There are many other density populated coastal cities around the world where off-shore LISA islands with lower population density would be attractive – San Francisco, Los Angeles, Shanghai, Hong Kong, London, Mumbai, to name just a few.

Fast, low cost access to and from the LISA islands could be provided by Maglev vehicles traveling in underwater ice tunnels. At speeds of 100 mph, a 10 mile trip would take only 6 minutes, a 20 mile trip only 12 minutes – much faster than subways and commuter rail lines.

Another very promising application for LISA ice structures is off-shore airports and seaports. There are many advantages for off-shore location – greater security and safety, lower pollution levels for people living in coastal cities and towns served by airports and seaports, less noise and distraction, reduced traffic congestion, etc. Floating off-shore LISA ice structures can provide highly effective, environmentally acceptable off-shore airports and seaports for coastal cities and towns all over the world, at much lower cost than using conventional construction approaches. Before describing the features and capabilities of off-shore LISA airports, we first address the following issues:

1. Why do we need new airports?
2. Why build new airports off-shore?
3. Where are existing and proposed off-shore airports?
4. Why off-shore LISA airports?

Why new airports? World population is growing. Now at 7 billion people. 9 billion is the projected population in 2050 AD. World GDP per capita is also growing and as living standards improve, people travel more. China and India, with 1/3rd of the world's population, are rapidly industrializing. As their economies grow, more and better transportation is needed for moving people and goods. China, for example, started a major program in 2008 to construct a national network of high speed rail lines. In only 6 years, China has built thousands of miles of new lines. Chinese air travel is also rapidly growing. Global Maglev (Chapter 2) will provide much of the future passenger travel around the world, but new airports will still be needed. In addition to providing new ports, off-shore LISA airports can replace existing ones that are unsafe, socially disruptive and environmentally harmful.

In 2012 airlines carried about 3 billion passengers and millions of tons of cargo. World annual airline revenues were 598 billion dollars (36), 1% of the world's annual GDP. Air passenger and cargo movements are increasing on the order of 5 percent per year. The world will definitely need new airports. In a recent (Jan. 26, 2014) article in the *Times of India* (37), S. Aiyar states, "As living standards rise towards western levels, India will need dozens of major new airports. These will face the hurdles of land scarcity, rising population, sky-high land prices and opposition to land acquisition. One way out in coastal locations is to build off-shore airports. This should become standard practice in India as is already the case in other countries like Japan."

Aiyar's article answers part of the second issue, why off-shore airports? Land scarcity, rising population, sky-high land prices and opposition to land acquisition are major reasons for locating airports off-shore. There are additional major reasons, however, Off-shore airports are much more secure, and much less vulnerable to terrorist attacks. They are also safer for the public, by reducing the chance of an airplane failure or accident causing deaths and damage to persons that live near a land-based airport. Off-shore airports also greatly reduce airplane noise and annoyance to people that live near airports. Many residents experience high noise levels that are distracting and wake them up at night. With off-shore airports, they could sleep peacefully. Also, there would be much less local emissions of pollutants from airplane engines. Finally, off-shore airports could reduce traffic congestion. With land-based airports in dense urban areas, travelers to and from the airport using cars, taxis, and buses converge on a small area, often causing traffic jams and delays on the highways that lead to it. With an off-shore airport that has multiple distributed points of access on the coastal region that it serves, congestion problems would be much less.

Where are existing and proposed off-shore airports? The answer, mostly in Asia. There a few proposed off-shore airports in other parts of the world, but their lower population density and less rapidly growing economies have not yet brought them to the point where they are ready to build. However, the much lower cost and greater capabilities of LISA off-shore airports should bring them to build off-shore. There are many locations outside of Asia that really would benefit from off-shore airports. On the order of one-third of the total world population live within 50 miles of a coastline (38). In the US, almost half of the population live within 50 miles of the coast many in dense populated cities. There is no shortage of potential locations for LISA off-shore airports. A few examples of existing off-shore airports are given below. They range from very large to small in size and daily traffic.

Hong Kong International Airport (HKIA) is the main airport for Hong Kong (39). It is located on Chek Lap Kok, an artificial island built island using reclaimed material from 9.83 km² of adjacent seabed. Figure 7.18 shows a view of the airport from the Ngong Ping 360 Cable Car (39). The statistics for the HKIA airport are very impressive. It handles 56 million passengers and 4 million tonnes of cargo annually, making it the 12th busiest airport in the world for passengers and the busiest for cargo. Its 2 parallel runways are 3800 meters long and 60 meters wide. Together they can handle more than 60 aircraft movements per hour. The airport cost 20 billion dollars to build, the world's most expensive air project, according to the Guinness Book of World Records.

The Chek Lap Kok off-shore airport replaced the old land-based Hong Kong Kai Tak airport, which was too small for the required passenger and cargo traffic, and was located in a densely populated urban area. One in every 3 flights experienced delays, and the noise levels in the surrounding urban area were extremely high (105 dB), affecting an estimated total of at least 340,000 people whenever an airplane landed or took off (39). HKIA airport opened in July 1998 after 6 years of construction. There were initial operating problems, but after 6 months, it achieved normal operation. Traffic keeps increasing, however. As a result, the Hong Kong government plans to build a 3d runway, adding 1606 acres of new reclaimed land. With the 3d runway, HKIA will be able to handle over 100 flights per day to meet a projected traffic in 2030 of 97 million passengers and 8.9 million tonnes of cargo per year. The estimated cost is 136.2 billion Hong Kong dollars (18 billion US dollars).

Japan has 5 off-shore airports, Kansai, Kobe, Kita Kyushu, Nagasaki, and Chubu. The Kansai off-shore airport (Figure 7.19) was an amazing construction project. The artificial island is 4 kilometers long and 2.5 kilometers wide. Three mountains were excavated for 21 million cubic meters of landfill (40). A

30 meter (100 feet) layer of earth was laid down over the sea floor.

Figure 7.18. Hong Kong International Airport (HKIA)

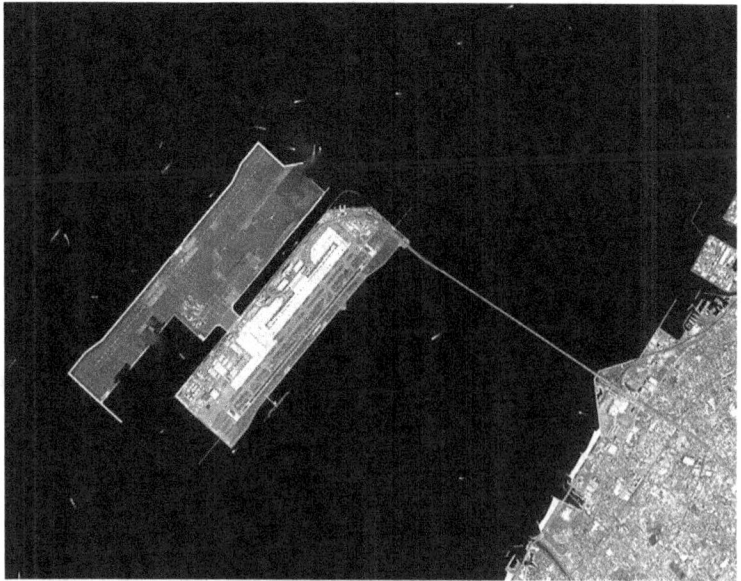

Figure 7.19. Satellite Image of Kansai International Airport in Osaka Bay, Japan.

The artificial island was predicted to sink by 5.7 meters due to the compression of the soft seabed silts by the weight of the landfill, but it sank considerably more, over 8 meters. It was still sinking in 2008, but at a much lower rate, 7 centimeters per year. Sinking is a significant concern for all off-shore artificial islands. Total construction cost for the Kansai airport was 20 billion dollars, including land reclamations, two runways, terminals and facilities.

Japan has very large air passenger traffic. In 2010, Japan's 50 largest airports had a total of 230 million passenger movements. Of the total, the two Tokyo inland airports, Haneda and Narita handled 95 million passengers (41). Of the remaining, 35 million passengers handled by the other 48 airports, 30 million were at the 5 off-shore airports, resulting in the off-shore airports taking 22% of the non-Tokyo air passenger movements in Japan – a significant fraction. Incheon International Airport is an off-shore artificial island airport that serves Seoul, the capital of South Korea (42). Built by land filling the sea that separated Yeongjong and Yongyu islands, it handled 41 million passengers in 2013, 40 percent of the total 100 million air passengers for all of South Korea. Incheon airport currently has three runways. It is being expanded to 5 runways, with a completion date of 2020 AD when it will be able to handle 100 million passengers annually. The airport is connected to the mainland by an expressway for cars and buses, along with ferry and rail service. A Maglev link to the airport is under construction. Together, Japan's five off-shore airports, the Incheon off-shore airport, and the Hong Kong off-shore airport handle 130 million passengers annually over 4 percent of the world's total of 3 billion passengers. With LISA off-shore airports along the east and west coasts of the US, plus India, new airports in Asia and Europe, the percentage of passenger movements handled by off-shore airports will grow substantially.

We turn now to 3 examples proposed off-shore airports, located off-shore from the coastal cities of San Diego in the US, Mumbai in India, and London in England. Over the years, there has been a lot of interest in an off-shore airport for San Diego (43). Figure 7.20 shows an illustration of the proposed airport which would be located 10 miles off-shore from Point Loma. It would be a floating airport anchored to the seabed. The proposed site is too far off-shore and the ocean too deep to build the airport on an artificial island using reclaimed land (43).

If built, it would be the world's first floating airport. The Ocean Works International Airport would have an area of 2000 acres (8.2 km²). Lindbergh Field, San Diego's present airport is 675 acres in extent and in a city of 3 million people. It has just 1 runway. It can't expand, and there is no land for a new airport. An off-shore airport is the only option. The projected cost of 20 billion dollars is probably too low. San Diego is a prime location for a LISA off-shore airport.

Figure 7.20. Illustration of San Diego Floating Off-Shore Airport

Other off-shore airports have been proposed. Marine Consulting Engineers Beckett Rankine has proposed to construct an airport at Goodwin Sands, 3 kilometers off-shore from Kent, to serve London (44). The estimated cost of the 4 runway airport is 39.2 billion pounds (65 billion US dollars).

London Heathrow airport is the 3rd busiest in the world with 72 million passengers per year in 2013. The only airports with greater passenger traffic are Atlanta-Hartsfield in the US (94 million passengers per year) and Beijing Capital in China (83 million per year).

The Indian city of Mumbai has 3 existing airports, but needs a 4th to handle future traffic. A land-based site has been proposed, but is very expensive and would cause severe disruption and displacement to local residents. As an alternative, an off-shore site at Madh Island, on land reclaimed from the sea is being considered (37). The cost of the land based site is estimated at 14,000 Crore (2.3 billion US dollars) considerably more than the off-shore airport, estimated at 6000 Crore (1 billion US dollars). In the future, as India industrializes, it will need dozens of new airports (37). It makes sense to build off-shore airports where possible to avoid the problems of land scarcity and expense, environmental impacts, and opposition from populations that would be adversely affected.

Fourth Issue: Why LISA off-shore airports? It can now be addressed by comparing them with existing off-shore airports. LISA off-shore airports would have the following major advantages:

1. Much cheaper, lower construction cost.
2. No sinking, settling, or movement of airport runways, terminals, and facilities.
3. Can be located off-shore in deep water and further from the coast than landfill airports.
4. Height of the airport surface above sea level is much greater than landfill airports, ensuring safety against severe storms and tsunamis.
5. Reduced environmental impact – does not require dredging and depositing many millions of cubic meters of landfill, disrupting habitats, and releasing toxic materials to the environment.

6. Reduced noise pollution and visual obstruction to urban population served by the off-shore airport, because of increased distance from the coastline.

Where would LISA airports be built? We look at 2 very promising US sites, off-shore from New York City and off-shore from Los Angeles. Both have very important and busy airports. The New York metropolitan region is served by 3 major airports: JFK, LaGuardia, and Newark. The airspace in the New York region is very congested, with frequent flight delays. These delays ripple across the US, causing an estimated 60 percent of the total flight delays in the country. A new off-shore LISA airport would considerably reduce airways congestion and flight delays.

Taking as an example of potential off-shore airports, we consider a LISA airport off-shore for Los Angeles. Figure 7.21 shows the layout of the present Los Angeles airport (LAX).

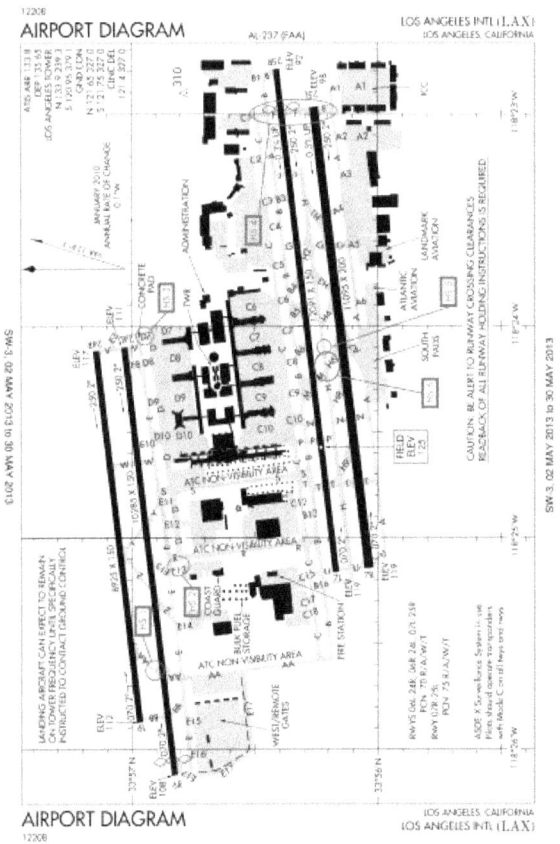

Figure 7.21. LAX Airport Diagram.

LAX has four parallel runways (Figure 7.21) oriented in the same direction, with lengths of 2720 meters, 3135 m, 3685 m, and 3382 m.(45) The two shorter runways are on one side of the airport, with the 2 longer runways on the other side. The distance between the outer runways on the left side of the airport from the outer runway is about 2 kilometers (2000 meters).

Taking the LAX layout as a guide lengthening the four runways to 3800 meters each, and making them parallel with their ends at the same starting and finishing line, the total length of the LISA airport would be 4 kilometers. The width of the LISA airport would be the same as LAX, i.e., 2 kilometers, for a total surface area of 8 square kilometers. LISA's terminals and facilities would be between the two runways on one side of the airport and the two runways on the other side, same as for LAX. Figure 7.22 shows this arrangement.

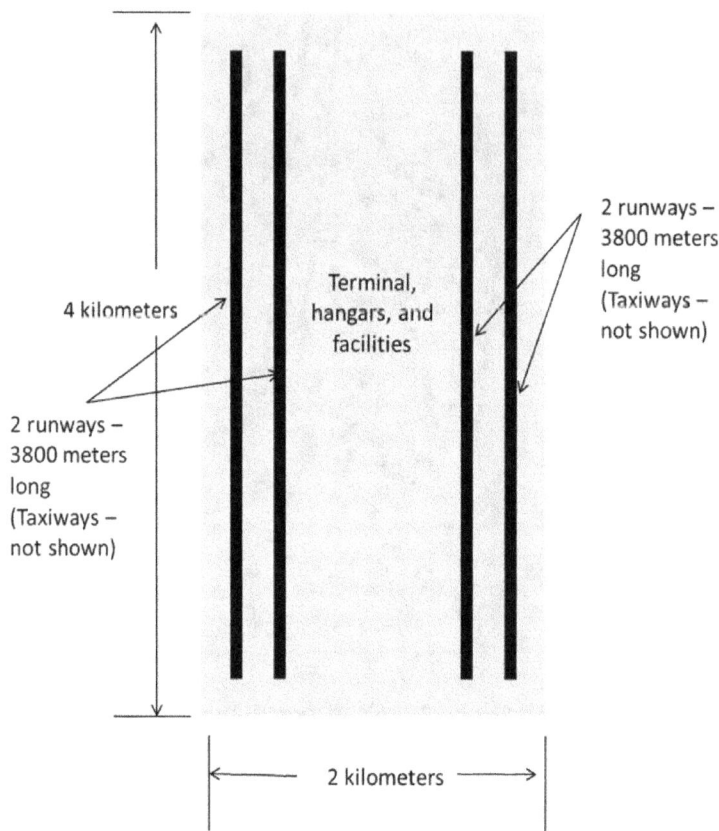

4 kilometers

Terminal, hangars, and facilities

2 runways – 3800 meters long (Taxiways – not shown)

2 runways – 3800 meters long (Taxiways – not shown)

2 kilometers

Figure 7.22. Layout of Floating LISA Airport. Basis: 8 km² of Ice Island Surface Area.

This parallel runway arrangement appears more efficient in the use of space than the 90 degree arrangement used by JFK, which may reflect constraints on land availability. However, there is no "one size fits all" model for off-shore LISA airports. The LISA airport structure would be manufactured as a number of smaller segments, on the order of 0.5 to 1 km² in area, and assembled to give whatever final shape is desired.

What is the cost of a LISA airport structure? For the example described above, a rectangular LISA airport 4 kilometers long and 2 kilometers wide, the cost can be estimated based on the values given in chapter 6. Deep cold water is assumed to be available, either at LISA's operating site, or through a relatively short pipe leading to water depths of 500 meters or more further off-shore.

Summarizing the capital costs for an 8 km² LISA airport

Component	Cost, Million Dollars
Insulation	
Top surface	124
Bottom surface	120
Side armored surface	92
Total insulation	336
Freezing and towing	340
Total capital cost	676 million dollars

The 676-million-dollar capital cost only covers the LISA structure, not the runway pavement, terminals, facilities, etc. Still it is a tiny fraction of the construction costs for artificial island off-shore airports which range from 20 billion dollars for already built airports, up to the proposed 65 billion dollars for a British airport at Goodwin Sands. Even if LISA costs double or triple the above projected 676 million dollars, it will still be a bargain compared to 20 billion or more. As described earlier, the refrigeration power cost is very reasonable, 13 million dollars annually, amounting to about 400 million dollars over a 30 year operating period – only a few cents per traveling passenger.

What R&D is needed? As with other potential LISA applications, R&D is needed to prove that they are practical in terms of construction, insulation, and refrigeration power. Adequate commercial refrigeration systems and insulation materials already exist. R&D can be carried out on a subscale, low cost basis to establish that the construction and insulation methods proposed here are practical.

For example, one could begin with a 50 meter (166 feet) wide coastal LISA ice barrier to test construction methods, distribution of coolant tubes, and comparing the pure water versus ice/water slush approaches. Relative rates of freezing and energy consumption could be measured and evaluated for the different test units to determine the best freezing approach.

Furthermore, individual test units do not have to be the full 1 kilometer length of a coastal ice barrier, just in the range of 100 to 200 meters long. They also do not have to be carried out in the ocean, but can be done in a fresh water lake.

As part of the R&D program, various types of insulation or different sections of a frozen unit could be tested and evaluated to determine their performance and relative simplicity of installation and operation.

The cost of such a subscale test program would be low, on the order of 10 to 20 million dollars. Following successful testing, the next stop would be scale up tests to approach operational size for the various LISA applications.

For the coastal ice barrier application, this would mean extending the length of the test unit to 1 kilometer. For the floating airport application, it would mean extending its length to 1 kilometer and its width to 400 meters. Assembling 20 such segments would produce an 8 km² LISA airport structure.

When can LISA airports begin operation? The answer depends on when the LISA R&D program would start and its level of funding. With an aggressive will funded program starting in 2015, LISA off-shore airports could be in operation ten years later, by 2025 AD.

We now turn to off-shore LISA seaports. There are two important advantages of off-shore seaports – greater safety and security, and reduced air pollution from ship engines. Terrorist attacks using explosive materials on a ship would do much less damage if the attacks occur at an off-shore LISA seaport tens of miles distant from a coastal city instead of in a city's harbor. Furthermore, the containers unloaded from ships at an off-shore seaport can be thoroughly inspected before they are transported to the mainland. Second, ship engines presently burn very dirty bunker fuel that emits toxic gases and microparticulates. A NOAA study estimates that 60,000 people world-wide die prematurely each year from pollutants and microparticulates emitted by ships (46). There is increasing pressure for ships to burn cleaner fuel when operating at seaports, and minimize their fuel consumption.

Where would off-shore seaports be built? The two busiest seaports on the west coast are the Port of Los Angles (47), the busiest in the US, and the adjoining Port of Long Beach (48), the 2nd busiest in the US. On the east coast, the Port of New York and New Jersey (49) is the 3rd busiest in the US. Statistics for the Ports of Los Angeles, Long Beach, and New York and New Jersey are summarized below.

Parameter	Los Angles	Long Beach	NY & NJ
# of ship berths	270	80	NA
Millions of metric tonnes thru port in 2010	158	78	32
Value of 2010 shipments, billion $	236	57	208
# ships/year (2010)	2182	NA	4811
Millions of TEU's per year (2010)	7.8	3.6	3.2

Development of the Global Maglev System (Chapter 2) will significantly reduce the amount of ocean shipping, because high value freight presently transported by ships will be transported much faster by Maglev at costs comparable to or less than by ship. However, there still will be ocean shipping for bulk cargo and TEUs to destinations not served by Global Maglev, and there will still will be a need for seaports.

Accordingly, Figure 7.23 shows a conceptual layout of a floating LISA seaport. It has four parallel ice islands, each 12 kilometers in length and 300 meters wide. The two central islands have 17 ship berths on each side of the island, for a total of 34 berths for each island. The berths are each 600 meters in length, long enough to handle ships as big as Emma Maersk, as an example, with its length of 397 meters – that's a gap of 200 meters between 2 Emma Maersks berthed along the ice island.

The two outer ice islands each have 16 berths on their 12 kilometer long inside edge, resulting in a total of 16+34+34+16 = 100 ship berths for the LISA seaport. The four ice islands are separated by three ocean passage ways, each 600 meters in width with Emma Maersk type container ships, each with a 56 meter beam width, berthed on each side of the ice islands, that still leaves a wide ocean water passage, 488 meters (1610 feet), for ships going into or out of the LISA seaport – over eight times their own width.

Ships enter the LISA seaport at one end, dock at a berth to load or unload cargo, and then leave through the other end of the seaport. The openings at the end are narrowed down to about 100 to 150 meters from the 600 meter wide passages between the ice islands, to minimize wave motions inside the seaport. The openings can be angled if necessary to further reduce any wave motion. The outer two ice island would have heights above sea level of 20 meters on their outer sides, but their inside edges would be lower, on the order of 10 meters above sea level, to better accommodate loading and unloading of cargo. Emma Maersk has a depth of 30 meters, deck edge to keel, resulting in a 14 meter height of the deck edge above sea level. The top surface of the two central ice islands would have also heights of 10 meters above sea level. Only the outer side surfaces of the two outer ice islands would require armored insulation against wave action. Their inner surfaces would use lower cost insulation, as would the two central ice island. The total top surface area of the four ice islands is approximately 4 x 0.3x12 = 14.4 square kilometers, compared to the 8 km² for the LISA airport shown in Figure 7.22.

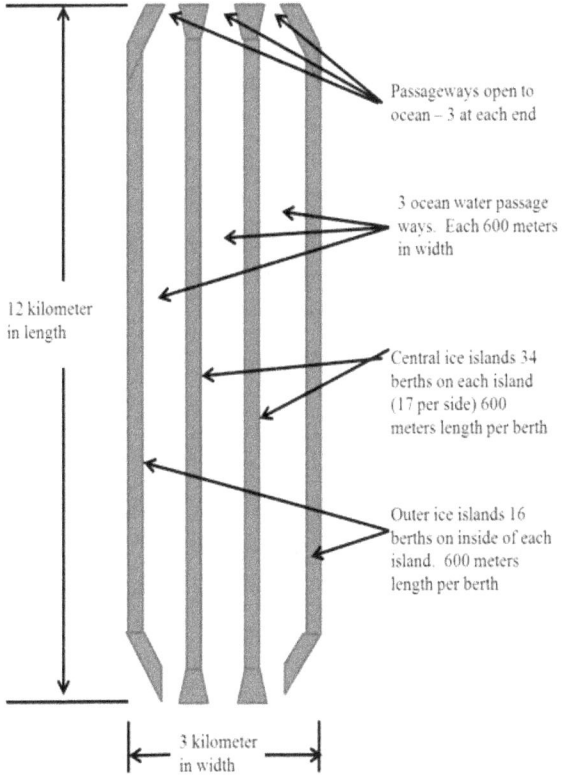

Passageways open to
ocean – 3 at each end

3 ocean water passage
ways. Each 600 meters
in width

12 kilometer
in length

Central ice islands 34
berths on each island
(17 per side) 600
meters length per berth

Outer ice islands 16
berths on inside of each
island. 600 meters
length per berth

3 kilometer
in width

Figure 7.23. Floating LISA Seaport. Basis: 100 total number of ship berths, 600 meter length per berth, 300 meter wide ice island, 14 km² total ice surface area.

The ice structures cost of the LISA airport was estimated as 676 million dollars. The cost of the LISA seaport ice structure with its greater surface area will be on the order of 1.5 billion dollars with additional costs for cargo handling facilities and a system that will transport cargo to and from land.

What would the LISA seaport ice structure cost be per ship docking at the seaport? Very little compared to the ships operating cost. At 3,000 ships per year – Los Angeles has 2182 ships per year – LISAs amortized capital cost of the ice structure per ship berthed would be 1.5 billion dollars divided by 30 x 3,000 = 90,000 ship berthings, equivalent to $16,000 per berthing. To put this cost in context, at full power the Emma Maersk burns 3,600 gallons of diesel fuel per hour (50) when transporting cargo across the ocean. At 3 dollars per gallon, this costs $10,800 per hour. The LISA amortized capital cost is then equivalent to 1 ½ hours of the fuel cost. A cargo ship traveling at a typical speed of 30 mph would take 150 hours for a 4500 mile trip, with a fuel cost 100 times greater than the LISA capital cost. Adding in the other operating and amortized capital costs for the cargo ship makes the LISA cost very small by comparison. LISAs ice structure capital cost will not be an important factor in deciding whether to build it.

On a final note, building off-shore LISA seaports will have major health benefits to populations that live near land based seaports. Ship, truck, and rail pollution due to the ports in southern California were the biggest source of air pollution there in 2006 (48). Regulators found that air pollution from the port causes 2000 cases of cancer per million population compared to an upper limit of 25 per million set by regulators, almost 100 times greater. Ships at the port emit 47 tons of nitrogen oxides daily, almost as much as the 350 largest factories and refineries in the region. Emissions are expected to grow by 70% by 2022 (48). Building off-shore LISA seaports will save many lives and improve people's health.

Conclusions: The Fork in the Road

James Powell and Gordon Danby

"When you come to a fork in the road, take it"
Yogi Berra

Yogi. You are absolutely right. Humanity is now at that Fork in the Road – do we continue massive consumption of fossil fuels or do we transition to other energy sources that do not pour carbon dioxide and other greenhouse gases into the atmosphere?

The Fossil Fuel Fork will lead to a collapse of modern civilization and environmental catastrophe – possibly even the extinction of humans. We've only begun to see the effects of climate change due to fossil fuel consumption – increasing global temperatures, stronger storms, rising sea levels and floods, much worse droughts and wildfires, melting glaciers, snowpacks, icecaps, and sea ice, acidifying oceans, and so on.

Things are going to get much worse, if we take the Fossil Fuel Fork. The damage and deaths from Hurricanes Sandy and Katrina will seem like small stuff compared to the bigger and more frequent devastating storms that will ensue. Millions of people around the World that live in low lying coastal regions will lose their homes and livelihoods as sea levels rise. As drought areas grow, food production will drop, further increasing the many millions of malnourished people in the World who now go hungry. As the oceans continue to acidify from the absorption of the increasing amount of carbon dioxide in the atmosphere, many oceanic species will not be able to form their shells, because the acidic ocean will dissolve them as fast as they form. Their extinction will destroy much of the oceanic food chains, turning the fertile seas into lifeless deserts.

At some point, even if humanity does stop consuming fossil fuels, the global warming process will become unstoppable. The Arctic permafrost is North America and Siberia is starting to emit carbon dioxide and methane from the organic carbon compounds stored in it, as it warms and the compounds oxidize. There is more carbon stored in the permafrost than there is carbon dioxide in the Earth's atmosphere.

Moreover, there are tremendous amounts of marginally stable methane hydrate material stored in the sea beds of Earth's oceans. Methane is a much more potent greenhouse gas than carbon dioxide. Oceanic methane is already starting to be released.

How much time until we reach the trigger point for unstoppable global warming from the realease of stored carbon in the permafrost and frozen oceanic methane hydrates? Nobody knows. We may have already passed it. If that happens, Earth will experience a mass extinction on the scale of the Permian-Triassic Extinction 250 million years ago, that wiped out more than 90 percent of Earth's species.

The trigger point and environmental catastrophe are inevitable if we take the Fossil Fuel Fork – even if we personally do not experience it, our children and grandchildren may. Babies born today in 2015, have a 90 year life expectancy, taking them to the end of the 21st Century. They are very likely to experience catastrophe if humanity continues down the Fossil Fuel Fork.

So, let's choose the other fork in the road, the Sustainable Life Fork, and transition from fossil fuels for energy to environmentally clean and sustainable energy sources and uses. That is the goal of this book, "*7 Big Projects for the World*". Each of the 7 chapters describes a practical way to help achieve the goal of clean sustainable energy sources and uses.

The proposed actions described here are critically important for a long-term sustainable future for humanity, but not the only ones, -- traveling the Sustainable Life Fork will require many other actions.

It is very important to realize that the 7 projects described have are practical to carry out – in fact, besides helping to prevent environmental catastrophe, they will have tremendous benefits for everybody with:

- More and better jobs
- Higher living standards and a better quality of life
- A stronger, more robust World economy
- Greatly reduced deaths, injuries, and damage to health and property from pollutions, accidents, storms, etc.
- New frontiers in space and the oceans

Implementing advanced 2nd generation Maglev networks in the US and Globally (chapters 1 and 2) will prevent millions of deaths and injuries from accidents on congested highways, greatly reduce health damage caused by pollution from cars and ships, substantially reduce travel cost and time for people and freight, and make travel much more comfortable and enjoyable for passengers.

With all of the benefits of Global Maglev – environmental, economic, social, and quality of life – it is difficult to understand why political leaders and investors are not already working to implement it. The answer of course, is the historic opposition from the vested interests in fossil fuels and existing transport technologies. If Global Maglev is to become a reality this opposition must be overcome.

Developing and implementing Maglev land and launch-to-space systems (Chapters 3 & 4), will also result in tremendous benefits for humanity – very low cost electric power beamed to Earth from space solar power satellites that is environmentally clean and sustainable – it will never run out. Other benefits are complete protection from asteroid and comet impact on Earth, much better and larger networks of satellites for Worldwide broadband internet and communications, much better weather prediction and much more detailed monitoring of crop yields, droughts, floods, wildfires, etc.

Low cost electric power beamed from space solar power satellites will also enable the production of synthetic gasoline, diesel and jet fuels, using carbon dioxide extracted from the atmosphere and hydrogen generated by electrolysis of water, the Big Project described in Chapter 5. With this capability to manufacture synthetic hydrocarbon fuel, those transport systems that are not electrified can be sustained indefinitely without needing to consume fossil fuels. When consumed for transport, the synthetic fuels release carbon dioxide to the atmosphere, which then in turn is extracted to make more synthetic fuel, with no net increase in atmospheric carbon dioxide.

Finally, the economic benefits and the greatly reduced deaths and injuries that large LISA ice structures will enable are tremendous. The economic benefits include prevention of trillions of dollars in property damage from storm surges, tsunamis, and rising sea levels (Chapter 7), production of very low cost electric power and fresh water using OTEC Cycles (Chapter 6), low cost mining of ocean resources and the creation of many new jobs.

The social benefits for humanity enabled by LISA structures will also be enormous. Millions of lives will be saved that would otherwise be lost to storm surges, tsunamis, and floods from rising sea levels. Millions of people will not be displaced and lose their homes due to rising sea levels (Chapter 7). Hazardous and polluting industries can be located on offshore LISA islands, greatly reducing the deaths and injuries from industrial accidents and pollution (Chapter 6). Low cost, less crowded, lower population density housing on offshore LISA islands adjacent to large, high expense coastal cities will be extremely attractive to millions of people (Chapter 7).

All of the 7 Big Projects described in this Book are technically and economically practical, and very attractive, both economically and socially. Nothing prevents them from starting development and implementation, other than opposition from a variety of vested industrial interests, e.g., fossil fuels, existing transport and space launch systems, etc.

The existing vested industrial interests want to continue along the Fossil Fuel Fork. To them, their near-term benefits are more important than the long-term sustainability of modern civilization and humanity. "Après moi, the deluge" – after me, the deluge.

The Sustainable Life Fork is there and waiting. All humanity needs is the will to take it.

Think of it in terms of the old story about the Dutch boy and the leaking dike. The dike had a small hole in it, through which the outside ocean was leaking. But as the water kept leaking, the hole was getting bigger. Soon, it would grow to the point where the dike would collapse, flooding the entire village. So the little Dutch boy stuck his finger in the dike, stopping the leak and saving his village. We need the equivalent of the Dutch boy as we transition from fossil fuel.

Gordon Danby

James Powell

References

Chapter 1

1. Department of Transportation

2. http://www.eia.gov/toos/faqs/fag.ctm?id=24 et+10

3. http://www.eia.gov/toos/faqs/fag.ctm?id=24 et+10

4. History of the Electric Car,
 en.wikipedia.org.wiki/history_of_the_electric_vehicle

Chapter 2

1. Projections' of Population Growth,
en.wikipedia.org/wiki/projections_of_population_growth

2. List of IMF Ranked Countries by Past and Projected GDP (nominal)
http;//en.wikipedia.org/wiki/list_of_countries_by_future GDP

3.International Energy Outlook 2013, US Energy Information
Administration, http://www.eia.gov/forecasts/ieo/electricity.cfm

4. OECD International Transport Forum,
http://www.uic.edu/ipsnews.net/news.asp?idnews=55943

5. Transport Outlook 2011 Meeting the Needs of 9 Billion People
OECD/ITF 2011

6. Global Greenhouse Gas Emissions Data,
http:www.epa.gov/climatechange/ghgemissions/global.html

7. World Energy Consumption,
en.wikipedia.org/wiki/world_energy_resoources_and_consumption

8. List of Countries by GDP(PPP) per Capita,
en.wikipedia.org/wiki/list_of_countries_byGDP_(PPP)_per_capita

9. List of Countries by Vehicles per Capita,
en.wikipedia.org/wiki/list_of_countries_by_vehicles_per_capita

10. List of countries by Past and Future Population,
en.wikipedia.org/wiki/list_ofcountries_ by past_and_future population

11. Gibraltar Bridge,
http://www.opacengineers.com/projects/gilbraltar

12. Strait of Gibraltar crossing, en.wikipedia.org/wiki
strait_of_gilbraltar_crossing

13. Channel Tunnel, en.wikipedia.org.wiki/channel_tunnel

14. Bering Strait crossing, en.wikipedia.org/wiki/bering _strait_crossing

15. "Czar Authorizes American Syndicate to Begin Work", New York
Times, August 2, 1906.

16. Trans Siberian Railway,
en.wikipedia.org/wiki/Trans_SIberian_Railway

17. Pan American Highway,
en.wikipedia.org/wiki/Pan_American_Highway

18. List of Continents by GDP(nominal), en.wikipedia.org/wiki/list_of_Continents_by_GDP_(nominal)

19. List of Asian Countries by GDP, en.wikipedia.org.wiki/list_of_Asian_countries_by_GDP

20. List of North American Countries By Population,
en.wikipedia.org.wiki/List_of_North_American_countries_by_population

21. List of Countries and dependencies by area, en.wikipedia.org.wiki/list_of countries_and_dependencies_by_area

22. List of North American Countries by GDP(PPP),
en.wikipedia.org/wiki/List_of_North_American_Countries_by_GDP_PPP

23. List of South American Countries by Population,
en.wikipedia.org/wiki/List_of_South_American_countries_by_Population

24. List of South American Countries by GDP(PPP),
en.wikipedia.org/wiki?List_of_South_American_countries_by_GDP_PPP

25. List of Countries by rail transport network size, en.wikipedia.org.wiki, List_of_countries_by_rail_transport_network_size

26. Projection of population Growth,
en.wikipedia.org/wiki/projection_of_population_growth

27. Rail Transport in South Africa,
en.wikipedia.org/wiki/rail_transport_in_South_Africa

28. Narrow Gauge Railways in Africa,
en.wikipedia.org/wiki/narrow_gauge_railways_in_Africa

29. List of African Countries by Population,
en.wikipedia.org/wiki/List_of_African_Countries_by_population

30. Silk Road, en.wikipedia.org/wiki/silk_road

31. Vasco da Gama, en.wikipedia.org/wiki/Vasco_da)Gama

32. Trans-Asian Railway, en.wikipedia.org/wiki/Iron_Silk_Road

33. Trans-Asian Railway – Indian Ocean Community,
http://sites.google.com/site/Indianocean
community1/trans_asia_railway

34. International Shipping and World Trade=Facts and Figures IMO
Library Services, February 2006

35.http://wiki.answers.com/Q/How_much_fuel_does_a_container_ship_burn

36. Big Polluters, one massive Container ship equals 50 million cars, Paul Evans, Http: www.gizmag.com/shipping-pollution/1526

37. US warns of pollution from merchant ships off the Florida Coast, www.guardian.co.UK/environment/2009/Mar/31/noaa-pollution-Florida_freighters_tankers_cruise ships.

Chapter3

1. StarTram, The New Race to Space, James Powell, George Maise, and Charles Pellegrino, Amazon.com

2. Apollo Program, en.wikipedia.org/wiki/Apollo_program

3. International Space Station, en.wikipedia.org/wiki/International_Space_Station

4. Furon Corporation, "Space Transportation Costs: Trends in price per pount to Orbit. 1990 to 2000", September 6, 2002

5. Saturn V, en.wikipedia.org/wiki/Saturn_V

6. Space Shuttle, en.wikipedia.org/wiki/space_shuttle

7. Space Launch Report 2013 Launch stats www.space launch report.com/log2013.html

8. The Case for Space Solar Power, John C. Mankins, Virginia Edition Publishing, LLC (2014)

9. List of Countries by electricity production, en.wikipedia.org/wiki/list_of_countries_by_electricity_production

10. International Energy Outlook 2013, www.eia.gov/forecast/ieo/electricity.cfm

11. 99942 Apophis, en.wikipedia.org/wiki/99942_Apophis

12. Asteroid Apophis 2036, "Astronomers Predict Asteroid Apophis to Hit in 2036, http://asteroid Apophis.com

13. Krakatoa, en.wikipedia.org/Wikipedia.org/wiki/Tanguska_Event

14. Angel, R. "Feasibility of Cooling the Earth with a Cloud of Small Spacecraft Near the Inner LaGrange Point," LZ, 17184-17189, PMAS, November 14, 2006/Vol 103?No.46

15. Meet Your Maker, Michael Page, New Scientist, Volume 223, Issue 2982, August 16, 2014

Chapter 4

1. Nicola Tesla, en.wikipedia.org/wiki/Nicola_Tesla
2. The Case for Space Solar Power, John C. Mankins, Virginia Edition Publishing (2014)
3. Space Based Solar Power, en.wikipedia.org/wiki/space_solar_power (2014)
4. Power From The Sun, Its Future, Peter E. Glaser, Science Magazine 62(3856), 857-861 (November 22, 1968)
5. StarTram: The New Race to Space, James Powell, George Maise, and Charles Pellegrino, Shoebox Press (2013)
6. List of Countries by Electricity Production, en.wikipedia.org/wiki/list_of_countries_by_electricity_productlon8.
7. International Energy Outlook 2013, EIA (US Energy Information Administration), http: www.eia.gov/forecasts/ieo/electricity.cfm
8. Cost of electricity by source, en.wikipedia.org/wiki/cost_of_electricity_by_source
9. List of IMF ranked countries by past and projected GDP(nominal), en.wikipedia.org/wiki/list_of_countries_by_ future_GDP

Chapter 5

1. Carbon Dioxide, en.wikipedia.org/wiki/Carbon_Dioxide
2. Carbon Dioxide From Gasoline and Diesel Fuels
http://www.eia.gov/facgs/fag.cfm?id=307Bpt=10
3. Total World CO2 Emissions, http://www.iea.org/newsroom and events/news/2012/may/name
4. US energy-related Carbon Dioxide Emissions by Sector, 2011, http://www.eia.gov/energy _in_brief/images/charts/energy_relations
5. Energy in the United States en.wikipedia.org/wiki/energy_in_the_United States
6. Short Term Energy Outlook, US Energy Information Administration, http://www.eia.gov/forecasts/steo/report/global
7. Transportation and Energy, Dr. Jean Paul Rodrique and Dr. Claude Contois, https//people.hofstra.edu/geotrans/eng/ch8en/conc:en?ch8c2en.html
8. List of IMF Ranked Countries by Past and Projected GDP (nominal) http;//en.wikipedia.org/wiki/list_of_countries_by_future GDP
9. http://www.eia.gov/toos/faqs/fag.ctm?id=24 et+10

10. History of the Electric Car,
 en.wikipedia.org.wiki/history_of_the_electric_vehicle

11. Electric Car, en.wikipedia.org./wiki/electric_car

12. List of Motor Vehicles Deaths in the US by year,
en.wikipedia.org/wiki/list_of_motor_vehicle_deaths_by_year

13. Lithium-USGS Mineral Resources Programs
 Minerals.usgs.gov/minerals/pubs/commodity/lithiums/mcs—
2010_lithi.pdf

14. World Lithium Supply, Eric Eason,
https//large.Stanford.edu/courses/2010/ph240/eason2/

15. Fischer-Tropsch process-
en.wikipedia.org/wiki/Fischer_Tropsch_Process

16. Synthetic Fuel, en.wikipedia.org/wiki/synthetic_gasoline

17. Water-gas shift reaction,
en.wikipedia.org/wiki/water_gas_shift_reaction

18. Mississippi River, en.wikipedia.org/wiki/Mississippi_river

19. Carbon Dioxide Removal,
en.wikipedia.org/wiki/carbon_dioxide_removal

20. Technology Developed for Extracting Carbon Dioxide from
Air,efficiently Maria Reyes:
 http://www.green optimist.com/2012/07/25/technology_developed_

21. 400 ppm: can artificial trees help pull CO2 from the Air?, David Biello,
http://www.scientific American.com/article/prospects-for-direct-
air...(May 16, 2013

22. Sucking CO2 from the skies with artificial trees, Gala Vinee, (October
4, 2012), http://www.bbc.com/future/story/c0121004-fake-trees-
to-clean-...

23. Direct Removal of Carbon Dioxide From Air Likely Not Viable, Report
suggests, Science Daily (May 9, 2011),
http://www.sciencedaily.com/releases/2011/05/110509114200.h
tm

24. Technical Specs of Common Wind Turbine Models(AWEO.org),
http:awes.org/windmodels.html

25. Statue of Liberty, en.wikipedia.org/wiki/statue_of_liberty

26. Wind Power, en.wikipedia.org/wiki/wind_power

27. Highest and cheapest Gas Prices by Country, Mark J Perry (May 15,2012) Wall Street pit.com/02107_highest_and_cheapest_gas_prices_...

Introducing LISA References

1. Fear of Juan de Fuca MegaThrust Earth Quake, http://www.live science.com/3775_tsunami_generating_earthquake

2. Coastal Flood Damage and Adaptation Costs Under 21st Century Sea—level rise. Jocken Hinkel, et al, www.pnas.org/cgi/doi/10.1073/pnas.1222469111

3. Rising Sea levels Seen As Threat To Coastal US, Justin Gillis, http://nytimes.com/2012/03/14/science/earth/study-rising...

4. Florida Faces Drastic Change From Sea Level Rise, Greg Allen, http://www.npr.org/templates/story/story.php?storyID=120498442

5. Pobeda Ice Island, en.wikipedia.org/wiki/pobeda_ice_island

6. RMS Titanic, en.wikipedia.org.wiki/RMS_Titanic

7. Ice Hotel, en.wikipedia.org.wiki/Ice-Hotel

8. Ice Hotel (Quebec), en.wikipedia.org/wiki/Ice_Hotel_(Quebec)

9. Best Ice Hotels in the World – The World Roamer, http://www.the worldroamer.com/best-ice-hotels-in-the-world

10. Harbin International Ice and Snow Sculpture Festival, en.wikipedia.org/wiki/Harbin_International_Ice_and_Snow_Festival

11. Incredible Pictures of Harbin Ice Festival, http:/edweb.tusd.kiz.az.us/sandre/harbin_ice.htm

12. Images of Harbin Ice Festival, bing.com/images

13. http:/en.wikipedia.org/wiki/File:Tower_at_Harbin_Ice_And_Snow_Festival_2012.jpg, Author, Shanghai Killer Whale, 5 January, 2013

14. http://en.wikipedia.org/wiki/File:Harbin_Ice_Festival.jg, Author, LiYan at zh Wikipedia(zh.wikipedia.org) 2003-08-09

15. Ice, The Nature, the history, and the uses of an Astonishing Substance, Marianna Gosnell, University of Chicago Press (2005)

16. Vattensagspan, en.wikipedia.org/wiki/File:Vattensagspan2.jpg, Author, Kr-Val, 8 May 2010

17. Image ODIO, en.wikipedia.org/wiki/File:Image_0010.jpg, Author, Charles Nichols.

Chapter 6

1. Manhattan en.wikipedia.org/wiki/Manhattan

2. Seasteading, en.wikipedia.org/wiki/seasteading.

3. Ocean Thermal Energy Conversion, Luis A. Vega, Encyclopedia of Sustainability Science and Technology, Springer, August 2012, pp 7296-7328, hinmrec.hnei.hawaii.edu/wp...Article-OTEC-by-Vega-Aug-2012.pdf.

4. First Generation 50 MW OTEC Plantship for the Production of Electricity and Desalinated Water, Luis A. Vega and Dominic Michaelis, OTC 20957, 2010 Off-shore Technology Conference, Houston, Texas, USA, 3-6 May 2010.

5. OTEC Overview, L.A. Vega, http://www.otecnews.org/portal/otec-articles/ocean-thermal-energy.conversion.

6. Ocean Thermal Energy Conversion, en.wikipedia.org/wiki/OTEC.

7. International Energy Outlook 2013, US Energy Information Administration, http://www.eia.gov/forecasts/ieo/electricity.cfm.

Chapter 7

1. 1900 Galveston Hurricane, en.wikipedia.org/wiki/1900_Galveston_Hurricane

2. Storm Surge Overview, http://www.nhe.noaa.gov/surge

3. 1938 New England Hurricane, en.wikipedia.org/wiki/1938_Hurricane

4. Hurricane Katrina, en.wikipedia.org/wiki/Hurricane_Katrina

5. Hurricane Sandy, en.wikipedia.org/wiki/Hurricane_Katrina

6. Tracking Storm Sandy Recovery, [http://live.reuters.com/Event/Tracking_Storm_Sandy/54277687]

7. Typhoon Haiyan, en.wikipedia.org/wiki/typhoon_Haiyan

8. Tsunami, en.wikipedia.org/wiki/Tsunami_wave_train

9. List of Historic Tsunamis, en.wikipedia.org/wiki/Historic_tsunami#Deadliest

10. 1755 Lisbon Earthquake, en.wikipedia.org/wiki/1755_Lisbon_earthquake

11. Megatsunami, en.wikipedia.org/wiki/Megatsunami

12. 99942.Apophis, http://en.wikipedia.org/wiki/99942_Apophis

13. 2004 Indian Ocean earthquake and tsunami, en.wikipedia.org/wiki/Indian_Ocean_eartghquake_a...

14. Fukushima Daiichi Nuclear Disaster, en.wikipedia.org/wiki/Fukushima_daiichi_nuclear_disaster

15. Cost and Consequences of the Fukushima Daiichi Nuclear Disaster, http://truth-out.org/news/item12832-costs-and -consequences-of...

16. Juan de Fuca Plate, http://en.wikipedia.org/wiki/Juan -de-Fuca_Plate

17. Tsunami Generating Earthquake Near US Possibly Imminent, Robin Lloyd, http://www.livescience.com/3775-tsunami-generating-earthquake

18. Fear of Juan de Fuca Mega Thrust Earthquake, http:///the watchers.adorraeli.com/2011/04/26/fear-of-juan-de-fuca

19. Did North American Quake Cause 1700 Japanese Tsunami?, Stefan Lovgren, http://news.nationalgeographic.com/news/pf/16098991.html

20. Coastal Flood Damage and Adaptation Costs Under 21st Century Sea-Level rise, Jochen Hinkel, etal, www.pnas.org/cgi/doi/10.1073/pnas.1222469111

21. Flood Control in the Netherlands, http:// en.wikipedia.org/wiki/Flood_Control_in_the_Netherlands

22. Rising Sea Levels Seen as Threat To Coastal US, Justin Gillis, http://www.nytimes.com/2012/03/14/science/earth/study-rising...

23. Florida Faces Drastic Change from Sea Level Rise, Greg Allen, http://www.npr.org/templates/story/story.php?storyId=120498442

24. What Percentage of the American Population Lives Near the Coast?, http://oceanservice.noaa.gov/facts/population.html

25. Living Ocean – NASA Science, http://science.nasa.gov/earth-science/oceanography/living-ocean/

26. Study: 634 Million People at Risk from Rising Seas, http://www.npr.org/templates/story/story.php?storyId=9162438

27. Carol Kellermann: Cleaning Up NY's Garbage Disposal, http://www.crainsnewyork.com/article/20120603/SUB/306039993

28. City Renters Rocked as Average Hits $3000, New York Post, Jennifer Gould Keil, July 11, 2013.

29. Amid Housing Scarcity, Many Buyers Are Going Home Empty Handed, Elizabeth Harris, NY Time, July 8, 2013.

30. Manhattan, en.wikipedia.org/wiki/Manhattan.

31. Table PL-H1 NYC, Total Housing Units and Vacancy Status New York City and Boroughs 2010, US Census Bureau 2010 Census Public Law 94-171 Redistricting File Population Division – New York City Department of City Planning (March 2011).

32. Housing Gap, Ken Schachter and Maura McDermott, Newsday, March 11, 2013.

33. En.wikipedia.org/wiki/Anatomical_Male_Figure.
34. Mumbai Population 2014 – World Population Review, http://world populationreview.com/world-cities/Mumbai-population/
35. The World (archipelago), en.wikipedia.org/wiki/The_World_(archipelago).
36. Annual Passenger Total Approaches 3 Billion According to JCAO, International Civil Aviation Organization, http://www.cao.int/Newsroom/Pages/annual-passenger-total-ap...
37. The Case for Offshore Airpots, SAiyar, http:// economictimes.indiatimes.com/opinions/comments/analysis...
38. Study: 634 million people at Risk from Rising Seas, http://www.npr.org/templates/story/story.php? Story Id = 9162438
39. Hong Kong International Airport, en.wikipedia.org/wiki/Hong_Kong_International_Airport
40. Kansai International Airport in Japan, en.wikipedia.org/wiki/list_of_the_busiest_airports_in_Japan
41. List of the Busiest Airports in Japan, en.wikipedia.org/wiki/list_of_the_busiest_airports_in_Japan
42. Inchon International Airport, en.wikipedia.org/wiki/Inchon_International_Airport
43. Ocean Works International: San Diego Offshore Airport?, http://www.fastcompany.com/1418997/oceanworks -International...
44. UK airport debate continues, http://world architecture news.com/index.php?fuseaction=...
45. Los Angeles International Airport, en.wikipedia.org/wiki/Los_Angeles_International_Airport
46. US warns of pollution from merchant ships off the Florida Coast,www.guardian.co.UK/environment/2009/Mar/31/noaa-pollution-Florida-freighters-tankers-cruise ships
47. Port of Los Angeles, en.wikipedia.org/wiki/Port_of_Los_Angeles
48. Port of Long Beach, en.wikipedia.org/wiki/Port_of_Long_Beach
49. Port of New York and New Jersey, en.wikipedia.org/wiki/Port_of_New_York_and _New_Jersey
50. The Emma Maersk, en.wikipedia.org/wiki/Emma_Maersk

Figure Credits

Preface

Figure 1 Louis XV-Rigaud1,

http://commons.wikipedia.org/wiki/file:LouisXV_Rigaud.jpg,
author Hyacinthe Rigaud

Figure 2 Anonymous – Prise de la Bastille,
http:/commons.wikipedia.org/wiki/file:anonymous_prise_de_la_bastille.
jpg, author, unknown

Prologue

Figure 1. US National Maglev Network, Maglev America, Author, Powell
and Jordan

Figure 2. Pan American Highway, http:
commons.wikimedia.org/wiki/file:Pan American Hwy.png, author,
en.user.seeweege

Figure 3. View of StarTram existing launch Tube, Author John Rather

Figure 4. Sun Tower Solar Power Satellite,

http://en.wikipedia.org/wiki/File:SunTower.jpg, author is NASA

Figure 5. George Schlegal-George Degen- New York 1873_,
http:/commons.wikipedia.org/wiki:file:
George_Schlegal_George_Degan_New_York_1873, author is George
Schlegal (artist), George Degen (publisher), Adam Cuerden (restoration)

Figure 6. Twin Towers_NYC, http://commons.wikimedia.org/wiki/file:
twin towers-NYC.jpg, Author, Carol M. Highsmith

Figure 7. Cumulative Height of Fossil Fuels Consumed by World If they
were laid on Manhattan Island, Author, Powell

Figure 8. Atmospheric CO2 Concentration from Fossil Fuel Consumption
as a Function of Time for 3 Rates of World Consumption, Author, Powell

Figure 9. Years in which Fossil Fuel Reserves are Exhausted for 3 Different
Consumption scenarios, Author, Powell

Figure 10 NOAA Katrina NOLA 17th Street Breach, http://commons.
Wikimedia.org/wiki/File:NOAA_Katrina_NOLA_17th_Street_Breach_Aug_
31_2005.jpg

Figure 11 Wikipedia Pyramids and Sphinx in Egypt

Figure 12 en.wikipedia.org/wiki/Tlipamina Cohur Gorgo

Figure 13 Great Wall of China

Figure 14 Watercolor of Erie Canal by John William Hill, 1829

Figure 3.12. Apophis Pass-by of Earth in 2029 Far View, en.wikipedia.org/wiki/file:Apophis_pass.svg, Author, Marco Polo

Figure 3.13. Apophis Pass by of Earth in 2029-Close In View, en.wikipedia.org/wiki/File: Apophis_pass.svg, Author, Marco Polo

Figure 3.14. Apophis Path at Risk in 2036, en.wikipedia.org/wiki/File: 2036_Apophis_Path_of_Risk.jpg, Author, Mario Roberto, Divran Ortiz Marrodoro

Figure 3.15. Tunguska Event Falling Trees, en.wikipedia.org?wiki/File: Tanguska_even_fallen_trees.jpg, Author, unknown

Figure 3.16. How a Space Lens Would Mitigate Global Warming, Space Lens, en.wikipedia.org/wiki/File: space_lens.jpg, Author, Michael Haggstrom

Figure 3.17. Lunar Base Concept Drawing From NASA http://commons.wikipedia.org/wiki/File: Luna_base_concept_drawing_578_23532.jpg, Author, NASA

Figure 3.18. Hubble Space Telescope view of Mars, showing Polar Ice Cap, en.wikipedia.org/wiki/File: Mars_as_seen_by_the _Hubble_Telescope, Author, NASA

Figure 3.19. ALPH Construction of Manned Base Inside North Polar Ice Cap, Author, Powell and Maise

Figure 3.20. Exploration of subsurface Europa Ocean by the NEMO Probe, Author, Powell and Maise

Figure 3.21. Space Debris Cloud in Low Earth Orbit (LEO), http://earthobservatory.nasa.gov/IOTO/view.php?id=40173.Author is NASA

Chapter 4 List of Figure Credits

Figure 4.1. Tesla Broadcast Tower 1904 http://en.wikipedia.org/wiki/File:Tesla_Broadcast_Tower_1904.jpeg

Figure 4.2. Earthmap 1000x500 compac; (http://en.wikipedia.org/wiki/file:Earthmap 1000x500 compac.jpg); author, jimht of shaw tot ca, modification by Rodrigold.

Figure 4.3. Space to Ground Microwave Using Laser Pilot, (http://en.wikipedia.org/wiki/File:Space_to_ground_microwave_pilot-beam.png) Author is NASA

Figure 4.4. SunTower Solar Power Satellite

(http://en.wikipedia.org.wiki/file: SunTower.jpg), author is NASA

Figure 4.5. Solar Power Satellite Sandwich or Abacus Concept;

(http://en.wikipedia.org/wiki/file: Solar_power_satellite_sandwich_ or _ abacus_concept.jpg), author is NASA

Figure 4.6. Solar Electric Generator Using MIC Structure Thin Film Authors are Powell and Maise.

Figure 4.7. Solar Electric Generator Using MIC Structure Solar Concentrator Authors are Powell and Maise.

Chapter 5 List of Figure Credits

Figure 5.1. US Energy – Related Carbon Dioxide Emissions by Sector, 2011, http://www.eia.gov/energy_in _brief/images/charts/energy/relat..., Author US energy Information Administration, Monthly Energy Review (May 2012).

Figure 5.2. US Energy 2009, http://wikipedia.org/wiki/file:USEenrgy 2009.jpg, Author, US Energy Information Administration

Figure 5.3. Automobiles per 1000 persons, Author, Powell and Jordan

Figure 5.4. Thomas Parker Electric Car, http://commons.wikimedia.org/wiki/File:Thomas_Parker_Electric _car, Author, Unknown

Figure 5.5. Edison Electric Car 1913, https//commons.wikimedia.org/wiki/file: Edison Electric Car 1913.jpg Author, Smithsonian

Figure 5.6. 1960 Henney Kilowatt, http://commons.wikimedia.org/wiki/file:kilowatt.jpg, Author D. Roberson.

Figure 5.7. Roadster Goodwood, http://commons/wikimedia.org/wiki/file:Roadster_Goodwood.jp g, author, Tesla Motors, Inc.

Figure 5.8 Lithium in Brine Pools or the Pacific Coast of South America, http://large.stanford.edu.courses/2010/pk240/season2/image/f2 101, Author USGS

Figure 5.9. NREL FT Diesel vs Conventional Diesel photo, http://commons. Wikimedia.org/wiki/File. NREL_FT_diesel_vs_conventianal_diesel_photo.jpg, Author USDOE

Figure 5.10. OSD Clean Fuel Initiative FT Diesel Emissions Presentation, http://wikipedia.org/wiki/File:OSD_Clean_Fuel_Initiative_FT_Diesel_Emi ssions_presentation, Author, Dr. Theodore K. Bornu, Edward Sheridan, William E. Harrison

Figure 5.11. Denali (Mt. McKinley),

http://commons,Wikimedia.org/wiki/File:Denali_Mt_McKinley.jpg,
Author, National Park Service
Figure 5.12. View of Mississippi River from Fire Point on Effigy Mounds
National Monument,
http://commons.org/wiki/file:Efme_view_from_fire_point Author,
National Park Service
Figure 5.13. Wind Mill Farm,
http:www.fiokr.com/photos/57904403@No5/5532750017/?rb+1,
Author, unknown
Figure 5.14. Statue of Liberty,
http://commons.wikimedia.org/wiki/file:Detroit_photographic,au
thor, unknown

Introducing LISA List of Figure Credits

Figure 1. Typical Large Ice Island Structure, Author, Powell
Figure 2. Features of On-Shore and Off-Shore Coastal Ice Barriers, Author,
Powell
Figure 3. Katrina New Orleans Flooded,
http:en.wikipedia.org/wiki/file:Katrina New Orleans Flooded_e…, Author
US Coast Guard
Figure 4. Great wave off Kanagawa,
http://en.wikipedia.org/wiki/File:Great_Wave_Off_Kanagawa2.jpg,
Author is Hokusai
Figure 5. The Netherlands compared to sea level,
http://en.wikipedia.org/wiki/file:the
Netherlands_compared_to_sea_level, author, Jan Arkeseijn
Figure 6. Iceberg A22A, South Atlantic Ocean (2002),
en.wikipedia.org/wiki/File:Iceberg_A22A_South_Atlantic, Author, ISS
Crew Earth Observations Experiment and the Image Science & Analysis
Laboratory, Johnson Space Center
Figure 7. The Iceberg Suspected of Having Sunk the RMS Titanic,
en.wikipedia.org/wiki/File:Titanic_iceberg, Author, Chief Steward of the
Liner Prinz Adelbert
Figures 8 & 9. Hotel de Glace, Quebec, Canada,
www.ice_hotel_Canada.com, author, unknown.
Figure 10.
http://en.wikipedia.org/wiki/File:Tower_at_Harbin_Ice_and_Snow
Festival_2012.jpg, Author, Shanghai Killer Whale, 5 January 2013
Figure 11. http://en.wikipedia.org/wiki/file:Harbin_la_Festival Author,
LiYan at zh.wikipedia (zH.wikipedia.org, 2003-08-09
Figure 12. http://en.wikipedia, org/wiki/org/File:vattensagspan2.jpg,
Author, Kr-Val, 8 May 2010

Figure 7.2. Sea Wall From West of Rapid Fire Battery, Fort Crockett, http://en.wikipedia.org/wiki/File:No._3._sea_wall._From_Wall..., author is National Archives and Records Administration.

Figure 7.3. NOAA-Hurricane-Katrina-Aug-28-05 2145UTC, http://en.wikipedia.org/wiki/File:NOAA_Hurricane_Katrina-Au..., author is NOAA.

Figure 7.4. Katrina New Orleans Flooded, http://en.wikipedia.org/wiki/File:KatrinaNewOrleansFlooded_e..., author is US Coast Guard.

Figure 7.5. Katrina Bayou La Batre 2005 boats ashore, http://en.wikipedia.org/wiki/File:Katrina_Bayou_La_Batre_20..., author is NASA.

Figure 7.6. Sandy Oct 28 2012 16.00 (UTC), http://en.wikipedia.org/wiki/File:Sandy_Oct_28_2012_1600(..., author is NASA.

Figure 7.7. 1755 Lisbon Earthquake, http://en.wikipedia.org/wiki/File:1755_Lisbon_earthquake.jpg, author unknown.

Figure 7.8. 2004-tsunami, http://en.wikipedia.org/wiki/File:-tsunami.jpg, author is David Rydevik.

Figure 7.9. Tsunami Size Scale 26 Dec 2004, http://en.wikipedia.org/wiki/File:Tsunami_size_scale_26Dec20..., author is Dcarrera.

Figure 7.10. Juan de Fuca Plate, http://en.wikipedia.org/wiki/File:Juan_de_Fuca_Plate.png, author is US Geologic Survey.

Figure 7.11. The Netherlands Compared to Sea Level, http://en.wikipedia.org/wiki/File:TheNetherlands_compared_to..., author, Jan Arkesteijn.

Figure 7.12. Features of On-Shore and Off-Shore Coastal Ice Barriers, author, J. Powell.

Figure 7.13. Potential Options for Insulation of Bottom Surfaces of LISA-2A Structures Resting on the Sea Bed, author, Powell.

Figure 7.14. Temperature Inside a Sea Bed Underneath a Large Ice Structure Resting on the Sea Floor, as a Function of Depth and Time After Structure is Formed, author, J. Powell.

Figure 7.15A. Average Distance Between Inhabitants of Manhattan Island, United States, author, Powell.

Figure 7.15B. Average Distance Between Inhabitants of Central Mumbai, India, author, Powell.

Figure 7.16. Satellite Photo of Artifical Archipelagos, Dubai, United Arab Emirates, author, NASA, en.wikipedia.org/wiki/File:Artifical_Archipelagoes,_Dubai...

Appendix

Bibliography of papers and reports on:

- StarTram (Maglev Launch)
- MITEE (Nuclear Thermal Propulsion Engine)
- SUSEE (Nuclear Space Power Reactor)
- ALPH (Nuclear Robotic Probe and Factory)
- MIC (Magnetically Inflated Cable Space Structures)

Detailed Papers and Reports on StarTram

1. StarTram: A New Concept for Very Low Cost Earth-to-Orbit Transport Using Ultra High Velocity Magnetic Launch, James Powell, George Maise, and John Paniagua, Paper IAF-01-S.6.04, 52nd International Astronautical Congress, 1-5 October 2001, Toulouse, France (36 pages).

2. Powell, J., and Maise, G., "Space Tram" US Patent No. 6,311,926B1, November 6, 2001.

3. StarTram: A New Approach for Low-Cost Earth-to-Orbit Transport, James Powell, George Maise, and John Paniagua, 2002 IEEE Space Conference, March 2002, Big Sky, Montana (17 pages).

4. StarTram C—A Maglev System for Ultra Low Cost Launch of Cargo to LEO, GEO, and the Moon, James Powell, George Maise, and John Paniagua, Paper IAC-03-IAA13.1.04, 54th International Astronautical Congress, October 2003, Bremen, Germany (18 pages).

5. StarTram: The Key to a Robust, Low Cost Earth/Lunar Transport System, James Powell, George Maise, and John Paniagua, International Lunar Conference 2003, November 16-22, 2003, Hawaii (23 pages).

6. StarTram: Ultra Low Cost Launch for Large Space Architectures, James Powell, George Maise, and John Paniagua, STAIF 2004 Conference, February 2004, Albuquerque, New Mexico (12 pages).

7. StarTram: The Key to Low Cost Lunar Bases and Human Exploration of Space, James Powell, George Maise, and John Paniagua, AIAA Space 2004, September 28-30, 2004, San Diego, California (12 pages).

8. StarTram: An Ultra Low Cost Launch System for Large Scale Exploration and Commercialization of Space, James Powell, George Maise, and John Paniagua, Paper IAC-04-V.05.07, 55th International Astronautical Congress, October 2-8, 2004, Vancouver, Canada (17 pages).

9. Ibid, StarTram viewgraphs presented at the 55th IAC meeting, Vancouver, Canada, (26 pages).

10. StarTram: An Ultra Low Cost Launch System to Enable Large Scale Exploration of the Solar System, James Powell, George Maise, and John Paniagua, STAIF 2006 Conference, February 12-15, 2006, Albuquerque, New Mexico (12 pages).

StarTram: An International Facility to Magnetically Launch Payloads at Ultra Low Unit Cost, George Maise, James Powell, John Paniagua, and James Jordan, Paper IAC-06-D3.2.7, 57th International Astronautical Congress, October 2-5, 2006, Valencia, Spain (14 pages).

11. Ibid: StarTram viewgraphs presented at the 57th IAC Meeting, Valencia, Spain (25 pages).

12. StarTram: The Maglev Launch Path to Very Low Cost, Very High Volume Launch to Space; presented at the 14th International EML Symposium, Victoria, Canada, June 10-13, 2008.

13. The Gen-1 Maglev Launch System for Ultra Low Cost Access to Space; James Powell, George Maise, and John Paniagua; presented at the 59th International Astronautical Congress (IAC), Glasgow, Scotland, September 29—October 3, 2008 (10 pages).

14. Ibid: Viewgraphs presented at the 59th IAC Meeting, Glasgow, Scotland (21 pages).

15. Maglev Launch—An Ultra Low Cost Way to Deploy Space Solar Power Systems; presented at the From the Sun to the Earth International Conference on solar Energy from Space, Ontario Science Center, Toronto, Canada, September 8-10, 2009 (17 pages).

16. Ibid: Viewgraphs presented at the From the Sun to the Earth Conference, Toronto, Canada (33 pages).

17. Maglev Launch: Ultra Low Cost, Ultra/High Volume Access to Space for Cargo and Humans; James Powell, George Maise, and John Rather, presented at SPESIF-2010—Space, Propulsion, and Energy Sciences International Forum, February 23-26, 2010, Johns Hopkins Applied Physics Laboratory, Baltimore, Maryland (15 pages).

18. Ibid: viewgraphs presented at SPESIF-2010 meeting, Baltimore, Maryland (33 pages).

19. A Development and Test Program for the Generation-1 Maglev Launch System, James Powell, George Maise, and John Rather, presented at SPESIF-2011 Meeting, Baltimore Maryland (16 pages).

20. Ibid: Viewgraphs presented at SPESIF-2011 Meeting, Baltimore, Maryland.

Detailed Papers and Reports on MITEE

1. MITEE: An Ultra Lightweight Nuclear Engine for New and Unique Planetary Science and Exploration Missions. James Powell, John Paniagua, George Maise, Hans Ludewig and Michael Todosow, Paper IAF-98-R.1.01, 49th International Astronautical Congress, Sept. 28—Oct. 2, 1998, Melbourne, Australia [27 pages].

2. Europa Sample Return Mission Utilizing MITEE Technologies. John Paniagua, James Powell, George Maise, Hans Ludewig, Michael Todosow, Paper IAF-98-Q.2.03, 49th International Astronautical Congress, Sept. 28—Oct. 2, 1998, Melbourne, Australia [15 pages].
3. Exploration of Jovian Atmosphere Using Nuclear Ramjet Flyer. George Maise, James Powell, John Paniagua, Hans Ludewig, Michael Todosow, Paper IAF-98-S.6.08, 49th International Astronautical Congress, Sept. 28—Oct. 2, 1998, Melbourne, Australia [11 pages].
4. High Performance Nuclear Thermal Propulsion System for Near Term Exploration Missions to 100 AU and Beyond. James Powell, John Paniagua, George Maise, Hans Ludewig, and Michael Todosow, Acta Astronautica, 44 No. 2-4, pp 159-166, Jan—Feb. 1999 [8 pages].
5. The Liquid Annular Reactor System (LARS) for Deep Space Exploration. George Maise, John Paniagua, James Powell, Hans Ludewig, and Michael Todosow, 2nd IAA Symposium on Realistic Near-Term Advanced Scientific Space Missions. June 29—July 1, 1998, Aosta, Italy; also Acta Astronautica, 44 No. 2-4, pp 167-174, Jan—Feb 1999 [13 pages].
6. New Approaches for the Exploration and Colonization of the Solar System: Road Map for the Next 30 Years in Space. James Powell, George Maise, and John Paniagua, Report PUR-7, Nov. 10, 1998 [18 pages].
7. The MITEE Family of Compact, Ultra Lightweight Nuclear Thermal Propulsion Engines for Planetary Space Exploration. James Powell, George Maise, and John Paniagua, Paper IAF 99-5.6.03, 50th International Astronautical Congress, October 4-8, 1999, Amsterdam, the Netherlands [28 pages].
8. SunBurn: A Concept Enabling Ultra High Spacecraft Velocities for Extra Solar System Exploration. George Maise, James Powell, and John Paniagua, Paper IAA-99-IAA.4.1.07, 50th International Astronautical Congress, October 4-8, 1999, Amsterdam, the Netherlands [18 pages].
9. A Cost Effective Space Infrastructure for Retrieval of Helium-3 from Uranus for Earth-Based Fusion Power Systems Utilizing the MITEE Nuclear Propulsion System. John Paniagua, James Powell, and George Maise, Paper IAA-99-R.3.10, 50th International Astronautical Congress, October 4-8, 1999, Amsterdam, the Netherlands [17 pages].
10. Compact, Ultra Lightweight Nuclear Thermal Propulsion Engines for Planetary Science Missions. James Powell, George Maise, John Paniagua, Hans Ludewig, and Michael Todosow, 10th Annual NASA/JPL/MFSC/AIAA Advanced Propulsion Research Workshop, Huntsville, Alabama, April 6-8, 1999 [25 pages].

11. Phase 1 Final Report: Lightweight High Specific Impulse (1000 sec) Space Propulsion Systems. James Powell, George Maise, John Paniagua, Jon Longtin, John Metzger, and Hui Zhang, PUR-12, October 1999 [221 pages].

12. Phase 1 NIAC Final Report: Exploration of Jovian Atmosphere Using Nuclear Ramjet Flyer. George Maise, James Powell, John Paniagua, and Robert Lecat, PUR-16, Nov. 30, 2000 [14 pages].

13. MITEE-B: A Compact Lightweight Bi-Modal Nuclear Engine to Deliver Both High Propulsive Thrust and High Electric Power. James Powell, George Maise, John Paniagua, and Stan Borowski, Paper IAF-01-S.6.05, 52nd International Astronautical Congress, Oct. 1-5, 2001, Toulouse, France [24 pages].

14. Europa One—A Manned Base for Exploration of the Outer Solar System and Near Interstellar Space. John Paniagua, James Powell, and George Maise, 52nd International Astronautical Congress, October 1-5, 2001, Toulouse, France [26 pages].

15. Phase 1 NIAC Final Report: Europa Sample Return Mission Utilizing High Specific Impulse Propulsion Refueled with Indigenous Resources. John Paniagua, James Powell and George Maise, November 30, 2001, Report PUR-21 [114 pages].

16. Compact MITEE-B: Bi-Modal Nuclear Engine for Unique New Planetary Science Missions. James Powell, George Maise, John Paniagua, and Stanley Borowski, AIAA 2002-3652, AIAA/ASME/SAE/ASEE Joint Propulsion Conference and Exhibit, July 2002, Indianapolis, Indiana [22 pages].

17. Phase 2 NIAC Interim Report: Exploration of Jovian Atmosphere Using Nuclear ramjet Flyer. George Maise, et al, PUR-26, Jan. 31, 2002 [125 pages].

18. Europa Sample Return Mission Utilizing High Specific Impulse Refueled with Indigenous Resources. John Paniagua, James Powell, and George Maise, Paper IAC-02-Q.2.05, 53rd International Astronautical Congress, October 10-19, 2002, Houston, Texas [14 pages].

19. Missions Possible: How Humanity Can Really Explore the Solar System Using Nuclear Propulsion, Report PUR-27, James Powell, George Maise, and John Paniagua, April 15, 2002 [22 pages].

20. Bi-Modal MITEE Engine for Nuclear Thermal/Nuclear Electric Propulsion. James Powell, George Maise, and John Paniagua, Advanced Space Propulsion Workshop, June 4-6, Pasadena, California [31 pages].

21. Exploration of Jovian Atmosphere Using Nuclear Ramjet Flyer. James Powell, George Maise, and John Paniagua, Advanced Space Propulsion Workshop, June 4-6, Pasadena, California [31 pages].

22. NEMO: Exploration of Europa's Subsurface Ocean and Return of Samples to Earth Using Nuclear Propulsion. James Paniagua, James Powell, and George Maise, Advanced Space Propulsion Workshop, Huntsville, Alabama, April 15-17, 2003 [35 pages].

23. MITEE-B: A Compact Ultra Lightweight Bi-Modal Nuclear Propulsion Engine for Robotic Planetary Science Missions. James Powell, George Maise, John Paniagua, and Stanley Borowski, STAIF 2003 Meeting, February 2003, Albuquerque, New Mexico [9 pages].

24. Pluto Orbiter/Lander/Sample Return Missions Using the MITEE Nuclear Engine. James Powell, George Maise and John Paniagua, 2003 IEEE Aerospace Conference, Big Sky, Montana, March 2003 [24 pages].

25. Exploration of Jovian Atmosphere Using Nuclear Ramjet Flyer. George Maise, et al., Phase II Final Report, March 1, 2003, NIAC Phase II Grant 07600-061 [163 pages].

26. HIP: A Hybrid NTP/NEP Propulsion System for Ultra Fast Robotic Orbiter/Lander Missions to the Outer Solar System. James Powell, George Maise, and John Paniagua, 54th International Astronautical Congress, October 2003, Bremen, Germany [17 pages].

27. MITEE and SUSEE: Compact Ultra Lightweight Nuclear Power Systems for Robotic and Human Exploration Mission. James Powell, George Maise, and John Paniagua, Paper IAC-04-IAA R.4/S.7-04, 55th International Astronautical Congress, Vancouver, Canada, October 2-8, 2004 [15 pages].

28. NEMO: A Mission to Explore and Return Samples from Europa's Oceans. James Powell, John Paniagua, and George Maise, STAIF 2004 Conference, February 2004, Albuquerque, New Mexico [7 pages].

29. Nuclear Propulsion and Power Systems for Near Term Exploration of the Solar System. James Powell, George Maise, and John Paniagua, AIAA 1st Space Exploration Conference, Jan 30—Feb. 1, 2005, Orlando, Florida [17 pages].

30. NEMO: A Mission to Search for and Return to Earth Possible Life Forms on Europa, Jesse Powell, James Powell, George Maise, and John Paniagua, Acta Astronautica, 57 pp 579-593, 2005 [15 pages].

31. Mini-MITEE: Ultra Small, Ultra Light NTP Engines for Robotic Science and Manned Exploration Missions. James Powell, George Maise, and John Paniagua, STAIF 2006 Conference, February 12-16, 2006, Albuquerque, New Mexico [10 pages].

32. MITEE: A Compact Near Term NTP Engine for New and Unique Robotic and Manned Exploration Missions. James Powell, George Maise, and John Paniagua, American Nuclear Space Conference, June 5-9, 2005, San Diego, California [9 pages].

33. The MITEE Hopper: A Compact NTP Spacecraft to Explore Multiple Surface Sites Using In-Situ Propellants. James Powell, George Maise, and John Paniagua, Paper IAC-06-D2.8/C3.5/C4.7/D3.5.06, 57th International Astronautical Congress, October 2-6, 2006, Valencia, Spain [12 pages].
34. A New Mission for the International Space Station (ISS) Enabled by Nuclear Thermal Propulsion—Cyclic Transport of Personnel and Supplies Between the Earth and the Moon, John Paniagua, James Powell, and George Maise, STAIF 2008 Conference, February 2008, Albequerque, New Mexico [9 pages].
35. Design and Development of the MITEE-B Bi-modal Nuclear Propulsion Engine, John Paniagua, James R. Powell, and George Maise.
36. Application of the MITEE Nuclear Ramjet for Ultra Long Range Flyer Missions in the Atmospheres of Jupiter and Other Giant Planets, George Maise, James Powell, John Paniagua, Edward Kush, Pasquale Sforza, and Hans Ludewig.

Detailed Papers and Reports on SUSEE
1. SUSEE: Ultra Light Nuclear Space Power Using the Steam Cycle, James Powell, George Maise, and John Paniagua, IEEE 2002 Space Conference, Big Sky, Montana, 2002 [17 pages].
2. SASSE: A Lightweight, High Efficiency Solar Thermal Steam Cycle For Satellites, James Powell, George Maise, and John Paniagua, Paper IAC-03-R.2.07, 54th International Congress, Bremen, Germany, 2003 [16 pages].
3. SUSEE—An Ultra Lightweight Nuclear Electric Propulsion System Based on Existing Water and Steam Cycle Technology, James Powell, George Maise, and John Paniagua, Advanced Space Propulsion Workshop, Huntsville, Alabama, April 15-17, 2003 [31 pages].
4. Compact Ultra Light Nuclear Electric Power Systems for Future Moon Bases and Colonies, James Powell, George Maise, and John Paniagua, International Lunar Conference 2003, Hawaii, November 16-22, 2003 [13 pages].
5. MITEE and SUSEE: Compact Ultra Lightweight Nuclear Power Systems for Robotic and Human Exploration Missions, James Powell, George Maise, and John Paniagua, Paper IAC-04-IAA-R.4/S.7-04, 55th International Astronautical Congress, October 2-8, 2004, Vancouver, Canada [15 pages].
6. Ibid: Viewgraph presentation [29 pages].
7. SUSEE: An Ultra Light Space Nuclear Power System Based on Conventional Water Reactor Technology, George Maise, James

Powell, and John Paniagua, American Nuclear Society, June 5-9, 2005, San Diego, California [9 pages].

8. SUSEE: A Compact, Lightweight Space Nuclear Power System Using Present Water Reactor Technology, George Maise, James Powell, and John Paniagua, STAIF 2006 Conference, February 12-16, 2006, Albuquerque, New Mexico [12 pages].

Detailed Papers and Reports on ALPH

1. ALPH—A Robotic Precursor to Produce Large Amounts of Supplies for Manned Outposts on Mars, James Powell, John Paniagua, and George Maise. Presented at 49th International Astronautical Congress, Melbourne, Australia, September 28—October 2, 1998, Paper IAF-98Q.3.08 [38 pages].

2. MICE: A Compact, Light Near Term Mobile Robot for Exploration of the Martian Polar Ice Cap, James Powell, George Maise, John Paniagua, Hans Ludewig, and Michael Todosow, Presented at 50th International Astronautical Congress, Amsterdam, the Netherlands, October 4-8, 1999, Paper IAF 99-Q.3.08 [18 pages].

3. Development of Self-Sustaining Mars Colonies Utilizing North Polar Cap and the Martian Atmosphere, James Powell, George Maise, John Paniagua, and Jesse Powell, Final Report, NIAC Research Grant 07600-053, November 20, 2000 [184 pages].

4. Self-Sustaining Mars Colonies Utilizing the North Polar Cap and Martian Atmosphere, James Powell, George Maise, and John Paniagua, Presented at 51st International Astronautical Congress, Rio de Janeiro, Brazil, Oct. 2-6, 2000, also published in Acta Astronautica, 48, No. 5-12, pp. 737-765 (2001) [27 pages].

5. The Mars Hopper—A Mobile Lightweight Probe to Explore and Return Samples from Many Widely Separated Locations on Mars, James Powell, George Maise, and John Paniagua, Presented at 52nd International Astronautical Congress, Toulouse, France, Oct. 1-5, 2001, Paper IAA-01-IAA13.3.08 [26 pages].

6. Fast Track Route to Mars Colony Using Nuclear Propulsion and Power, James Powell, George Maise, and John Paniagua, Presented at 40th Aerospace Sciences Meeting and Exhibit, Reno, Nevada, Jan. 14-17, 2002, Paper AIAA 2002-0996 [32 pages].

7. CADMUS—A Robotic Mars Factory Returning Supplies to Earth Orbit, James Powell, George Maise, and John Paniagua, Presented at 2003 IEEE Aerospace Conference, Big Sky, Montana, March 2003 [18 pages].

8. Xanadu: A Polar Base for Manufacturing Supplies on Mars, James Powell, George Maise, and John Paniagua, Presented at STAIF 2004 Conference, Albuquerque, New Mexico, February 2004 [9 pages].

9. Multi-MICE: A Network of Interacting Nuclear Cryoprobes to Explore Ice Sheets on Mars and Europa, Jesse Powell, James Powell, George Maise, and John Paniagua, Presented at Space 2005 Conference, Long Beach, California, Sept. 2005 [14 pages].
10. MERIT: A New Approach for a Large Scale Space Infrastructure Based on Mars, James Powell, George Maise, and John Paniagua, Presented at 2005 STAIF Conference, February 13-17, 2005, Albuquerque, New Mexico [10 pages].
11. ALPH: A Low Risk, Cost Effective Approach for Establishing Manned Bases and Colonies on Mars, James Powell, George Maise, John Paniagua, and Jesse Powell, Presented at AIAA Space 2005 Conference, Long Beach, California, August 30—Sept. 1, 2005 [17 pages].
12. ALPH: A compact Robotic Nuclear Powered Factory to Build and Supply Bases on Mars Prior to Manned Landing, James Powell, George Maise, and John Paniagua, Present at American Nuclear Society Space Nuclear Conference, San Diego, California, June 5-9, 2005 [17 pages].
13. Multi-MICE: A Network of Interactive Nuclear Cryoprobes to Explore Ice Sheets on Mars and Europa, George Maise, James Powell, Jesse Powell, John Paniagua and Hans Ludewig, NASA Institute of Advanced Concepts, Phase 1 Report, NIAC Subaward No. 07605-003-047, May 1, 2006 [145 pages].
14. Multi-MICE: Nuclear Powered Mobile Probes to Explore Deep Interiors of the Ice Sheet on Mars and the Jovian Moons, George Maise, James Powell, Jesse Powell, John Paniagua, and Hans Ludewig. Presented at STAIF 2007 Conference, Albuquerque, New Mexico, February 11-15, 2007 [10 pages].
15. MICE: A System of Compact Mobile Nuclear Probes to Explore the Deep Interior of Mars North Polar Cap, George Maise, James Powell, John Paniagua, Jesse Powell, and Hans Ludewig, presented at the 57th International Congress, Valencia, Spain, October 2-5, 2006, paper IAC-06-A3.P3.5.

Detailed Papers and Reports on MIC
1. MIC—A Self Deploying Magnetically Inflated Cable System for Large Scale Space Structures, James Powell, George Maise, and John Paniagua. Acta Astronautica 48, No 5-12, pp 331-352, 2001 [21 pages].
2. Deployment of Large Structures in Space Using the Magnetically Inflated Cable (MIC) System, James Powell, George Maise, John Paniagua, and John Rather, Paper IAC-06-D1.2.09; delivered at the 57th International Astronautical Congress, Valencia, Spain, October 2-5, 2006 [12 pages].

3. Ibid, viewgraphs of oral presentation at 57th International Astronautical Congress, October 2-5, 2006 [28 pages].

4. MIC: Magnetically Deployable Structures for Power, Propulsion, Processing, Habitats, and Energy Storage at Manned Lunar Bases, James Powell, George Maise, John Paniagua, and John Rather, to be delivered at STAIF-2007 Conference, Albuquerque, New Mexico, February 11-15, 2007 [9 pages].

5. Magnetically Inflated Cable (MIC) System for Large Scale Space Structures, James Powell, George Maise, John Paniagua, and John Rather, NIAC (NASA Institute of Advanced Concepts) Phase 1 Report, May 1, 2006 [162 pages].

6. MIC—Large Scale Magnetically Inflated Cable Structures for Space Power, Propulsion, Communications and Observational Applications, James Powell, George Maise, and John Rather, delivered at SPESIR-2010 International Forum, John Hopkins Applied Physics Laboratory, February 23-26, 2010 (12 pages).

7. Ibid: Viewgraphs of oral presentation at SPESIF International Forum, February 23-26, 2010 (31 pages).

8. A Development and Test Program for the Magnetically Inflated Cable (MIC) Large Space Structures System, James Powell, George Maise, and John Rather, delivered at SPESIF-2011 International Forum, Johns Hopkins Applied Physics Laboratory, March 15-17, 2011 (14 pages).

9. Ibid: Viewgraphs of oral presentation at SPESIF International Forum, March 15-17, 2011 (26 pages).

List of Maglev Reports

1. Powell, J. and Danby, G., 1966. "High Speed Transport By Magnetically Suspended Trains", Paper 66WA/RR-5, ASME Winter Annual Meeting, New York, NY. Also, Powell, J. and Danby, G., 1967 "A 300 mph Magnetically Suspended Train, Mech Eng 89, p. 30-35

2. Powell, J. and Danby, G., 1969. "Electromagnetic Inductive Suspension and Stabilization System For A Ground Vehicle"; US Patent 3,470,828

3. Powell, J. and Danby, G., 1969, "Magnetically Suspended Trains: The Application of Superconductors to High Speed Transport", Cryogenics and Industrial Gases, 4 (10), p.19

4. Powell, J. and Danby, G., (3. 1970. "Dynamically Stable Cryogenic Magnetic. Suspensions for Vehicles in Very High Velocity Transport Systems". Recent Advances in Engineering Science, Gordon and Breach, Vol 5: p. 159-182

5. Powell, J: and Danby, G., 1971. "The Linear Synchronous Motor and High Speed Transport" Proc Intersociety Energy Conversion Eng. Conference, Boston, MA, p. 11 8-131

6. Powell, J. and Danby, G., 1971. "Magnetic Suspension For Levitated Tracked Vehicles" Cryogenics 11: p. 192-204

7. Danby, G. and Powell, J., "The Central Role of Cryogenics in Magnetically Levitated High Speed Trains" Proc. Of XIII International Conference on Refrigeration, Washington, DC

8, Powell, J. and Danby, G., 1971. "Cryogenic Suspension and Propulsion systems for 200 —2000 mph Ground Transport" Proc. Cryogenic Society of America Conf. on Applications of Cryogenic Technology, Vol 4, p. 299-332

9. Danby G. and Powell, J., 1972. "Integrated Systems for Magnetic Suspension and Propulsion of Vehicles" Proc. 1972. Applied Superconductivity Conf., Annapolis, p.120-126

10. Powell, J. and Danby, G., 1972. "Integrated Magnetic Suspension and Propulsion Systems" Proc. IEEE Meeting of Industrial Applications Society. Philadelphia, PA (10 pages)

11. Danby, G. , Jackson, J., and Powell, J. 1974. "Force Calculations for Hybrid (Ferro — Null Flux) Low Drag Systems, IEEE Trans on Magnetic Mag 10, p. 443.446

12. Danby, G. and Powell, .1., 1974. "Hybrid Superconducting Magnetic Suspensions for Very Efficient High Speed Ground Transport" Proc. 1974, Applied Superconductivity Conference

13. Danby, G. and Powell, J., 1988. "Design Approaches and Parameters for Magnetically Levitated Transport Systems" Proc. 2nd Annual Conference on Superconductivity and its Applications, Elsevier Science Publishing

14. Powell, J,. 1992. "Large Scale Implementation of Maglev in the United States" AAAS Symposium on Maglev Transport, Chicago, IL (10 pages)

15. Powell, J. and Danby, G., 1995. "Passenger and Freight Maglev for the US'-, Proc of the Future Transportation Technology Conf., Costa Mesa, CA. SAE Paper 95-1921

16. Powell, J., 1995. "The Application of Maglev Technology to Intermodal Transportation". Proc of National Aviation and Transportation Center, 4th Annual Symposium on Global Intermodalism and Economic Development, July 24-27, 1995

17, Powell, J. and Danby, G., 1996. "Integrating Passenger and Freight Service: Maglev Technology Approach", Proc of 514 Annual Symposium on Intermodal Transportation. Bordeaux, France (41 pages)

18. Powell, J. and Danby, G., 1998. "Transport by Magnetic Levitation", Encyclopedia of Applied Physics, Vol 22, p 233-261

19, Powell, J. and Danby, G., 1989. Co-Chairman, Maglev Technology Advisory Committee for US Senate Committee on Environment and Public Works, "Benefits of Magnetically Levitated Transport for the United States", Volume 1, Executive Summary, (30 pages) and Volume 2, Technical Report (238 pages)

20. Powell, J. and Danby, G., 2000. "Magnetic Levitation: A New Mode of Transport for the 21th Century", Lecture Given at the Award of the 2000 Franklin Medal for Engineering to Powell and Danby by the Franklin Institute (50 Pages)

21. Powell, J. and Danby, G., 2002. "Maglev 2000 'Transportation Technology", Final Report Vol 1 (347 pages) and Vol 2, (452 pages) Federal Railroad Administration and Florida Department of Transportation (849 total pages)

22. "Florida Maglev Deployment Program Kennedy Space Center Circulator National Demonstration Project", Tilden, Lobnitz & Cooper, 2002, (238 Pages)

23. Powell, J. and Danby, G., 2005. "Final Report, Federal Transit Administration, Maglev 2000 Project FL-26-7023", Maglev 2000 of Florida Corporation (100 pages)

24. Powell, J. "Maglev Presentation", ETA Low Speed Urban Maglev Workshop. W.Kulyk, Director, Office of Mobility Innovation, September 8.9, 2005, Washington, DC (30 pages)

25. Powell, J. and Danby, G. "Integration of Maglev Guideways with Railroad Track: The MERRI System", Report DPMT-1, October 5, 1996 (70 Pages)

26. Powell, J. and Danby, G. "Maglev Vehicles", IEEE Potentials, p.7-12, October/November, 1996

27. Powell, J. and Danby, G., "The Development of Maglev-Yamanashi and Beyond", Invited Talk at Dedication of Japan Railways Yamanashi Maglev Test Line, April 4, 1997) (23 Pages)/

28. Powell, J. and Danby, G., "The M-2000 Maglev System for the United States", Presentation to the FRA Maglev Advisory Committee, March 24, 1997 (28 Pages)

29. Powell, J. and Danby, G., "Maglev Technologies for Combined Freight and Passenger Movement — Application to Industrialized and Rapidly Industrializing Nations", 6th International Symposium in Intermodal Transportation, Mexico City June 18-20, 1997 (29 Pages)

30. Powell, J, and Danby, G., and Bragden, C., "Developing A High Speed Maglev Land Bridge in Central America" 66 International Symposium on Intermodal Transportation, Mexico City, June 18-20 1997 (28 pages)

31. Powell, J. and Danby, G., "The Water Train: Long Distance Transport of Water by Maglev", Report DPMT-3, December 13, 1997 (34 pages)

32. Powell, J., "Electrical Power Storage Using Maglev — The Maglev Power Storage System", Report DPMT-14, November 1, 2000 (126 Pages)

33. Powell. J. and Danby, G., "Low Speed Application of Superconducting Maglev", presentation to Transportation Research Board, January 14, 1999, Washington, DC (30 pages)

34. Powell, J., "Cost Projections for the M-2000 Maglev System" Report M-2000/002, presentation to the FRA Workshop, August 24, 1999, Washington, DC (30 pages)

35. Danby, G., "The M-2000 Maglev System" Report 2000/001, Presentation to the FRA Workshop, August 24, 1999, Washington, DC (26 Pages).

36. Danby G., and Powell, J., "Progress in Design and Testing of Maglev 2000 Technology", Report No. M-2001, Presentation to Transportation Research Board 2000 Annual Meeting, Washington, DC, January 9-13, 2000 (29 pages1

37. Powell, and Danby, G., "Maglev: An Evolutionary Technology for the Transport of Freight, People, and Resources", Presented at International Congress on the Implementation Follow Up of Habitat Agenda in Islamic Cities, Tunis, Tunisia, March 24-26, 2000 (46 pages)

38. Powell, J. and Danby, G., Morena, J, Wagner, T, and Smith, C., "The Maglev 2000 Urban Transit Systems", July 31, 2002 (15 Pages).

39. Powell, James, and Danby, Gordon, "The 2nd Generation Maglev 2000 Transport System: Design, Technology, Status, and Future Applications, Maglev 2000 of Florida, Sept. 2006 (166 pages)

40. Jordan, James, and Powell, James " Maglev Transport – A Necessity in the Age of No Oil", presented Capital Science 2008, Arlington, VA., National Science Foundation, March 29-30, 2008 (40 pages)

41. Ibid, PowerPoint presentation, March 29-30, 2008 (25 pages)

42. Powell, James and Gordon Danby, "Energy Efficiency and Economics of Maglev Transport", Presented at 2008 Advanced Energy Conference, Stony Brook University, NY, November 19-20, 2008 (22 pages)

43. Ibid, PowerPoint presentation, November 19-20, 2008 (18 pages)

44. Powell, James, et al, "A National Maglev Network for the US – Design and Capabilities", Presented at Maglev 2008, 20th International Conference on Magnetically Levitated Systems and Linear Drives. San Diego, California, December 15-18, 2008 (9 pages)

45. Danby, Gordon, et al, "Fabrication and Testing of Full-Scale Components For the 2nd Generation Maglev 2000 System, 20th International Conference on Magnetically Levitated Systems and

Linear Drives, San Diego, California, December 15-18, 2008 (10 pages)

46. Powell, James, et al, "Adaptation of the LIRR System to Maglev for Faster, More Convenient, and Lower Cost Service", Presented at 2nd Advanced Energy Conference, Stony Brook, Long Island, New York, November 18-19, 2009 (18 pages)

47. Ibid, PowerPoint Presentation, Nov 18-19, 2009 (29 pages)

48. Griffis, F.H. (Bud), et al, "How Maglev Can Enable Stewart Airport to become the 4th Major Airport for the NYC Region". Presented at the 2nd Advanced Energy Conference, Stony Brook, Long Island, New York, November 18-19, 2009 (21 Pages)

49. Ibid, PowerPoint Presentation, Nov 18-19, 2009 (23 Pages)

50. Powell, James, et al, "The West Coast Maglev Network Transport for the 21st Century, Maglev 2000, January 20, 2010 (39 pages)

51. Powell, James, et al, "The New York Maglev Network – A New Transport System for the 21st Century, Maglev 2000, January, 2011 (12 pages)

52. Powell, James, et al, "Maglev Energy Storage and the Grid", Presented at the 2010 Advanced Energy Conference, New York, NY, November 8-9, 2010 (11 pages)

53. Griffis, F.H, (Bud), et al, "Feasibility Study for New Danby Powell Maglev System & Preliminary Route Study (New York City to Stewart International Airport), Preliminary Guideway Plans and Updated Cost Estimates, Polytechnic Institute of NY, January 2011 (144 pages)

54. Powell, James, et al, "The Maglev America Project: A 29,000 National Maglev Network for the United States", Presented at Maglev 2011, 21st International Conference on Magnetically Levitated Systems and Linear Drives, Daejeon, Korea, Oct 10-13, 2011 (15 pages)

55. Ibid, PowerPoint Presentation, October 10-13, 2011 (17 pages).

56. Griffis, F.H.(Bud), et al, "Adaptation of Existing Railroad Trackage for Levitated Maglev Vehicles", presented at Maglev 2011, 21st International Conference on Magnetically Levitated Systems and Linear Drives, Daejeon, Korea, Oct. 10-13, 2011 (6 pages)

57. Powell, James, et al, "Large Scale Storage of Electrical Energy Using Maglev", presented at Maglev 2011, 21st International

Conference on Magnetically Levitated Systems and Linear Drives, Daejeon, Korea, Oct 10-13, 2011 (19 pages)

Maglev 2000 Proprietary Reports

1. Powell, J., "Geometry and Magnetic Forces on the Electromagnetic Loops for the M¬2000 Narrow Beam Guideway", Memo M-2000-JP-3-01-6/14/96, June 14, 1996 (61 pages).

2. Powell, J., "Fringe Fields From SC Magnets", Memo M-2000-JP-3-02-7/26/96, July 26, 1996 (13 pages)

3. Powell, J., "Computer Analysis of Magnetic Field Distributions From M-2000 SC Quadrupole Arrays: Nature and Requirements for Phase A Studies", Memo M-2000-JP¬03-8/03/96 (32 pages).

4. Powell, J., "Refrigeration Systems for High Temperature Superconducting Vehicle Magnets", Memo M-2000-JP-4-01-8/17/96, August 17, 1996 (6 pages).

5. Powell, J., and Maise, G, "Vehicle Design and Mass Budget", Memo M-2000-JP/GM-01-1-8/24/96 August 24, 1996 (14 pages)

6. Maise, G., "Potential Damage to Sides of Vehicles from Windborne Debris", Memo M-2000=GM-1-01-09/26/96 September 26, 1996 (4 pages)

7. Maise, G., "Aerodynamic Lift to Augment Magnetic Levitation", Memo M-2000-GM¬1-03-12/11/96 December 11, 1996 (5 pages)

8. Powell, J., "Optimization of Placement of LSM Winding", Memo M-2000-JP-3-06¬12/31/96, December 31, 1996 (5 pages).

9. Maise, G., "Vehicle Drag Coefficient", Memo M-2000-GM-1-02-10/31/96 October 31, 1996 (9 pages)

10. Maise, G., "Design of M-2000 Vehicle", Memo M-2000-GM-1-04-1/21/97, January 31, 1997 (6 pages).

11. Powell, J., "Vehicle Speeds and Accelerations on Proposed Brevard County Route", Memo M-2000-JP-01-02-1/10/97, January 10, 1997 (11 pages)

12. Maise, G., "Tilting of M-2000 Guideway and/or Vehicles to Eliminate Lateral Forces on Passengers" Memo M-2000-GM-1-05-07/11/97, July 11,1997 (6 pages)

13. M-2000 Program Review; Presentation to C. Smith, Florida Department of Transportation, August 1, 1947 (33 pages)

14. Maise, G., "Aerodynamic Loads on the Guideway Structure Due to Hurricane — Level Wind Forces", Memo M-2000 — GM-1-07/12/31/97, December 31, 1997 (10 pages)

15. Powell, J., "Design of Guideway Loops and Panels, Memo M-2000-JP-3-04¬11/15/96, November 15, 1997 (24 pages)

16. Powell, J., "Reduction of Fringe Fields in the Passenger Cabin by Use of Asymmetric Quadrupoles", Report DPMT-6 December 15, 1997 (7 pages)

17. Powell, J., "The NOVI Ride Control System: A Method for Eliminating Vibration and Maximizing Ride Comfort in Maglev Transportation Systems", Report DPMT-4, December 15, 1997 (23 pages)

18. Powell, J., "On Surface Maglev Guideways", Report DPMT-2, September 15, 1997 (11 pages)

19. Powell, J., "Magnetic Anchoring of Maglev Guideway Beams", Report DPMT-5, December 15, 1997 (22 pages).

20. Powell, J., "Morena, G., Powell, J., and Danby, G., "Transport of Bulk Cargo for Above and Underground Mining by Low-Cost, High-Speed Magnetically Levitated Vehicles", Report M-205, National Mining Associates 21" Annual Transportation and Distribution Seminar, January, 1998 (15 pages)

21. Powell, J., ed., "Cost Projections for the M-2000 Maglev System", Report DPMT-20, May 15, 1999 (192 pages).

22. Powell, J., "The Matrushka Magnet – A Low Cost Ultra-Low Refrigeration Load Magnet System", Report DPMT-9, April 1998 (113 Pages)

23. Powell, J., "Non-conventional Methods for Large Scale Manufacture of Maglev Guideway Loops", Report DPMT-7, February 15, 1999 (31 pages)

24. Powell, J., "Analysis of Eddy Current Heating in Guideway Conductors and Determination of Allowable Limits on Conductor Size", Report DPMT-8, February 15, 1999 (37 pages)

25. Lazareth, O., Skaritka, J., and Powell, J. -- "Comparison of Analytical Calculations of Magnetic Forces for the M-2000 Maglev Systems with Experimental Measurements", Report DPMT-11, August 11, 1999 (35 pages)

26. Powell, J., "Power Transmission and Distribution Architecture for the M-2000 Maglev System" Report DPMT-12, December 15, 1999 (76 pages)

27. Powell, J., "Levitation, Propulsion, Power, and Braking Systems for Maglev 2000 Vehicles", Report DPMT-22, May 18, 2000 (42 pages)

28. Powell, J., "Communications, Control, and Safety for the M-2000 Maglev System', Report DPMT-23, May 18, 2000 (13 pages).

29. Maise, G., "Design of the M-2000 Maglev Passenger Vehicle, Report DPMT-21, May 30, 2000 (27 pages)

30. Powell, J., "The IRT Levitation Demonstration", Report DPMT-26, May 2001 (20 Pages)

31. Skaritka, J., "Superconducting Magnet Design', Report DPMT-24, May 2001 (23 Pages)

32. Harmer, E, Danby, G., Lazareth, O., and Powell, J., "Experimental Measurements of the Magnetic Forces Between the Maglev 2000 Superconducting Quadrupole and Powered Guideway Loops, with Comparison to Values Predicted Using the Maglev 2000 Computer Code", October 15, 2002 (57 pages)

33. Lazareth, O. and Powell, J., "Computer Analyses of Magnetic Forces Between the M-2000 Vehicle and Powered Guideway Loops", Report DPMT-25 (59 pages)

34. Lazareth, O. and Powell, J. "Planar Guideway Performance of Compact Urban M-2000 Revenue Vehicle Part L Levitation and Stability Performance on Non-Powered Guideway", Report M-2000 PG-2-1 (73 pages)

35. Lazareth, O. and Powell, J., "Planar Guideway Performance of Urban/Suburban M¬2000 Maglev Revenue Vehicle, Part 1: Levitation and Stability Performance on Non-Powered Guideway" Report M2000 PG-1-1 (172 pages)

36. Lazareth, O. and Powell, J., "Propulsion and Power Performance of Maglev 2000 Vehicles on Planar Guideway", Report M-2000 PG-3-1, June 1, 2003 (114 pages)

About the Authors

James R. Powell, Ph.D. is a Director of the MAGLEV 2000 of Florida  **Corporation** Dr. Powell and his colleague, Dr. Gordon Danby are the recipients of the 2000 Benjamin Franklin Medal in Engineering for their invention of superconducting Maglev trasnsport . The medal was awarded to Drs. Powell and Danby by The Franklin Institute "for their invention of a magnetically-levitated transport system using super conducting magnets and subsequent work in the field." The Franklin Institute awards medals annually in recognition of the recipients' genius and civic spirit and in memory of the Institute's namesake, Benjamin Franklin, who exhibited those same qualities. Some noted past recipients of the Franklin Institute medals include Alexander Graham Bell, Thomas Edison, Neils Bohr, Max Planck, Albert Einstein, and Stephen Hawking.

He was a senior scientist at Brookhaven National Laboratory (BNL) from 1956 through 1996. His experiences have led to significant advances in the design and analysis of advanced reactor systems, cryogenic and super-conducting power transmission, plasma physics, mine safety, fusion reactor technology, electronuclear (accelerator) breeder systems, transmutation of nuclear wastes, space nuclear thermal propulsion, electromagnetic hypervelocity guns, hydrogen and synthetic fuels, and transportation infrastructure.

He holds patents for the Particle Bed Reactor (PBR) for nuclear rocket propulsion, the use of aluminum structure in fusion reactors; blankets employing solid lithium ceramics and alloys for tritium breeding; and, demountable super conducting magnet systems and the Advanced Vitrification System (AVS) for high-level nuclear and toxic wastes. He and Dr. Danby are the holders of the first patent for superconducting Maglev in 1968, as well as many recent patents on their 2nd generation advanced Maglev system.

Dr. Powell holds a Bachelor of Science in Chemical Engineering from the Carnegie Institute of Technology and a Doctor of Science in Nuclear Engineering earned in 1958 from the Massachusetts Institute of

Technology. Dr. Powell has published almost 500 professional papers and reports. He is a member of the American Nuclear Society.

Gordon Danby, Ph.D. is a Director of the MAGLEV 2000 of Florida Corporation, together with Dr. Powell, he was awarded the Franklin  Institute Medal 2000 for Engineering for their Maglev inventions. He retired from Brookhaven National Laboratory where he worked on the theory and experimental development of accelerators and magnetic detectors for the study of basic properties of matter. Dr. Danby, together with Dr. Powell, is directing the development of advanced 2nd generation Maglev by their company Maglev 2000 of Florida.

Gordon Danby is widely respected for his contribution to the practical application of theoretical science to technology. His achievements are recognized by his peers as changing Magnetic Resonance Imaging and Transportation Industries.

From the Franklin Award citation, *"Danby's pioneering research efforts in magnetic technology led to the production of open Magnetic Resonance Imaging (MRI) machines that are better, faster and more patient friendly than their tunnel-style predecessors. Danby, along with James Powell, also invented the Superconducting Maglev, a magnetically levitated, high speed train system. The practical and efficient design of the Maglev provides mixed freight and passenger service and interfaces easily with other transport modes."*

Dr. Danby received his B.S. in physics and math from Carleton University in Ottawa, Canada, and his Ph.D. in nuclear physics from McGill University in Montreal, Canada. He is a fellow of the American Physical Society. In 1983, the New York Academy of Sciences honored Danby with the Boris Pregel Award for Applied Science and Technology.

Contributors

Robert J. Coullahan, CEM, CPP, CBCP is an industry leader in national preparedness, critical infrastructure protection, command, control,  communications, intelligence, surveillance and reconnaissance (C4ISR) systems and advanced technology development and integration. After completing active US Army duty in the 1970s, he relocated to Washington, DC area and for over 35 years he has supported and led programs in support to government and commercial operations, grown and created corporations. He is a recognized subject matter expert in emergency management and security. He served in management roles for over 20 years with Science Applications International Corporation (SAIC) where he was Senior Vice President. For 5 years he served as Vice President for Government & International Programs leading an award-winning university consortium that created the *World Data Center-A for Human Interactions in the Environment* and addressed *in situ* and remotely sensed scientific and socioeconomic data products for the US Global Change Research Program. He is the President and Founder of Readiness Resource Group Incorporated (RRG), a veteran-owned business supporting professional services and technology integration for homeland security, infrastructure resilience and energy and transport systems (reference www.readinessresource.net). Currently based in Las Vegas, Nevada, his company supports emergency planning, training, exercises, test and evaluation, studies and analysis, systems integration and advanced technology development for the U.S. Government. These programs range from technical assistance for public safety applications of unmanned aircraft systems (UAS) to fielding of radio-frequency (RF) visualization tools for military and border security applications. He is board certified in emergency management, security management, homeland security, and continuity management. He attended Rutgers University and earned his Bachelor's degree from the University of California. He also holds an M.S. in Telecommunications and an M.A. in Security Management (Forensic Sciences) from The George Washington University.

Ernest Fazio served as an enlisted man in the US Coast Guard Reserve as

both an Engineman and an Electronics Technician.

After the Coast Guard he worked for the New York Telephone Company for 8 years building and testing the outside plant equipment.

Fazio enjoyed a successful career in the financial services industry until his retirement in 2000. His reputation as a knowledgeable, constructive critic in the Long island community has won him respect and recognition. He has been awarded the "Front Page Award" from Long Island Business News and "The Spirit of Long Island" award from the Long Island Association.

He is a self-taught advocate of alternate energy and energy conservation systems. In the early 90's he helped bring together investors that launched a technology idea that was born at Brookhaven National Laboratories. The invention enhances the value of Solar Photovoltaic systems. In 1982 he designed and built a state-of-the-art energy efficient house, where he still lives.

Fazio hosted two radio shows over a span of eight years 1988-1996 on WHPC in Nassau County "Tomorrow's Technology Today" and "Community Issues" and presently hosts a radio interview show on WLIX, "Ernestly Speaking"

Two years after his retirement he was asked to lead Long Island MidSuffolk Business Action (LIMBA). This organization, under Fazio's direction, has become recognized as the most respected forum on Long Island. LIMBA produces 42 programs each year and engages speakers including congressional representatives, university presidents, inventors, authors, business leaders and leaders from the not-for-profit world.

In 2004 Fazio he was invited to join RPANY (Regional Plan Association of NY). Many of the goals of that organization are in concert with LIMBA. Since that affiliation LIMBA now works to accomplish goals that have importance to the whole Long Island Region. Including Brooklyn and Queens. It is now known as Long Island Metro Business Action.

F.H. (Bud) Griffis, PE, PhD is Professor of Construction Engineering and

Management in the Department of Civil and Urban Engineering at Polytechnic Institute of NYU. Until July 2006, he was Provost, Dean of Engineering and Applied Science and Vice President for Academic Affairs at Polytechnic University. He joined the faculty of the Department of Civil Engineering as a tenured professor in January 2000. He was also in charge of all capital construction for Polytechnic University. Professor Griffis is a Professor Emeritus in the Department of Civil Engineering and Engineering Mechanics at Columbia University, having taught there from July 1986 to December 1999. He was the head of the Construction Engineering Program and the Director of the Center for Infrastructure Studies. In addition, he was a Principal in the firm of Robbins, Pope and Griffis Engineers, P.C. of New York and now Chairman of GB Services International, LLC.

He retired with the rank of Colonel from the U.S. Army Corps of Engineers in 1986 after serving as Commander and District Engineer of the New York District. In that position he was responsible for all Army and Air Force construction from Southern New Jersey to Maine and included Greenland and Labrador. He was responsible for water resource development and regulation in the Northeast. Prior to coming to New York, he was Area Engineer, Construction Manager, and Contracting Officer for the Ramon Airbase construction in Israel. For the 20 years preceding his Israeli assignment he commanded Corps of Engineers Construction and Combat Units in the U.S., Korea, Viet Nam and Germany.

He is a registered Professional Engineer in the State of New York. He holds a BS from the United States Military Academy at West Point and MSCE, MSIEOR, and PhD from Oklahoma State University and is a graduate of the US Army War College. He is a Fellow of the American Society of Civil Engineers and the Society of American Military Engineers and an elected member of the National Academy of Construction. He is the author of two textbooks and numerous technical publications.

James Jordan is the founder and President of the Interstate Maglev Project

and Executive Vice President of Maglev 2000 and a co-author with Powell and Danby of *"The Fight For Maglev"*. He is also the managing editor of *"Maglev America"* and *"7 Projects for a Better World"*. He can be reached at

james.jordan@magneticglide.com

The energy crises of the 1970s focused the Navy career of James Jordan. The new era of scarce oil and rapid increases in oil prices dramatically introduced Commander Jordan to the national military and economic security consequences of America's growing dependence on oil. Commander Jordan served as the director of the Navy Energy R&D program office in the Pentagon. As director, he developed strategies and technologies aimed at sustaining military and national economic security in the new oil reality.

In 1979, Mr. Jordan retired from the Navy and became a senior policy advisor to the late Senator John C. Stennis, Chairman, Armed Services Committee and Defense Appropriations Committee. In this capacity, Mr. Jordan was a Senate staff leader in energy, transportation, environment, and agricultural policy.

In 1988, after leaving the U.S. Senate, Mr. Jordan founded several entrepreneurial ventures directed toward development of environmentally sustainable energy and economic growth: efficient all-electric Maglev (**mag**netic **lev**itation) transportation, carbon capture and storage, nuclear waste isolation, hydrogen and electric power co-generation, advanced nuclear power generation and earth science data management.[2]

Education: MBA, Harvard Business School, Cambridge, MA; Distinguished Graduate, Industrial College of the Armed Forces at the National Defense University, Washington, DC; BA, University of North Carolina, Chapel Hill, NC, Student Body President and Graduate of Senior High School, Greensboro, NC.

Consortium International Earth Science Information Network, (www.ciesin.org), now located at Columbia University. In 1992, Mr. Jordan introduced CIESIN to the U.N. Conference on Environmental Development (UNCED) in Rio.

George Maise, Ph.D. is the vice president and treasurer of StarTram, Inc.  His areas of expertise include aerodynamics, thermodynamics, heat transfer, rocket propulsion and orbital mechanics.

Dr. Maise received his B.S. (cum laude) in mechanical engineering from the University of California, Berkeley, and his M.A. and Ph.D. in aerospace sciences from Princeton University. At Princeton, he did his research under the guidance of Prof. John B. Fenn (Nobel Laureate, Chemistry 2002).

Dr. Maise has worked for several aerospace companies (Northrop Aircraft, General Dynamics, Grumman) and also for two aerospace research laboratories (United Aircraft Research Labs and AeroChem Research Laboratory). He has extensive experience—both analytical and experimental—on a wide range of thermal/hydraulic problems, e.g. aerodynamic heating, reentry heating, aircraft and rocket propulsion, etc. While at Grumman, Dr. Maise taught fluid mechanics courses at Hofstra University.

In 1974 he joined the Scientific Staff at Brookhaven National Laboratory (BNL). Dr. Maise supervised a group of BNL scientists and outside consultants providing advice to NRC (Nuclear Regulatory Commission) on thermal/hydraulic loads during postulated accident scenarios. When the Reagan administration launched its Strategic Defense Initiative (SDI), commonly called "Star Wars," Dr. Maise transferred to work on a "black program" to develop an ultra high-speed, nuclear-powered rocket engine. He conducted analytical and experimental programs to characterize the thermal-hydraulic performance of the nuclear-powered rocket. When the SDI program ended due to the end of the Cold War, Dr. Maise switched gears again and joined the International Safeguards Projects Office at BNL, where he provided advice to the U.S. State Department on IAEA requests for technical assistance.

In 1997 he left BNL to work as a private consultant to several high-technology firms. During this period Drs. Powell & Maise patented the StarTram invention (2001), & formed a company called StarTram, Inc. (2001).

Dr. Maise lives with his wife Vita Joy in Head of the Harbor, Long Island, New York. They have two grown children.

Charles Pellegrino, Ph.D.⍰ is the author of more than a dozen books, including the bestselling Titanic trilogy, covering the history of all three expedition groups to the lost liner (on which he participated as deep-ocean  invertebrate zoologist, microbiologist, and forensic archaeologist).

Pellegrino has contributed to numerous technical and popular-readership scientific publications, based on his work in paleobiology, forensic archaeology (ranging from Pompeii to Hiroshima and Nagasaki), and nuclear propulsion systems for space exploration.

He is probably best known as the scientist whose "dinosaur biomorph recipe" became the basis for the Jurassic Park series. Pellegrino served as a scientific consultant on James Cameron's Titanic and hydrothermal vent expeditions—during which deep-penetrating robotic probes were designed and used as "practice" for exploration of new oceans turning up on (and under) Titan, Enceladus, Europa, and other icebound worlds of the Outer Solar System. He also served as a consultant on Cameron's fictional film series, Avatar.

Dr. Jesse Powell, Ph.D. is Founder and President of Maglev Strategies, in which capacity he works to identify new markets and opportunities for maglev technologies. He coordinates between Maglev 2000, Inc. and third party companies in the scoping of new projects, and manages technology transfer issues. Currently, he is focused on maglev space launch, maglev energy storage, and maglev water transport as the areas most likely to attract funding in the United States.

From 2002 to 2013, Dr. Powell has worked in the field of Oceanography. He worked at Scripps Institution of Oceanography, where he studied the impact of ocean fronts and mesoscale ocean structures on plankton distributions. During this time, he also worked on a range of technology projects spanning from mission designs for the exploration of Mars and Europa, to the use of autonomous underwater vehicles to map ocean life, to machine vision systems for plankton identification and the automatic classification of fish eggs of important species for habitat mapping.

Jesse Powell at Brookhaven National Lab on the Occasion of the 50th Anniversary of the invention of Superconducting Maglev by Gordon Danby and his father, James Powell https://www.bnl.gov/video/index.php?v=514

Dr. Powell holds a BS in Biology and BA in French Literature from University of California, San Diego, a MS in Molecular Biology from San Diego State University, and a PhD from Scripps Institution of Oceanography.

John Powell, 1957   B.S. in Mechanical engineering from Carnegie Institute
of Technology

1957 to 1965 Aerospace Engineer: Douglas
Aircraft, General Dynamics
(Atlas ICBM propulsion systems)
Registered Professional Engineer California
since 1965.
1965 to 1993 Scripps Institution of
Oceanography 1965 to 1993:
Principal Development Engineer
Design of oceanographic equipment and large
scale laboratory facilities designed for coastal
research and wave mechanics studies.

Conducted numerous hydraulic experiments for oceanographic problems
and research.

1982 to 1993 Scripps Institution of Oceanography: Head of Hydraulics
Laboratory

Principal Investigator for experimental hydraulic projects such as:

Innovative studies of dredging equipment for the Army Corps of Engineers

Sand fluidization,sand bypass systems, and beach erosion

Ocean wave studies.

Floating island projects, ship modeling

Private design consultant to commercial Water Parks (wave pools, surf
generators, innovative rides) and Civic Aquariums, Disney World.

1993 Retired from Scripps,but continued part time at the Hydraulics
Laboratory and as Lecturer and Adjunct Professor, University of
California, San Diego. Taught Engineering Design courses and
supervised teams of engineering students solving industrial sponsored
projects, 1990 to 1996

1996 to 2006 Moved to England. continued private consulting practice

2006 Returned to California,private consulting practice to present.

www.ingramcontent.com/pod-product-compliance
Lightning Source LLC
Chambersburg PA
CBHW071246220526
45468CB00001B/19